Glimpses in Plant Physiology

Dr. Chetana Mandavia
Professor
Department of Agril. Botany
College of Agriculture
Junagadh Agricultural University
Junagadh - 362 001

Dr. S.V.Patel
Professor
Department of Biochemistry
Junagadh Agricultural University
Junagadh - 362 001

Dr. (Ms). M.K. Mandavia
Professor
Department of Biochemistry
Junagadh Agricultural University
Junagadh - 362 001

Dr. B.A.Golakiya
Professor and Head
Department of Biochemistry
Junagadh Agricultural University
Junagadh - 362 001

Dr. V.P. Chovatia
Professor and Head
Department of Botany
Junagadh Agricultural University
Junagadh - 362 001

IBDC

International Book Distributing Co.
(Publishing Division)

Published by
INTERNATIONAL BOOK DISTRIBUTING CO.
(Publishing Division)
Khushnuma Complex Basement
7, Meerabai Marg (Behind Jawahar Bhawan)
Lucknow 226 001 U.P. (INDIA)
Tel. : 91-522-2209542, 2209543, 2209544, 2209545
Fax : 0522-4045308
E-Mail : ibdco@airtelmail.in

First Edition 2009

Price: Rs. 275/-
ISBN 81-8189-270-4 (P.B.)

Composed & Designed at :
Panacea Computers
3rd Floor, Agrawal Sabha Bhawan
Subhash Mohal, Sadar Cantt. Lucknow-226 002
Phone : 0522-2483312, 9452295008, 9335927082
E-mail : prasgupt@rediffmail.com

Printed at:
Salasar Imaging Systems
C-7/5, Lawrence Road Industrial Area
Delhi - 110 035
Tel. : 011-27185653, 9810064311

Dr. B.K.Kikani
Vice-chancellor

Junagadh Agricultural University
Junagadh-362 001
(Gujarat)

FOREWORD

The last decade has seen a rapid expansion in the knowledge of plant physiology, and it has evolved as a multidisciplinary science, encompassing new frontiers in biochemistry, biophysics and molecular biology. Due to the growth of these frontier areas, voluminous information has been added to this discipline, enriching it greatly. And that is precisely the reason, it is imperative to choose and summarize new findings that are most representative of the status of plant physiology during the last decade. The present book fulfills this need, as it assembles and explains in a narrative format, the underlying concepts of plant physiology within a framework of modern approaches with the help of extensive illustrations based on up-to-date literature.

The general aim of the book is to present to senior students and investigators the results of recent work in a few of those branches of plant physiology which are at present attracting attention. By such a presentation of portion of the science which are still in a state of rapid growth, it is hoped that further investigations will be stimulated, and students will get insight into the basic themes based on which future line of work can be formulated. I wish to congratulate my scientists for their laudable effort of condensing, organizing and synthesizing all the recent findings in a clear comprehensive manner. I hope, this book would provide to students a foundation in physiological principles that will enable them to grow with the discipline as it moves forward and its frontiers expanded.

B.K.Kikani

Dr. D.B. Kuchhadia
Director of Research and Dean
P.G Studies

Junagadh Agricultural University
Junagadh-362 001
(Gujarat)

PREFACE

It is both a pleasure and a duty to write a few words about this timely publication, in which the authors have attempted to consider both the senior students and teachers interested alike in the fundamentals of Plant Physiology. Students majoring in Plant Breeding, Genetics, Horticulture, Floriculture and Biochemistry opt for Plant Physiology as the supporting discipline. A need was felt for such a text book that is moderate in size, adequately full in detail and gives due justice to fundamentals and recent advances equally. I think that this need is fulfilled with this book entitled " Glimpses in Plant Physiology " which gives glimpses into the existing and new perspectives of Plant Physiology. The choice of the topics is exhaustive, selected with meticulous carve; the available information has been well organized and presented in a systematic format. The figures and drawings used in the book make the text more clear. I am sure, this book will be of immense value to teachers, researchers and students and will open a new roads to them. It will give them opportunity to encounter the excitement, mystery and challenge of Plant Physiology.

D.B.Kuchhadia

Contents

Chapter 1

The Plant Kingdom and Tissue System

1.1 The Plant Kingdom

Since the time of Aristotle (384–322 B.C.E.), biologists have sought to classify organisms. At first the purpose was ease of identification ("artificial" classification schemes). Carolus Linnaeus (1707–1778), arguably the greatest of the pre-modern Natural Historians, sought to classify plants and other organisms according to affinity groups that reflected the mind of the Creator. Later, after Darwin, the goal of classification was to derive evolutionary relationships ("natural" classification schemes). For the past 150 years, biologists have emphasized natural systems of classification and have attempted to define morphological criteria that best reflect evolutionary relationships.

We now know that morphology, the form and structure of organisms, is the end product of the action of genes. Virtually all of the information needed to form a complete organism is encoded in its DNA sequences. DNA sequence analysis has thus provided evolutionary biologists with a powerful new tool for arriving at a truly natural classification system.

On the basis of phylogenetic analyses of highly conserved DNA sequences, living organisms have been divided into three major domains: Bacteria, Archaea, and Eucarya (Figure 1.1).

The **Eucarya** include the **eukaryotes,** organisms whose cells contain a true nucleus. The **Bacteria,** or **eubacteria,** which include the cyanobacteria, lack a true nucleus and are therefore prokaryotic. The **Archaea,** or **archaebacteria,** are also prokaryotic, but they differ from the Bacteria: Besides their morphological and biochemical differences, they are often adapted to extreme environments, such as sulfur hot springs or saline ponds. Phylogenetic studies have indicated that the Archaea and Eucarya split after the Bacteria separated from the common ancestor. Thus Archaea and Eucarya represent sister groups. This closer relation between Archaea and Eucarya is reflected in their similar promoter structures and RNA polymerases, the presence of histones, and many other characteristics.

Before the three domains of life were recognized, the five-kingdom classification scheme of living organisms was widely accepted. According to this scheme, all

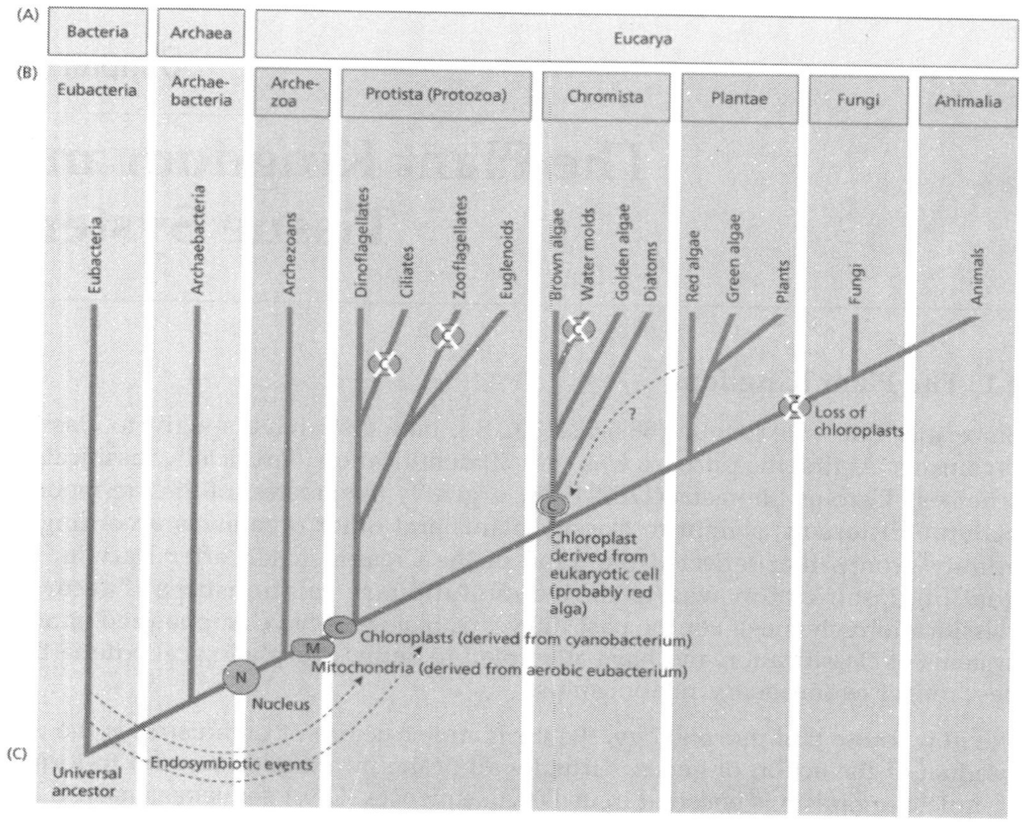

Figure 1.1. Natural classification scheme and phylogeny of living organisms. (A) The three domains of life. (B) Division of organisms into two prokaryotic kingdoms and six eukaryotic kingdoms. (C) Hypothetical phylogeny showing the origin of the major groups of eukaryotes. Proponents of a natural classification scheme divide the protists into three separate kingdoms according to their relatedness: Archezoa, Protozoa, and Chromista. The appearance of the nucleus and the acquisition of mitochondria and chloroplasts as endosymbiotic organelles are indicated by the dashed lines and arrows.(Campbell,1996,*Biology* 4th ed. Benjamin/Cummings, Menlo Park, CA)

living organisms were divided into five kingdoms: Monera, Protista, Fungi, Plantae, and Animalia.

The **Monera** include all prokaryotic organisms: the eubacteria and the archaebacteria. Some biologists still adhere to this grouping, which is based on the dichotomy between prokaryotes and eukaryotes. However, the discovery that the archaebacteria and eukaryotes are sister groups suggests that the archaebacteria do not belong in the same kingdom with the eubacteria. Hence, biologists who prefer a natural classification scheme reject the Monera grouping, while those who prefer an artificial classification scheme (i.e., one based on criteria other than evolutionary relationships) continue to use it. Whereas in the natural system the distinction between prokaryote and eukaryote is ignored, in the artificial system this distinction becomes the primary criterion for dividing the organisms.

The "kingdom" **Protista** presents a similar dilemma. Protists are usually defined as unicellular eukaryotes, although the classification also includes multicellular algae with relatively simple structures. Proponents of a natural classification scheme divide the protists into three separate kingdoms according to their relatedness: Archezoa, Protozoa, and Chromista (Figure 1.1).

The **Archezoa** are extremely primitive eukaryotic cells, such as the intestinal parasite *Giardia,* which lacks both mitochondria and chloroplasts. *Giardia* is thought to have branched off before the endosymbiotic bacteria that gave rise to these organelles were incorporated in the eukaryotes.

Fungi were formerly classified with the plants. However, fungi are sufficiently different from plants and the other eukaryotes to be placed in a separate kingdom. Having lost their chloroplasts, fungi are heterotrophic; that is, they depend on other organisms for their food, and they satisfy their nutritional needs by absorbing inorganic ions and organic molecules from the external environment. Most fungal species are filamentous and possess cell walls made of chitin, the same substance that is found in insect exoskeletons.

Bryophytes are small (rarely more than 4 cm in height), very simple land plants, and the least abundant in terms of number of species and overall population. Bryophytes do not appear to be in the direct line of evolution leading to the vascular plants; rather they seem to constitute a separate minor branch. Bryophytes include mosses, liverworts, and hornworts. These small plants have life cycles that depend on water during the sexual phase. Water facilitates **fertilization,** the fusion of gametes to produce a diploid **zygote,** a feature also seen in the algal precursors of these plants. Bryophytes are like algae in other respects as well: They have neither true roots nor true leaves, they lack a vascular system, and they produce no hard tissues for structural support. The absence of these structures that are important for growth on land greatly restricts the potential size of bryophytes, which, unlike algae, are terrestrial rather than aquatic.

The **ferns** represent the largest group of spore-bearing **vascular plants.** In contrast to the bryophytes, ferns have true roots, leaves, and vascular tissues, and they produce hard tissues for support. These architectural features enable ferns to grow to the size of small trees. Although ferns are better adapted to the drying conditions of terrestrial life than bryophytes are, they still depend on water as a medium for the movement of sperm to the egg. This dependence on water during a critical stage of their life cycle restricts the ecological range of ferns to relatively moist habitats.

The most successful terrestrial plants are the **seed plants**. Seed plants have been able to adapt to an extraordinary range of habitats. The embryo, protected and nourished inside the seed, is able to survive in a dormant state during unfavorable growing conditions such as drought. Seed dispersal also facilitates the dissemination of the embryos away from the parent plant.

Another important feature of seed plants is their mode of fertilization. Fertilization in seed plants is brought about by wind- or insect-mediated transfer of **pollen**, the gamete-producing structure of the male, to the sexual structure of the female, the **pistil**. Pollination is independent of external water, a distinct advantage in terrestrial environments. Many seed plants produce massive amounts of woody tissues, which enable them to grow to extraordinary heights. These features of seed plants have contributed to their success and account for their wide range.

There are two categories of seed plants: gymnosperms (from the Greek for "naked seed") and angiosperms (based on the Greek for "vessel seed," or seeds contained in a vessel). **Gymnosperms** are the less advanced type; about 700 species of gymnosperms are known. The largest group of gymnosperms is the conifers ("cone-bearers"), which include such commercially important forest trees as pine, fir, spruce, and redwood.

Two types of cones are present: male cones, which produce pollen, and female cones, which bear ovules. The ovules are located on the surfaces of specialized structures called **cone scales**. After wind-mediated pollination, the sperm reaches the egg via a **pollen tube**, and the fertilized egg develops into an embryo. Upon maturation, the cone scales, which are appressed during early development, separate from each other, allowing the naked seeds to fall to the ground.

Angiosperms, the more advanced type of seed plant, first became abundant during the Cretaceous period, about 100 million years ago. Today, they dominate the landscape, easily outcompeting their cousins, the gymnosperms. About 250,000 species are known, but many more remain to be characterized. A typical angiosperm life cycle, that of *Zea mays* (corn), is shown in Figure 1.2. The major innovation of the angiosperms is the flower; hence they are referred to as *flowering plants*. There are other anatomical differences between angiosperms and gymnosperms, but none so crucial and far-reaching as the mode of reproduction.

1.2 Flower Structure and the Angiosperm Life Cycle

The flower consists of several leaflike structures attached to a specialized region of the stem called the **receptacle** (Figure 1.3). **Sepals** and **petals** are the most leaflike. Petals have the primary function of attracting insects to serve as pollinators, accounting for their often showy and brightly colored appearance. The **stamen** is the male sexual structure, and the **pistil** is the female sexual structure. The pistil is composed of one or more united **carpels**; the pistil, or in some flowers a whorl of pistils, is sometimes referred to as the **gynoecium**. The stamen consists of a

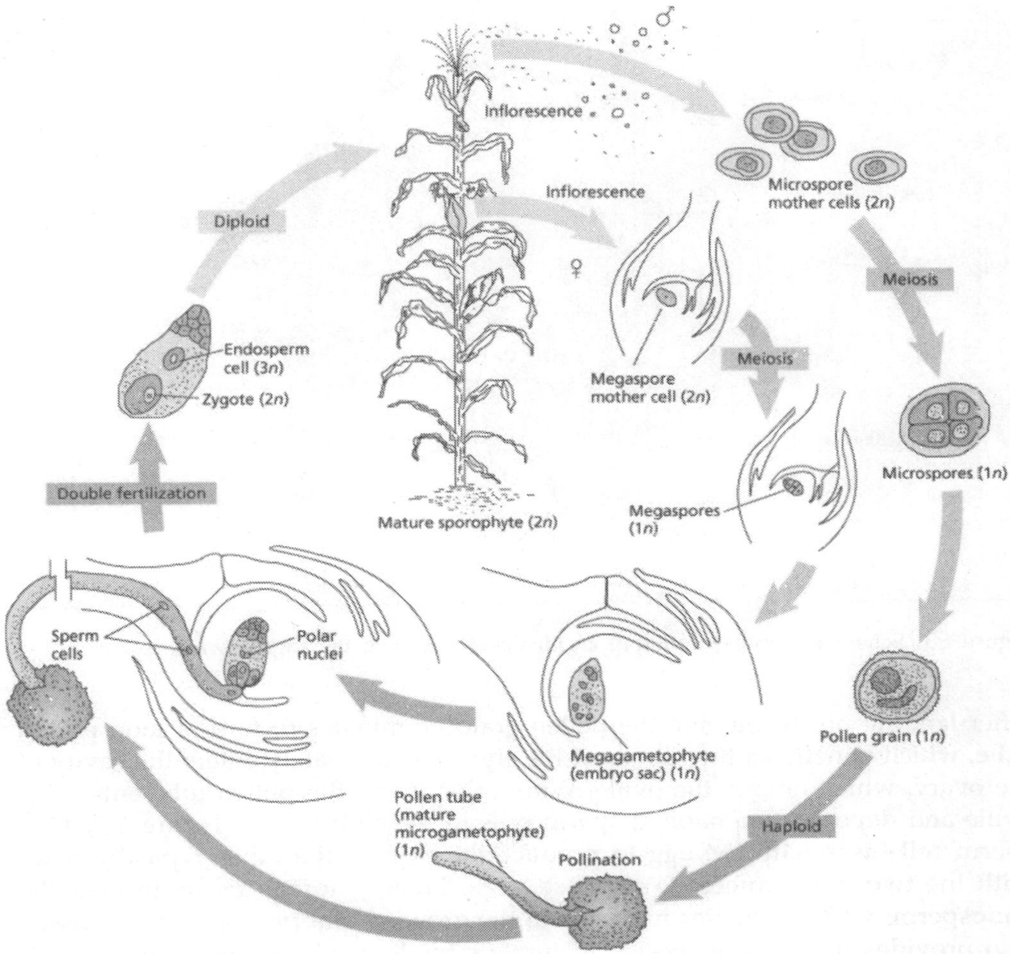

Figure 1.2 Life cycle of corn (*Zea mays*), a monocot. The vegetative plant represents the diploid sporophyte generation. Meiosis occurs in the male and female flowers, represented by the tassels and ears, respectively. The haploid microspores (male spores) develop into pollen grains, and the single surviving haploid megaspore (female spore) divides mitotically to form the embryo sac (megagametophyte). The egg forms in the embryo sac. Pollination leads tc the formation of a pollen tube containing two sperm cells (the microgametophyte). Finally, double fertilization results in the formation of the diploid zygote, the first stage of the new sporophyte generation, and the triploid endosperm cell.

narrow stalk called the **filament** and a chambered structure called the **anther**. The anther contains tissue that gives rise to **pollen grains**. The pistil consists of the **stigma** (the tip where pollen lands during pollination), the **style** (an elongated structure), and the **ovary**. The ovary, the hollow basal portion of the pistil, completely encloses one or more **ovules**. Each ovule, in turn, contains an **embryo sac**, the structure that gives rise to the female gamete, the **egg**.

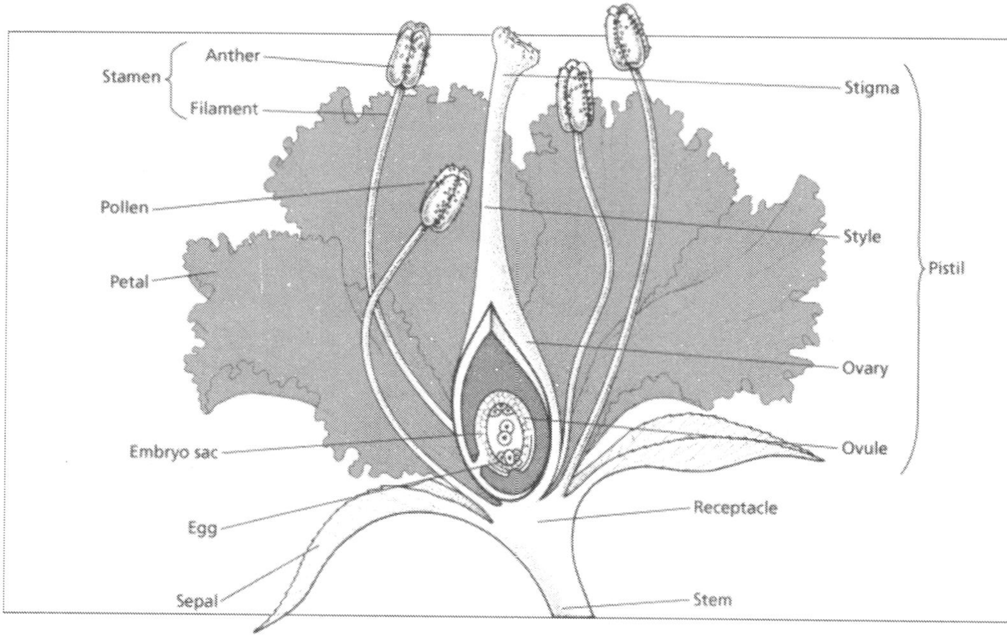

Figure 1.3 Schematic representation of an idealized flower of the angiosperms.

After landing on the stigma, the pollen grain germinates to form a long **pollen tube**, which penetrates the tissues of the style and ultimately enters the cavity of the ovary, which houses the ovule. Within the ovary, the pollen tube enters the ovule and deposits two haploid sperm cells in the embryo sac (Figure 1.3). One sperm cell fuses with the egg to produce the zygote; the other typically fuses with the two **polar nuclei** to produce a specialized storage tissue termed the **endosperm**, which provides nutrients to the growing embryo. Endosperm tissue also provides the bulk of the worldrquote s food supply in the form of cereal grains. As in conifers, in angiosperms the outer tissues of the ovule harden into a protective **seed coat**. Angiosperm seeds have a second layer of protective tissues, the fruit. The **fruit** consists of the ovary wall and, in some cases, receptacle tissue.

Angiosperms are divided into two major groups, **dicotyledons** (dicots) and **monocotyledons** (monocots). This distinction is based primarily on the number of **cotyledons**, or seed leaves. In addition, the two groups differ with respect to other anatomical features, such as the arrangement of their vascular tissues, and their floral structure.

As the dominant plant group on Earth, and because of their great economic and agricultural importance, angiosperms have been studied much more intensively than other types of plants, and they are discussed extensively in this book. Plant physiologists have focused on a relatively small number of species that represent convenient experimental systems for the study of specific phenomena. Therefore, while we focus on these famous few, it is important to keep in mind the tremendous diversity of form and function that exists within the angiosperms, and the even greater diversity of form and function that is found within the plant kingdom as a whole.

1.3 Plant Tissue Systems: Dermal, Ground, and Vascular

Dermal tissue. The **epidermis** is the **dermal tissue** of young plants undergoing primary growth . It is generally composed of specialized, flattened polygonal cells that occur on all plant surfaces. Shoot surfaces are usually coated with a waxy **cuticle** to prevent water loss and are often covered with hairs, or trichomes, which are epidermal cell extensions.

Pairs of specialized epidermal cells, the **guard cells**, are found surrounding microscopic pores in all leaves . The guard cells and pores are called **stomata** (singular stoma), and they permit gas exchange (water loss, CO_2 uptake, and O_2 release or uptake) between the atmosphere and the interior of the leaf.

The root epidermis is adapted for absorption of water and minerals, and its outer wall surface typically does not have a waxy cuticle. Extensions from the root epidermal cells, the **root hairs** , increase the surface area over which absorption can take place.

Ground tissue. Making up the bulk of the plant are cells termed the ground tissue. There are three types of ground tissue: parenchyma, collenchyma, and sclerenchyma.

- **Parenchyma**, the most abundant ground tissue, consists of thin-walled, metabolically active cells that carry out a variety of functions in the plant including photosynthesis and storage.
- **Collenchyma** tissue is composed of narrow, elongated cells with thick primary walls . Collenchyma cells provide structural support to the growing plant body, particularly shoots, and their thickened walls are nonlignified so they can stretch as the organ elongates. Collenchyma cells are typically arranged in bundles or layers near the periphery of stems or leaf petioles.
- **Sclerenchyma** consists of two types of cells, sclereids and fibers Both have thick secondary walls and are frequently dead at maturity. Sclereids occur in a variety of shapes, ranging from roughly spherical to branched, and are widely distributed throughout the plant. In contrast, fibers are narrow, elongated cells that are commonly associated with vascular tissues. The main function of sclerenchyma is to provide mechanical support particularly to parts of the plant that are no longer elongating.

In the stem, the pith and the cortex make up the ground tissue . The **pith** is located within the cylinder of vascular tissue, where it often exhibits a spongy texture because of the presence of large intercellular air spaces. If the growth of the pith fails to keep up with that of the surrounding tissues, the pith may degenerate, producing a hollow stem. In general, roots lack piths, although there are exceptions to this rule. In contrast, the **cortex** , which is located between the epidermis and the vascular cylinder, is present in both stems and roots.

At the boundary between the ground tissue and the vascular tissue in roots, and occasionally in stems, is a specialized layer of co rtex known as the **endodermis** . This single layer of cells originates from cortical tissue at the innermost layer of the root cortex and forms a cylinder that surrounds the central vascular tissue, or **stele**. Early in root development, a narrow band composed of the waxy substance **suberin** is formed in the cell walls circumscribing each endodermal cell. These suberin deposits, called **Casparian strips**, form a barrier in the endodermal walls to the intercellular movement of water, ions, and other water-soluble solutes to the vascular cells.

Leaves have two interior layers of ground tissue that are collectively known as the **mesophyll.** The **palisade parenchyma** consists of closely spaced, columnar cells located beneath the upper epidermis. There is usually one layer of palisade parenchyma in the leaf. Palisade parenchyma cells are rich in chloroplasts and are a primary site of photosynthesis in the leaf. Below the palisade parenchyma are irregularly shaped, widely spaced **spongy mesophyll** cells. The spongy mesophyll cells are also photosynthetic, and the large spaces between these cells allow diffusion of carbon dioxide. The spongy mesophyll also contributes to leaf flexibility in the wind and this flexibility facilitates the movement of gases within the leaf.

Vascular tissues: xylem and phloem. The vascular tissue is composed of two major conducting systems: the xylem and the phloem. The **xylem** transports water and mineral ions from the root to the rest of the plant. The **phloem** distributes the products of photosynthesis and a variety of other solutes throughout the plant .

The **tracheids** and **vessel elements** are the conducting cells of the xylem . Both of these cell types have elaborate secondary-wall thickenings and lose their cytoplasm at maturity; that is, they are dead when functional. Tracheids overlap each other, whereas vessel elements have open end walls and are arranged end to end to form a larger unit called a **vessel.** Other cell types present in the xylem include parenchyma cells, which are important for the storage of energy-rich molecules and phenolic compounds, and sclerenchyma fibers.

The **sieve elements** and **sieve cells** are responsible for sugar translocation in the phloem . The former are found in angiosperms; the latter perform the same function in gymnosperms. Like vessel elements, sieve elements are often stacked in vertical rows, forming larger units called **sieve tubes**, whereas sieve cells form overlapping arrays. Both types of conducting cells are living when functional, but they lack nuclei and central vacuoles and have relatively few cytoplasmic organelles.

Substances are translocated from sieve cell to sieve cell laterally through circular or oval zones containing enlarged pores, called **sieve areas**. In contrast, sieve tubes translocate substances through large pores in the end walls of the sieve elements, called **sieve plates**. Sugar movement through sieve tubes is more efficient and rapid than through sieve cells and represents a more evolutionarily advanced mechanism.

Sieve elements are associated with, and depend on, densely cytoplasmic parenchyma cells called **companion cells**. The analogous cells adjacent to the sieve cells of gymnosperms are called **albuminous cells**. Companion cells provide proteins and metabolites necessary for the functions of the sieve tube elements. In addition, the phloem frequently contains storage parenchyma and fibers that provide mechanical support.

Chapter **2**

Water Potential and Its Measurement

2.1 Components of Water Potential

Students planning further study of plant water relations should note that the components of water potential defined in the text are sometimes given different names and symbols. In particular, the following equation gives the water potential and its components

$$\psi_w = \psi_s + \psi_p$$

This Equation is often replaced by the following equivalent equation:

$$\psi_w = \pi + P$$

In this alternative convention, *P* is the same as ψ_p. It is the hydrostatic pressure of the solution, and may be positive, as in turgid cells, or negative, as in xylem water. The symbol π is called **osmotic pressure** and is the negative of ψs. That is, π has positive values, and ψs has negative values. "Osmotic pressure" is the term that physical chemists, zoologists, and many others use to denote the effect of dissolved solutes on the free energy of water. Most handbooks of physics and chemistry use the term "osmotic pressure" and the symbol π. The negative sign in front of π in the equation above accounts for the reduction in water potential (ψw) by dissolved solutes. Thus $\psi s = -\pi$.

Unfortunately, some authors have mixed the conventions for ψs and π, leading to unnecessary confusion about what is meant by the symbol π . Thus, π is sometimes incorrectly called osmotic potential instead of osmotic pressure and it may be used either as a positive quantity or as a negative quantity.

Regarding matric potential, it is usually designated by the symbol *t*.

2.2 Measuring Water Potential

Plant scientists have expended considerable effort in devising accurate and reliable methods for evaluating the water status of a plant. Four instruments that have been used extensively to measure ψ_w , ψ_s , and ψ_p are described here: psychrometer, pressure chamber, cryoscopic osmometer and pressure probe.

Psychrometer (ψ_w measurement)

Psychrometry (the prefix "psychro-" comes from the Greek word *psychein*, "to cool") is based on the fact that the vapor pressure of water is lowered as its water potential is reduced. Psychrometers measure the water vapor pressure of a solution or plant sample, on the basis of the principle that evaporation of water from a surface cools the surface.

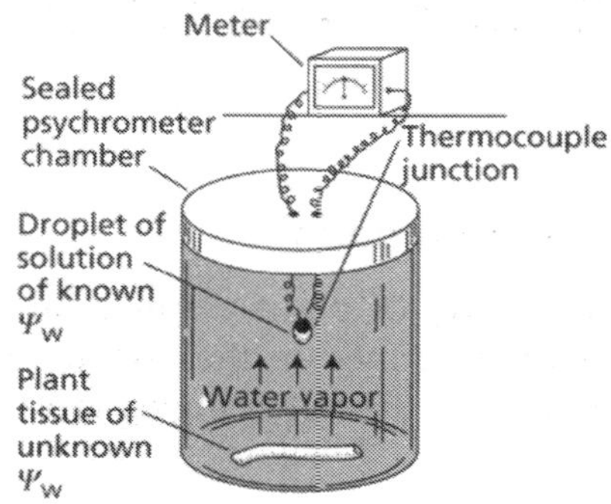

Figure 2.1. Diagram illustrating the use of isopiestic psychrometry to measure the water potential of a plant tissue.

One psychrometric technique, known as *isopiestic psychrometry*, has been used extensively by John Boyer and coworkers and is illustrated in Figure 2.1 . Investigators make a measurement by placing a piece of tissue sealed inside a small chamber that contains a temperature sensor (in this case, a thermocouple) in contact with a small droplet of a standard solution of known solute concentration (known ψ_s and thus known ψ_w). If the tissue has a lower water potential than that of the droplet, water evaporates from the droplet, diffuses through the air, and is absorbed by the tissue. This slight evaporation of water cools the drop. The larger the difference in water potential between the tissue and the droplet, the higher the rate of water transfer and hence the cooler the droplet. If the standard solution has a lower water potential than that of the sample to be measured, water will diffuse from the tissue to the droplet, causing warming of the droplet. Measuring the change in temperature of the droplet for several solutions of known ψ_w makes it possible to calculate the water potential of a solution for which the net movement of water between the droplet and the tissue would be zero (Figure 2.1) signifying that the droplet and the tissue have the same water potential.

Psychrometers can be used to measure the water potentials of both excised and intact plant tissue. Moreover, the method can be used to measure the ψ_s of solutions. This can be particularly useful with plant tissues. For example, the ψ_w of a tissue is measured with a psychrometer, and then the tissue is crushed and the ψ_s value of the expressed cell sap is measured with the same instrument. By combining the two measurements, researchers can estimate the turgor pressure that existed in the cells before the tissue was crushed ($\psi_p = \psi_w - \psi_s$).

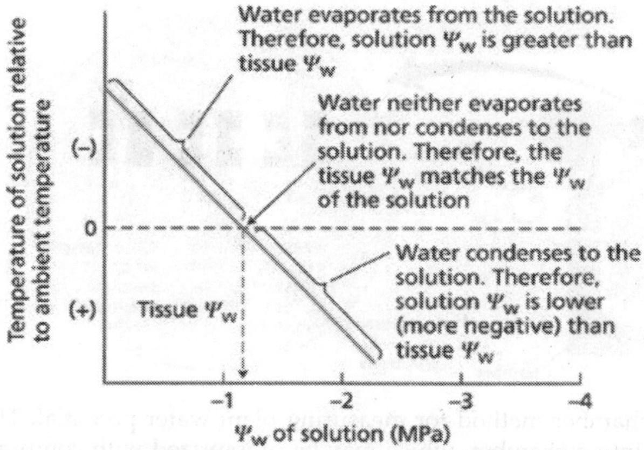

Figure 2.2. The temperature of the sensor in the psychrometer shown in this Figure depends on the water potential of the solution relative to the water potential of the tissue sample.

A major difficulty with this approach is the extreme sensitivity of the measurement to temperature fluctuations. For example, a change in temperature of 0.01°C corresponds to a change in water potential of about 0.1 MPa. Thus, psychrometers must be operated under constant temperature conditions. For this reason, the method is used primarily in laboratory settings. There are many variations in psychrometric technique; int erested readers should consult Brown and Van Haveren 1972, Slavik 1974, and Boyer 1995.

Pressure chamber (ψ_w measurement)

A relatively quick method for estimating the water potential of large pieces of tissues, such as leaves and small shoots, is by use of the **pressure chamber** . This method was pioneered by Henry Dixon at Trinity College, Dublin, at the beginning of the twentieth century, but it did not come into widespread use until P. Scholander and coworkers at the Scripps Institution of Oceanography improved the instrument design and showed its practical use.

In this technique, the organ to be measured is excised from the plant and is partly sealed in a pressure chamber (Figure 2.3). Before excision, the water column in the xylem is under tension. When the water column is broken by excision of the organ (i.e., its tension is relieved allowing its ψ_p to rise to zero), water is pulled rapidly from the xylem into the surrounding living cells by osmosis. The cut surface consequently appears dull and dry. To make a measurement, the investigator pressurizes the chamber with compressed gas until the distribution of water between the living cells and the xylem conduits is returned to its initial, pre-excision, state. This can be detected visually by observing when the water returns to the open ends of the xylem conduits that can be seen in the cut surface. The pressure needed to bring the water back to its initial distribution is called the *balance pressure* and is readily detected by the change in the appearance of the cut surface, which becomes wet and shiny when this pressure is attained.

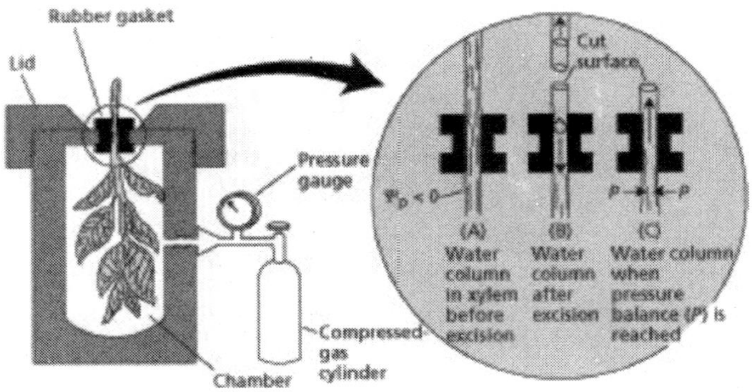

Figure 2.3 The pressure chamber method for measuring plant water potential. The diagram at left shows a shoot sealed into a chamber, which may be pressurized with compressed gas. The diagrams at right show the state of the water columns within the xylem at three points in time

The pressure chamber is often described as a tool to measure the tension in the xylem. However, this is only strictly true for measurements made on a non-transpiring leaf or shoot (for example, one that has been previously enclosed in a plastic bag). When there is no transpiration, the water potential of the leaf cells and the water potential in the xylem will come into equilibrium. The balancing pressure measured on such a non-transpiring shoot is equal in magnitude but opposite in sign to the pressure in the xylem (ψ_p). Because the water potential of our non-transpiring leaf is equal to the water potential of the xylem, one can calculate the water potential of the leaf by adding together ψ_p and ψ_s of the xylem, provided one collects a sample of xylem sap for determination of ψ_s. Luckily ψ_s of the xylem is usually small (> -0.1 MPa) compared to typical midday tensions in the xylem (ψ_p of –1 to –2 MPa). Thus, correction for the ψ_s of the xylem sap is frequently omitted.

Balancing pressure measurements of transpiring leaves are more difficult to interpret. The fact that water is flowing from the xylem to the leaf means that differences in water potential must exist. When the transpiring leaf or shoot is cut off, the tension in the xylem is instantly relieved and water is drawn into the leaf cells until the water potentials of the xylem and the leaf cells come into equilibrium. Because the total volume of the leaf cells is much larger than the volume of sap in the xylem, this equilibrium water potential will be heavily weighted towards that of the leaf. Thus, any measurement of the balancing pressure on such a leaf or shoot will result in a value that is approximately the water potential of the leaf, rather than the tension of the xylem. (To be exact, one would have to add the $Ø_s$ of the xylem sap to the negative of the balancing pressure to get the leaf water potential.) One can explore the differences between the water potential of the xylem and the water potential of a transpiring leaf by comparing balancing pressures measured on covered (i.e., non-transpiring) versus uncovered (transpiring) leaves.

Pressure chamber measurements provide a quick and accurate way of measuring leaf water potential. Because the pressure chamber method does not require delicate instrumentation or temperature control, it has been used extensively under field conditions. For a more complete description of the theory and operation of the pressure chamber.

Cryoscopic osmometer (ψ_s measurement)

The **cryoscopic osmometer** measures the osmotic potential of a solution by measuring its freezing point. Solutions have *colligative* properties that collectively depend on the number of dissolved particles and not on the nature of the solute. For example, solutes reduce the vapor pressure of a solution, raise its boiling point, and lower its freezing point. The specific nature of the solute does not matter. One of the colligative *properties* of solutions is the decrease in the freezing point as the solute concentration increases. For example, a solution containing 1 mol of solutes per kilogram of water has a freezing point of –1.86°C, compared with 0°C for pure water.

Various instruments can be used to measure the freezing-point depression of solutions. With a cryoscopic osmometer, solution samples as small as 1 nanoliter

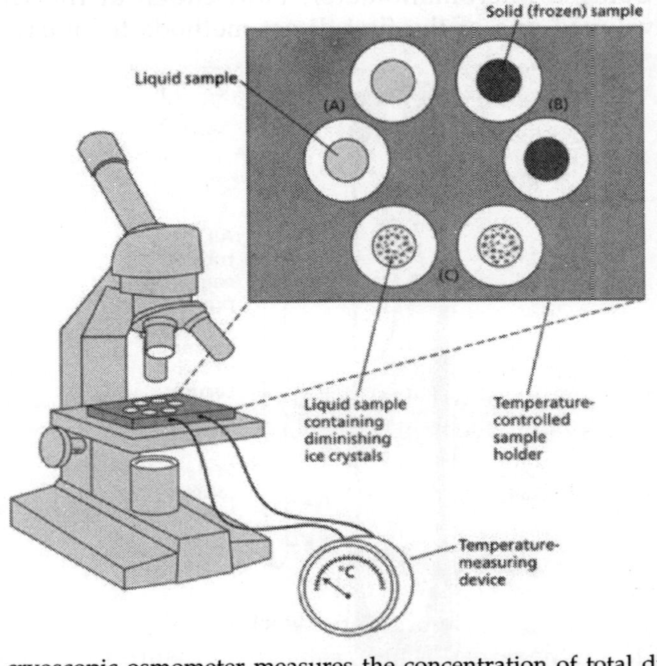

Figure 2.4. A cryoscopic osmometer measures the concentration of total dissolved solutes by measuring the freezing-point depression of a solution. (A) Very small liquid samples are loaded onto the temperature-controlled stage of a microscope. (B) When the temperature is quickly reduced, the samples supercool and freeze. (C) Slowly warming the stage causes the samples to thaw. The temperature at which the last ice crystal melts provides a measure of the melting point of the sample.

(10^{-9} L) are placed in an oil medium located on the temperature-controlled stage of a microscope (Figure 2.4.). The very small sample size allows sap from single cells to be measured and permits rapid thermal equilibration with the stage. To prevent evaporation, the investigator suspends the samples in oil-filled wells in a silver plate (silver has high thermal conductivity). The temperature of the stage is rapidly decreased to about -30° C, which causes the sample to freeze. The temperature is then raised very slowly, and the melting process in the sample is observed through the microscope. When the last ice crystal in the sample melts, the temperature of the stage is recorded (note that the melting and freezing points are the same). It is straightforward to calculate the solute concentration from the freezing-point depression; and from the solute concentration (c_s), ψ_s is calculated as $-RTc_s$ .This technique has been used to measure droplets extracted from single cells.

Pressure probe (ψ_p measurement)

If a cell were as large as a watermelon or even a grape, measuring its hydrostatic pressure would be a relatively easy task. Because of the small size of plant cells, however, the development of methods for direct measurement of turgor pressure has been slow. Using a micromanometer, Paul Green at the University of Pennsylvania developed one of the first direct methods for measuring turgor

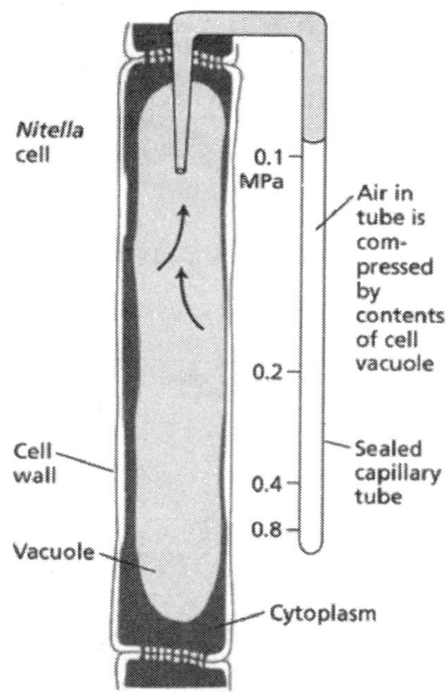

Figure 2.5 Use of the micromanometer, a pressure probe, to measure cell turgor pressure. *Nitella* cells (which are particularly large—about 100 mm in diameter and many centimeters long) were used for these measurements. (Green and Stanton, 1967, *Science* 155: 1675-1676)

pressure in plant cells . In this technique, an air-filled glass tube sealed at one end is inserted into a cell (Figure 2.5.). The high pressure in the cell compresses the trapped gas, and from the change in volume one can readily calculate the pressure of the cell from the ideal gas law (pressure × volume = constant). This method works only for cells of relatively large volume, such as the giant cell of the filamentous green alga *Nitella.* For smaller cells, the loss of cell sap into the glass tube is sufficient to deflate the cell and this yields artifactually low pressures.

For higher plant cells, which are several orders of magnitude smaller in volume than *Nitella,* a more sophisticated device, the **pressure probe**, was developed by Ernest Steudle, Ulrich Zimmermann, and their colleagues in German. This instrument is similar to a miniature syringe (Figure 2.6). A glass microcapillary tube is pulled to a fine point and is inserted into a cell. The microcapillary is filled with silicone oil, a relatively incompressible fluid that can be readily distinguished from cell sap under a microscope. When the tip of the microcapillary is first inserted into the cell, cell sap begins to flow into the capillary because of the initial low pressure of that region. Investigators can observe such movement of sap under the microscope and counteract it by pushing on the plunger of the device, thus building up a pressure. In such fashion the boundary between the oil and the cell sap can be pushed back to the tip of the microcapillary. When the boundary is

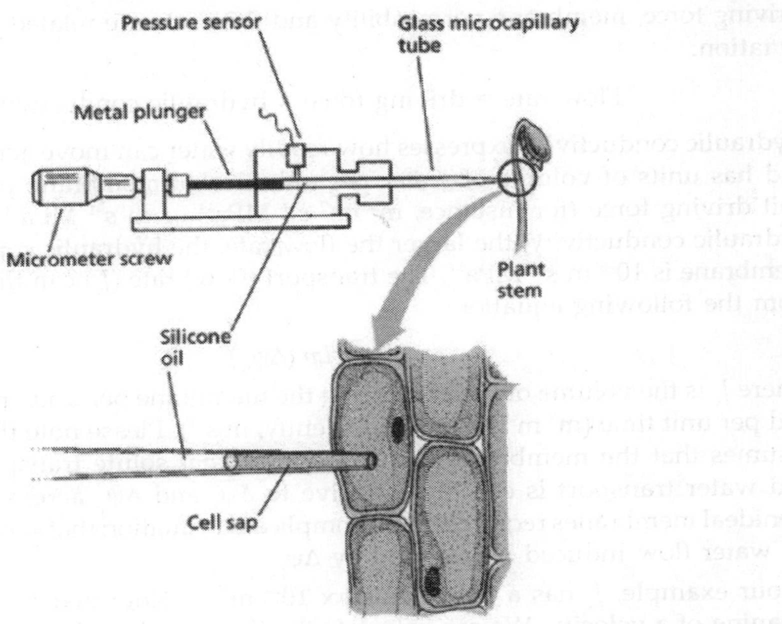

Figure 2.6 Diagram of the simplest pressure probe (not to scale). The primary advantage of this method over the one shown in Web Figure 3.5.E is that cell volume is minimally disturbed. Minimal disturbance is of great importance for the tiny cells that are typical of higher plants, in which loss of even a few picoliters (10^{-12} L) of fluid can substantially reduce turgor pressure.

returned to the tip and is held in a constant position, the initial volume of the cell is restored and the pressure inside the cell is exactly balanced by the pressure in the capillary. This pressure is measured by a pressure sensor in the device. Thus the hydrostatic pressure of individual cells may be measured directly.

This method has been used to measure ψ_p and other parameters of water relations in cells of both excised and intact tissues of a variety of plant species . The primary limitation of this method is that some cells are too small to measure. Furthermore, some cells tend to leak after being stabbed with the capillary and others plug up the tip of the capillary, thereby preventing valid measurements. The pressure probe has also been adapted to measure positive and negative values of ψ_p in the xylem . However, technical problems with cavitation limit the measurement of negative ψ_p by this technique.

2.3 Hydraulic Conductivity

Consider a cell with an initial water potential of –0.2 MPa, submerged in pure water. From this information we know that water will flow into the cell and that the driving force is $\Delta\psi_w$ = 0.2 MPa, but what is the initial rate of movement? The rate depends on permeability of the membrane to water, a property usually called the **hydraulic conductivity (*Lp*)** of the membrane .

Driving force, membrane permeability and flow rate are related by the following equation:

$$\text{Flow rate} = \text{driving force} \times \text{hydraulic conductivity}$$

Hydraulic conductivity expresses how readily water can move across a membrane and has units of volume of water per unit area of membrane per unit time per unit driving force (for instance, $m^3\ m^{-2}\ s^{-1}\ MPa^{-1}$ or $m\ s^{-1}\ MPa^{-1}$). The larger the hydraulic conductivity, the larger the flow rate. the hydraulic conductivity of the membrane is $10^{-6}\ m\ s^{-1}\ MPa^{-1}$. The transport (flow) rate (J_v) can then be calculated from the following equation:

$$J_v + Lp\ (\Delta\psi_w)$$

where J_v is the volume of water crossing the membrane per unit area of membrane and per unit time ($m^3\ m^{-2}\ s^{-1}$ or, equivalently, $m\ s^{-1}$). Please note that this equation assumes that the membrane is ideal—that is, that solute transport is negligible and water transport is equally sensitive to $\Delta\psi_s$ and $\Delta\psi_p$ across the membrane. Nonideal membranes require a more complicated equation that separately accounts for water flow induced by $\Delta\psi_s$ and by $\Delta\psi_p$.

In our example, J_v has a value of $0.2 \times 10^{-6}\ m\ s^{-1}$. Note that J_v has the physical meaning of a velocity. We can calculate the flow rate in volumetric terms ($m^3\ s^{-1}$) by multiplying J_v by the surface area of the cell.

The resulting value is the initial rate of water transport. As water is taken up, cell ψ_w increases and the driving force ($\Delta\psi_w$) decreases. As a result, water transport slows with time. As elaborated in p. 43 of the textbook, the rate approaches zero in an exponential manner.

Chapter **3**

Leaf Transpiration

3.1 Relationship between Water Potential and Soil Hydraulic Conductivity

Soil hydraulic conductivity is a function of the water potential of the soil. Conductivity measures the ease with which water moves through the soil. As water content (and hence the water potential) decreases, the hydraulic conductivity decreases drastically (Figure 3.1 note the logarithmic scales).

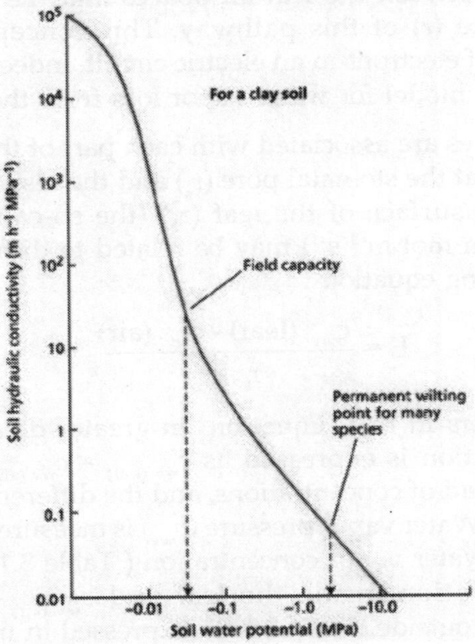

Figure 3.1. Soil hydraulic conductivity as a function of the water potential of the soil. Conductivity measures the ease with which water moves through the soil. The decrease in conductivity as the soil dries is due primarily to the movement of air into the soil to replace the water. As air moves in, the pathways for water flow between soil particles become smaller and more tortuous, and flow becomes more difficult. The overall shape of this curve is representative of many soils, but the shape for a particular soil may be influenced by the size distribution of its particles and by its organic matter content. The field capacity is the amount of water the soil is able to retain against gravitational forces. The permanent wilting point is the soil water potential value at which plants cannot regain turgor pressure even at night, in the absence of transpiration.(Lu *et al* ,1998, *J. Exp. Bot.* 49:453-460)

The decrease in conductivity as the soil dries is due primarily to the movement of air into the soil to replace the water. As air moves in, the pathways for water flow between soil particles become smaller and more tortuous, and flow becomes more difficult.

The overall shape of this curve is representative of many soils, but the shape for a particular soil may be influenced by the size distribution of its particles and by its organic matter content.

The field capacity (labeled on the curve) is the maximum amount of water the soil is able to retain against gravitational forces. The permanent wilting point (labeled on the curve) is the soil water potential value at which plants cannot regain turgor pressure even at night, in the absence of transpiration.

3.2 Leaf Transpiration

Transpiration from the leaf depends on two major factors: the **difference in water vapor concentration** between the leaf air spaces and the external air and, the **diffusional resisitance (*r*)** of this pathway. This concept of transpiration is analogous to the flow of electrons in an electric circuit. Indeed, an electrical analog is commonly used as a model for water vapor loss from the leaf .

In this analog, resistances are associated with each part of the pathway; the major ones are the resistance at the stomatal pore (r_s) and the resistance due to the layer of unstirred air at the surface of the leaf (r_b) (the so-called boundary layer). Transpiration rate (E, in mol m^{-2} s^{-1}) may be related to diffusional resistances (r, in s m^{-1}) by the following equation :

$$E = \frac{c_{wv}\,(\text{leaf}) - c_{wv}\,(\text{air})}{r_s + r_b}$$

Let's examine the factors in this Equation in greater detail. The difference in water vapor concentration is expressed as $c_{wv\,(leaf)} - c_{wv\,(air)}$. (Sometimes vapor pressures are used instead of concentrations, and the difference is called the **water vapor pressure deficit**. Water vapor pressure (p_{wv}) is measured in kilopascals (kPa) and is proportional to water vapor concentration (Table 3.1) where $c_{wv\,(leaf)}$ is the water vapor concentration inside the leaf and $c_{wv\,(air)}$ is the water vapor concentration of the air outside the leaf, both expressed in moles per cubic meter (mol m^{-3}). Resistance (r) is the inverse of conductance; that is, resistance = 1/ conductance. Thus, a high resistance is the same as a low conductance. Expression of this value in terms of resistances is preferred over expression in terms of conductances in some instances because resistances in series may be summed to calculate a total resistance, as in the above equation, whereas a similar calculation of conductances in series is more complicated. In the leaf, the total resistance is due mostly to the diffusion limitation imposed by the stomatal pore, but other parts of the pathway for water vapor loss, such as the boundary layer (which we will discuss shortly), may contribute significantly to r.

Table 3.1 : Relationships among water vapor concentration (C_{wv}), water vapor pressure (P_{wv}), relative humidity (RH), and water potential (ψ_w)

C_{wv} (mol m^{-3})	P_{wv} (kPa)	RH	ψ_w (MPa)[a]
0.961	2.34	1	0.00
0.957	2.33	0.996	-0.54
0.951	2.32	0.990	-1.36
0.923	2.25	0.960	-5.51
0.865	2.11	0.900	-14.2
0.480	1.17	0.500	-93.6
0	0	0	-infinity

Note : These data are for 20ºC

3.3 Stomatal Conductance and Yields of Irrigated Crops

In the last few decades, the breeding of agricultural crops for higher yields has been very successful. Breeders usually select for high-yielding genotypes empirically, without particular attention to specific plant traits that might be conducive to higher yields. However, comparing old, low-yielding lines of any crop with advanced, high-yielding lines shows clearly that many morphological, physiological, and biochemical traits have been altered by the intense selection pressures for higher yields. These changes indicate that selection for high-yielding genotypes has generated indirect selection pressures on the altered traits. If we exclude effects of genes that regulate the expression of two or more unrelated traits (pleiotropic genes), the study of high-yielding lines might reveal specific traits and genes associated with the higher yield. With this information, breeders could explicitly select for yield-enhancing traits to improve yields further.

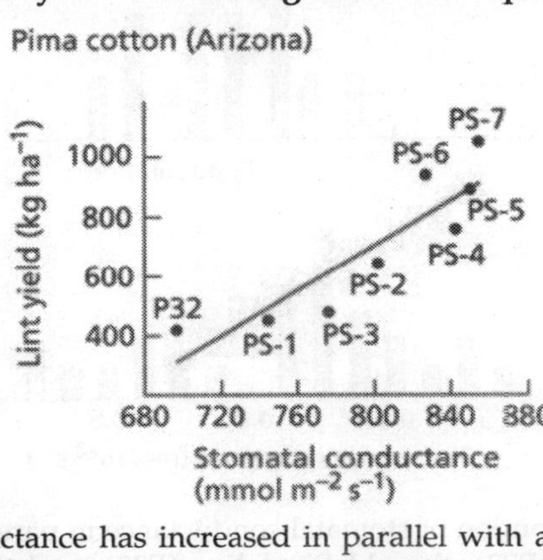

Figure 3.2. Stomatal conductance has increased in parallel with agronomic yields in irrigated Pima cotton (*Gossypium barbadense*) selected for higher yields at a high temperature. The figure shows the relationship between lint yield and stomatal conductance in a historical series of Pima cotton grown in Arizona. The abbreviations P32 and PS-1 through PS-7 designate successive commercial releases between 1949 and 1996. (Pl. Physiology, by Taiz & Zeigers, 2004)

Changes in plant traits associated with yield increases can be studied conveniently in a historical series of a crop. Historical series include successive commercial releases within a breeding program, with each member of the series representing an incremental increase in yield. In studies aimed at characterizing relationships between specific traits and higher yields, all the members of the series are grown together under the same conditions, and the traits of interest, such as plant architecture or photosynthetic rates, are measured in parallel with yields. Recent studies of this type focusing on historical series of Pima cotton (*Gossypium barbadense*) and bread wheat (*Triticum aestivum*) have shown a remarkable positive correlation between yield increases and increases in stomatal conductance (Fig. 3.1)

Pima cotton is a high-quality fiber cotton grown in hot environments under intensive irrigation. The historical series used in the Pima studies included eight members, released between 1949 and 1996, and encompassing a nearly threefold increase in lint yield. The yield increase attained in each commercial release was accompanied by a corresponding increase in stomatal conductance. Stomatal conductance increased by about 30 mmol m^{-2} s^{-1} for each 100 kg ha^{-1} increase in yield.

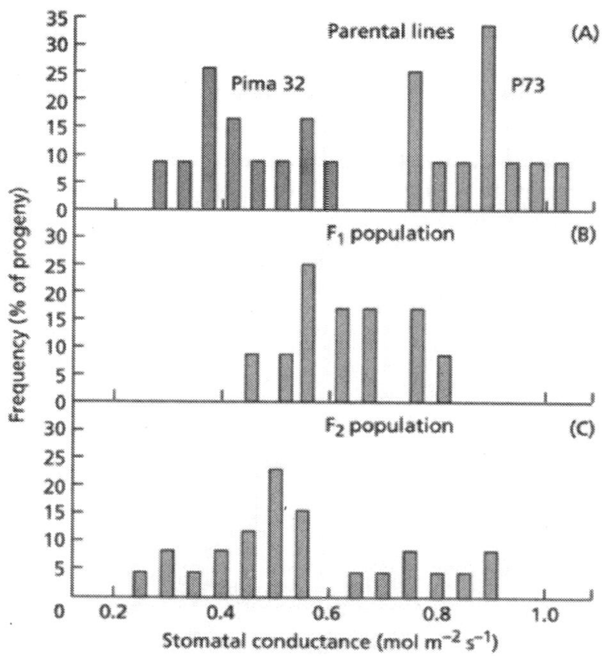

Figure 3.3. Frequency distribution of stomatal conductance in parental populations of the high-yielding Pima cotton line (P73), the old Pima line (P32), and their F_1 and F_2 progeny. (A) Nonoverlapping distribution of stomatal conductance in the two parental populations. (B) Distribution of stomatal conductance in an F_1 population from a cross between the two parents. (C) Distribution of stomatal conductance in an F_2 population derived from F_1 parents.(Percy *et al* ,1996, *Physiol. Plantarum*.96: 389-394)

Genetic crosses between low- and high-yielding Pima strains have shown that stomatal conductance in Pima cotton has a clear-cut genetic component (Figure 3.2.). The studies with stomata of Pima cotton are the first to show a clear-cut genetic regulation of stomatal conductance proper, of the stomatal response to temperature, and of proton pumping rates of isolated guard cells.

In the absence of explicit selection for higher stomatal conductance in the Pima breeding program, what could be the nature of the indirect selection pressures that caused the increases in conductance paralleling the increases in yield? Higher stomatal conductance increases CO_2 diffusion into the leaf and favors higher photosynthetic rates . Higher photosynthetic rates could in turn favor a higher biomass and higher crop yields. Advanced Pima lines show a higher photosynthetic capacity than older, low-yielding lines, but photosynthetic rates measured in the same leaves used to measure stomatal conductance were not positively correlated with yields . Thus, higher stomatal conductance appears to favor higher yields by a mechanism not directly related to photosynthesis.

Evapotranspiration at the leaf surface lowers leaf temperature , and higher stomatal conductance enhances this leaf cooling. Optimal daytime temperatures for growth, photosynthesis, and reproduction in Pima cotton are below 30°C, while afternoon air temperature in thè hot Pima-growing areas often exceeds 40°C. Evaporative cooling of the leaves thus reduces the gap between optimal growth temperatures and air temperature, and provides heat resistance. Studies with the Pima historical series showed that, because of leaf cooling, leaf temperatures were several degrees lower than air temperature, and leaf and canopy temperature were lower in the advanced, high-yielding lines than in low-yielding lines.

The same relationship among yields, stomatal conductance, and leaf temperature was found in a historical series of bread wheat grown in the warm Yaqui Valley

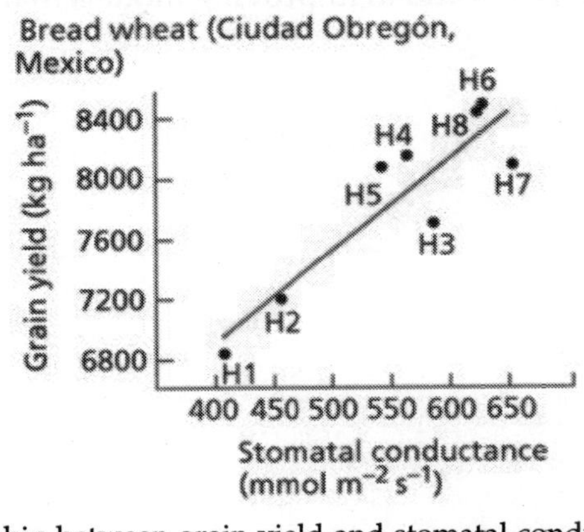

Figure 3.4 The relationship between grain yield and stomatal conductance in a historical series of semidwarf bread wheat grown in Ciudad Obregón, Mexico. The abbreviations H1 through H8 designate successive commercial lines released by the International Maize and Wheat Improvement Center between 1962 and 1988. (Lu *et al*, 1998, 5. *Exp Bot.* 49:453-460)

of northwestern Mexico (Figure 3.3) . These studies indicate that selection for higher yields in irrigated crops grown at high temperatures imposes indirect selection pressures for high stomatal conductance that lowers leaf temperature and appears to reduce deleterious effects of heat stress on critical flowering and fruiting stages, thus resulting in higher crop yields.

If selection for higher yields generates indirect selection pressures for higher stomatal conductance, can breeders explicitly select for higher conductance and thus obtain lines with higher yields? Studies with Pima cotton have shown that selection of high-conductance F_2 progeny from a cross between high- and low-conductance parents produces high-conductance F_4 lines with higher lint yields than low-conductance lines have . These experiments indicate that high stomatal conductance could be used as a selection trait for high yields in irrigated crops grown at high temperature

The observed increases in stomatal conductance in crops grown at high temperature also have valuable implications for models developed to predict global climate changes, which are expected to occur because of increases in atmospheric CO_2 . Under atmospheric CO_2 concentrations that are twice as high as the present ones, evapotranspiration would decrease over the continents and air temperatures would increase significantly over the tropical land masses, amplifying the changes resulting from atmospheric radiative effects . However, these models have yet to take into consideration the effect of higher temperatures on the stomatal control of evapotranspiration. Incorporation of the stomatal response to temperature in climate change models is likely to improve the accuracy of the predictions generated by the models. These studies illustrate both the complex interactions between physical and biological factors in the biosphere, and the advantage of a thorough understanding of the physiological properties of plants for efforts aimed at increasing agronomic yields and at improving models that predict global climate changes.

Chapter 4

Essential Minerals and their Deficiency Symptoms

4.1 Introduction

Visual nutrient deficiency symptoms can be a very powerful diagnostic tool for evaluating the nutrient status of plants. One should keep in mind, however, that a given individual visual symptom is seldom sufficient to make a definitive diagnosis of a plant's nutrient status. Many of the classic deficiency symptoms such as tip burn, chlorosis and necrosis are characteristically associated with more than one mineral deficiency and also with other stresses that by themselves are not diagnostic for any specific nutrient stress. However, their detection is extremely useful in making an evaluation of nutrient status. In addition to the actual observations of morphological and spectral symptoms, knowing the location and timing of these symptoms is a critical aspect of any nutrient status evaluation. Plants do not grow in isolation, they are part of the overall environment and as such they respond to environmental changes as that affect nutrient availability. Also, plants do influence their environment and can contribute to environmental changes, which in turn can affect the nutrient status of the plant.

4.2 Visual Symptoms

Stresses such as salinity, pathogens, and air pollution induce their own characteristic set of visual symptoms. Often, these symptoms closely resemble those of nutrient deficiency. Pathogens often produce an interveinal chlorosis, and air pollution and salinity stress can cause tip burn. Although at first these symptoms might seem similar in their general appearance to nutrient deficiency symptoms, they do differ in detail and/or in their overall developmental pattern. Pathological symptoms can often be separated from nutritional symptoms by their distribution in a population of affected plants. If the plants are under a nutrient stress, all plants of a given type and age in the same environment tend to develop similar symptoms at the same time. However if the stress is the result of pathology, the development of symptoms will have a tendency to vary between plants until a relatively advanced stage of the pathology is reached.

4.3 Effects of Environment

Plants remove substantial amounts of nutrients from the soil during their normal growth cycle and many long-term environmental changes occur as a result of this process. Effects on the soil go considerably beyond the straight removal or depletion of nutrients. Charge balance must be maintained in the plant-soil system during nutrient uptake. Charge balance is usually achieved by the excretion of proton and/or hydroxyl ions by the plant to replace the absorbed nutrient cations or anions. For example when plants are fertilized with ammonia, they acquire most of their nitrogen in the form of the ammonium cation, rather than from the usual nitrate anion. Because nitrate is the only anion used by the plant in large amounts, the net result of this change is that during normal nutrient uptake the proton excretion will far exceed that of hydroxyl ions. In the case of vigorously growing plants, the amount of excreted protons can be sufficiently large as to decrease the pH of the soil by several pH units. Changes in soil pH of such magnitude can have large implications for a number of soil processes such as soil structure, nutrient availability and leaching of nutrients. The immediate effect on the soil may be favorable for some plants, especially acid-loving plants, in that it tends to make iron more available. However, in the long run, lowering the soil pH can be deleterious to plants in that the availability of nutrients will change. A lower soil pH will allow micronutrients to be more readily leached from the soil profile, eventually resulting in deficiencies of nutrients such as Cu and Zn. Additionally, when the pH of the soil drops much below pH 5, the solubility of Al and Mn can increase to such an extent as to become toxic to most plant growth.

Plants are often thought of as passive in relation to the environment. However this is not always a valid assumption; for there are many plants that clearly manipulate their environment in a fashion that tends to makes certain nutrients more readily available. For example, iron is a limiting nutrient in many agricultural areas, but it comprises about 3% of the average soil which, if available, would be far in excess of the needs of the average plant. Some plants actively excrete protons, and the resulting decrease in pH increases the solubility of iron in their environment. In addition, other plants excrete phytosi-derophores that chelate the soil iron rendering it a more available form for the plants.

4.4 Acute and Chronic Deficiency Symptoms

At first glance, it would appear that the distinction of deficiency symptoms for the 13 known essential mineral nutrients should be relatively simple. But such an assumption is incorrect. In fact, the deficiency symptoms are quite complex because each nutrient has a number of different biological functions and each function may have an independent set of interactions with a wide range of environmental parameters. In addition, the expression of these symptoms varies for acute or chronic deficiency conditions. **Acute deficiency** occurs when a nutrient is suddenly no longer available to a rapidly growing plant. **Chronic deficiency** occurs when there is a limited but continuous supply of a nutrient, at a rate that is insufficient to meet the growth demands of the plant.

Most of the classic deficiency symptoms described in textbooks are characteristic of acute deficiencies. The most common symptoms of low-grade, chronic deficiencies are a tendency towards darker green leaves and stunted or slow growth. Typically most published descriptions of deficiency symptoms arise from experiments conducted in greenhouses or growth chambers where the plants are grown in hydroponics or in media where the nutrients are fully available. In these conditions, nutrients are readily available while present, but when a nutrient is depleted, the plant suddenly faces an acute deficiency. Thus, hydroponic studies favor the development of acute deficiencies.

In experiments designed to study **micronutrient** deficiency symptoms, micronutrients are usually omitted from the nutrient solution. Micronutrients are often present in the seed or as contaminants in the environment, so a plant of adequate size will exhaust these trace amounts of micronutrient and develop characteristic acute deficiency systems. When deficiency symptoms of **macronutrients** are sought, the macronutrient is removed suddenly from a suitable sized rapidly growing plant. Alternatively the plant can initially be given a one-time supply of the nutrient that is sufficient for a limited amount of growth. Because macronutrients are continuously required in relatively large amounts by rapidly growing plants, the available nutrients will be rapidly depleted, resulting in an acute deficiency.

In natural systems, the plant encounters many degrees and types of stresses that result in different types of symptoms occurring over time. Perhaps the most common nutrient deficiency in natural environments is the case of a limited nutrient supply that is continuously renewed at a low rate from soil weathering processes. In such cases, the limited nutrient availability results in chronic nutrient deficiency symptoms.

4.5 Symptom Development is Affected by Mobility

For very mobile nutrients such as nitrogen and potassium, deficiency symptoms develop predominantly in the older and mature leaves. This is a result of these nutrients being preferentially mobilized during times of nutrient stress from the older leaves to the newer leaves near the growing regions of the plant. Additionally, mobile nutrients newly acquired by the roots are also preferentially translocated to new leaves and the growing regions. Thus old and mature leaves are depleted of mobile nutrients during times of stress while the new leaves are maintained at a more favorable nutrient status.

The deficiency symptoms of very weakly mobile nutrients such as calcium, boron, and iron are first displayed in the growing regions and new leaves while the old leaves remain in a favorable nutrient status. (This assumes that these plants started with sufficient nutrient, but ran out of nutrient as they developed). In plants growing very slowly for reasons other than nutrition (such as low light) a normally limiting supply of a nutrient could, under these conditions, be sufficient for the plant to slowly develop, may be even without symptoms. This type of development is likely to occur in the case of weakly mobile nutrients because excess nutrients in the older leaves will eventually be mobilized to supply newly developing tissues. In contrast, a plant with a similar supply that is growing rapidly will develop severe deficiencies in the actively growing tissue such as leaf edges and the growing region of the plant. A classic example of this is calcium deficiency in vegetables such as lettuce where symptoms develop on the leaf margins (tip burn) and the growing region near the meristems. The maximal growth rate of lettuce is often limited by the internal translocation rate of calcium to the growing tissue rather than from a limited nutrient supply in the soil.

In case of mobile nutrients such as sulfur and magnesium deficiency symptoms are normally seen over the entire plant. However the growth rate and rate of nutrient availability can make a considerable difference on the locations at which the symptoms develop. If the nutrient supply is marginal compared to the growth rate, symptoms will appear on the older tissue, but if the nutrient supply is very low compared to the growth rate, or the nutrient is totally depleted, the younger tissue will become deficient first.

4.6 Induced Deficiencies

When the observed symptoms are the direct result of a nutrient deficiency, the actions needed for correction are relatively straight-forward. However symptoms are often the result of interactions with other environmental factors limiting the availability of the nutrient whose symptoms are expressed. The classic instance is that of iron deficiency induced by an excess of heavy metals in the environment. Transition metals such as Cu, Zn Cr and Ni compete with Fe and each other for plant uptake. Competition for uptake is not specific to Fe and heavy metals but is true for all mineral nutrients that are chemically similar and have similar uptake mechanisms. For example if the availability of Cu or Zn is relatively less than that of Fe, then excessive concentrations of some other metal such as Ni or Cr will induce a deficiency of one of these nutrients rather than Fe. In the case of the macronutrients, excessive amounts of Mg will compete with K for uptake and can possibly induce a K deficiency. The barrenness of serpentine soils is the result of such competition, with the high Mg of these soils inducing a Ca deficiency. The toxicity of a low pH soil is another example of a basic nutrient deficiency. Low pH has a two-fold effect on soil nutrients: It enhances the leaching of cations, reducing their availability in the soil, and the relatively abundant protons in the soil compete with Ca and other cations for uptake. Thus, nutrient deficiencies can be induced by a number of different mechanisms often working in concert to limit the availability of a nutrient.

4.7 Nutrient Use Efficiency

Although all plants of the same species respond similarly to nutrient stress, plants of similar species will often show significant differences in their nutrient use efficiency. This results from differences in growth rate, root distribution, phase of development, and efficiency of nutrient uptake and utilization. This implies that in any given location, plants from one species may become nutrient-deficient, while those from another species growing in the same environment right next to them, may not show any deficiency symptoms.

Growth rate also affects nutrient status. When the nutrient supply is barely inadequate for growth under existing environmental conditions, many plants adjust their growth rate to match that supported by the available nutrient supply without displaying typical visual deficiency symptoms.

Agricultural systems differ from natural systems in that crop plants have been selected primarily for rapid growth under low stress conditions. This rapid growth rate results in a high nutrient demand by these plants and a higher incidence of nutrient deficiency unless supplemental fertilizers are supplied. It is not uncommon to find agricultural crops showing severe signs of nutrient stress, with native plants growing in the same area showing little or no indication of nutrient stress. In agriculture systems chronic deficiency symptoms develop mostly in crops with little or limited fertilization. Acute nutrient deficiency symptoms most often occur when new crops with a higher nutrient demand are introduced, or less productive lands are brought under cultivation for the production of rapidly growing crop plants.

4.8 Nutrient Status

Not all tissues of a plant are at the same nutrient status during times of stress. Leaves on the same plant that are exposed to different environmental conditions, (such as light), or those of different ages may have considerable differences in nutrient status. Mineral nutrients are for the most part acquired by the roots and translocated throughout the plant. The distance of any part of the plant to the roots will influence nutrient availability, particularly in the case of the less mobile nutrients. In plants recovering from nutrient deficiency, the root and conductive tissues recover first. For example, in the case of recovery from Fe deficiency, it is common to see the veins re-green while the interveinal tissue remains chlorotic and Fe-deficient.

In order to maintain rapid, optimal growth, all plant tissues must have a favorable nutrient status. Although a plant may be marginally low in a number of nutrients, only one nutrient at a time will limit overall growth. However, if the supply of that limiting nutrient is increased even slightly, the resulting increase in growth will increase the demand for all other nutrients and another nutrient, the next lowest in availability, will become limiting.

4.9 Measurment of Nutrient Status

Although visual diagnostic symptoms are an extremely valuable tool for the rapid evaluation of the nutrient status of a plant, they are only some of the tools available.

Other major tools include microscopic studies, spectral analysis, and tissue and soil analysis. These methods all vary in their precision, rapidity and their ability to predict future nutrient status. Because of the close interaction between plant growth and the environment, all predictions of future nutrient status must make assumptions about how the environment will change in that time frame.

The principle advantage of visual diagnostic symptoms is that they are readily obtained and provide an immediate evaluation of nutrient status. Their main drawback is that the visual symptoms do not develop until after there has been a major effect on yield, growth and development.

Tissue analysis is nutrient-specific but relatively slow; tissues must be sampled, processed and analyzed before the nutrient status can be determined. An analysis of the mineral nutrient content of selected plants tissues, when compared against Critical Level values (which are available for most crop plants, can be used to evaluate the plant nutrient status at the time of sampling with a relatively high degree of confidence and can be extrapolated to project nutrient status at harvest. Soil analysis is similar to tissue analysis but evaluates the potential supplying power of the soil instead of plant nutrient status. Plant analysis provides information as to what the plant needs, while soil analysis provides information about the status of the nutrient supply.

Spectral analysis of nutrient status is still in its infancy and is presently used primarily in the inventory of global resources and in specialized studies. Microscopic studies are most valuable in looking at the physiological aspects of nutrient stress rather than the evaluation of plant nutrient status on a whole plant or crop basis.

4.10 Deficiency Symptom of Macro and Micronutrients

It is unusual to find any one leaf or even one plant that displays the full array of symptoms that are characteristic of a given deficiency. It is thus highly desirable to know how individual symptoms look, for it is possible for them to occur in many possible combinations on a single plant. Most of the terms used below in the description of deficiency symptoms are reasonably self evident; a few however have a distinct meaning in the nutrient deficiency field. For example, the term chlorotic, which is a general term for yellowing of leaves through the loss of chlorophyll, cannot be used without further qualification because there may be an overall chlorosis as in nitrogen deficiency, interveinal, as in iron deficiency, or marginal, as in calcium deficiency. Another term used frequently in the description of deficiency symptoms is necrotic, a general term for brown, dead tissue. This symptom can also appear in many varied forms, as is the case with chlorotic symptoms.

Nutrient deficiency symptoms for many plants are similar, but because of the large diversity found in plants and their environments there is a range of expression of symptoms. Because of their parallel veins, grasses and other monocots generally display the affects of chlorosis as a series of stripes rather than the netted interveinal chlorosis commonly found in dicots. The other major difference is that the marginal necrosis or chlorosis found in dicots is often expressed as tip burn in monocots.

Magnesium. The Mg-deficient leaves (Figure 4.1) show advanced interveinal chlorosis, with necrosis developing in the highly chlorotic tissue. In its advanced form, magnesium deficiency may superficially resemble potassium deficiency. In the case of magnesium deficiency the symptoms generally start with mottled chlorotic areas developing in the interveinal tissue. The interveinal laminae tissue tends to expand proportionately more than the other leaf tissues, producing a raised puckered surface, with the top of the puckers progressively going from chlorotic to necrotic tissue. In some plants such as the Brassica (The mustard family, which includes vegetables such as broccoli, brussel sprouts, cabbage, cauliflower, collards, kale, kohlrabi, mustard, rape, rutabaga and turnip.), tints of orange, yellow, and purple may also develop.

Figure 4.1 Magnesium deficiency symptoms in tomato. (Epstein and Bloom, 2004, *Plant Nutrition*, Sinauer Associates, Sunderland, MA)

Manganese. These leaves (Figure 4.2.) show a light interveinal chlorosis developed under a limited supply of Mn. The early stages of the chlorosis induced by manganese deficiency are somewhat similar to iron deficiency. They begin with a light chlorosis of the young leaves and netted veins of the mature leaves especially when they are viewed through transmitted light. As the stress increases, the leaves take on a gray metallic sheen and develop dark freckled and necrotic areas along the veins. A purplish luster may also develop on the upper surface of the leaves. Grains such as oats, wheat, and barley are extremely susceptible to manganese deficiency. They develop a light chlorosis along with gray specks which elongate and coalesce, and eventually the entire leaf withers and dies.

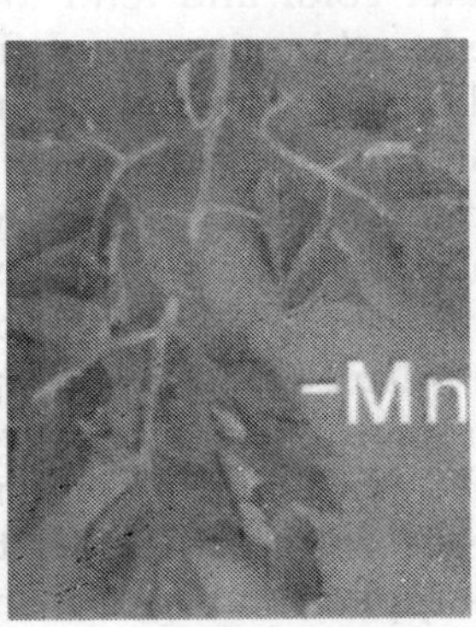

Figure 4.2 Manganese

Molybdenum. These leaves (Figure 4.3.) show some mottled spotting along with some interveinal chlorosis. An early symptom for molybdenum deficiency is a general overall chlorosis, similar to the symptom for nitrogen deficiency but generally without the reddish coloration on the undersides of the leaves. This results from the requirement for molybdenum in the reduction of nitrate, which needs to be reduced prior to its assimilation by the plant . Thus, the initial symptoms of molybdenum deficiency are in fact those of nitrogen deficiency. However, molybdenum has other metabolic functions within the plant, and hence there are deficiency symptoms even when reduced nitrogen is available. In the case of cauliflower, the lamina of the new leaves fail to develop, resulting in a characteristic whiptail appearance. In many plants there is an upward cupping of the leaves and mottled spots developing into large interveinal chlorotic areas under severe deficiency. At high concentrations, molybdenum has a very distinctive toxicity symptom in that the leaves turn a very brilliant orange.

Figure 4.3. Molybdenum deficiency symptoms in tomato. (Epstein and Bloom, 2004)

Nitrogen. The chlorotic symptoms (Figure 4.4) shown by this leaf resulted from nitrogen deficiency. A light red cast can also be seen on the veins and petioles. Under nitrogen deficiency, the older mature leaves gradually change from their normal characteristic green appearance to a much paler green. As the deficiency progresses these older leaves become uniformly yellow (chlorotic). Leaves approach a yellowish white color under extreme deficiency. The young leaves at the top of the plant maintain a green but paler color and tend to become smaller in size. Branching is reduced in nitrogen deficient plants resulting in short, spindly plants. The yellowing in nitrogen deficiency is uniform over the entire leaf including the veins. However in some instances, an interveinal necrosis replaces the chlorosis commonly found in many plants. In some plants the underside of the leaves and/or the petioles and midribs develop traces of a reddish or purple color. In some plants this coloration can be quite bright. As the deficiency progresses, the older leaves also show more of a tendency to wilt under mild water stress and become senescent much earlier than usual. Recovery of deficient plants to applied nitrogen is immediate (days) and spectacular.

Figure 4.4. Nitrogen deficiency symptoms in tomato.(Epstein and Bloom, 2004)

Phosphorus. These phosphorus-deficient leaves (Figure 4.5) show some necrotic spots. As a rule, phosphorus deficiency symptoms are not very distinct and thus difficult to identify. A major visual symptom is that the plants are dwarfed or stunted. Phosphorus deficient plants develop very slowly in relation to other plants growing under similar environmental conditions but without phosphorus deficiency. Phosphorus deficient plants are often mistaken for unstressed but much younger plants. Some species such as tomato, lettuce, corn and the brassicas develop

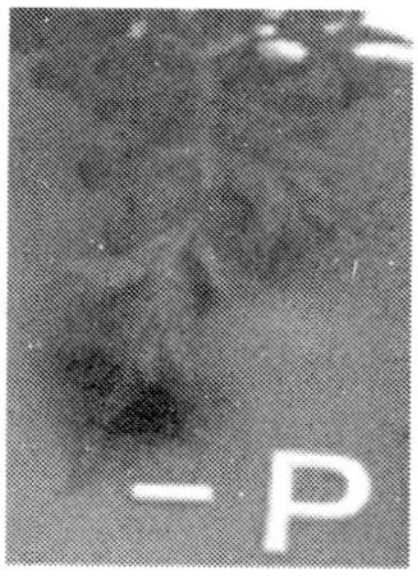

Figure 4.5. Phosphorus deficiency symptoms in tomato. (Epstein and Bloom, 2004)

a distinct purpling of the stem, petiole and the under sides of the leaves. Under severe deficiency conditions there is also a tendency for leaves to develop a blue-gray luster. In older leaves under very severe deficiency conditions a brown netted veining of the leaves may develop.

Figure 4.6 Sulfur deficiency symptoms in tomato. (Epstein and Bloom, 2004)

Sulfur. This leaf (Figure 4.6) shows a general overall chlorosis while still retaining some green color. The veins and petioles show a very distinct reddish color. The visual symptoms of sulfur deficiency are very similar to the chlorosis found in nitrogen deficiency. However, in sulfur deficiency the yellowing is much more uniform over the entire plant including young leaves. The reddish color often found on the underside of the leaves and the petioles has a more pinkish tone and is much less vivid than that found in nitrogen deficiency. With advanced sulfur deficiency brown lesions and/or necrotic spots often develop along the petiole, and the leaves tend to become more erect and often twisted and brittle.

Figure 4.7. Zinc deficiency symptoms in tomato. (Epstein and Bloom, 2004)

Zinc. This leaf (Figure 4.7) shows an advanced case of interveinal necrosis. In the early stages of zinc deficiency the younger leaves become yellow and pitting develops in the interveinal upper surfaces of the mature leaves. Guttation is also prevalent. As the deficiency progress these symptoms develop into an intense interveinal necrosis but the main veins remain green, as in the symptoms of recovering iron deficiency. In many plants, especially trees, the leaves become very small and the internodes shorten, producing a rosette like appearance.

Boron. These boron-deficient leaves (Figure 4.8) show a light general chlorosis. The tolerance of plants to boron varies greatly, to the extent that the boron concentrations necessary for the growth of plants having a high boron requirement may be toxic to plants sensitive to boron. Boron is poorly transported in the phloem of most plants, with the exception of those plants that utilize complex sugars, such as sorbitol, as transport metabolites. In a recent study, tobacco plants engineered to synthesize sorbitol were shown to have increased boron mobility, and to better tolerate boron deficiency in the soil.

In plants with poor boron mobility, boron deficiency results in necrosis of meristematic tissues in the growing region, leading to loss of apical dominance and the development of a rosette condition. These deficiency symptoms are similar to those caused by calcium deficiency. In plants in which boron is readily transported in the phloem, the deficiency symptoms localize in the mature tissues, similar to those of nitrogen and potassium. Both the pith and the epidermis of stems may be affected, often resulting in hollow or roughened stems along with necrotic spots on the fruit. The leaf blades develop a pronounced crinkling and there is a darkening and crackling of the petioles often with exudation of syrupy material from the leaf blade. The leaves are unusually brittle and tend to break easily. Also, there is often a wilting of the younger leaves even under an adequate water supply, pointing to a disruption of water transport caused by boron deficiency.

Figure 4.8. Boron deficiency symptoms in tomato. (Epstein and Bloom, 2004)

Calcium. These calcium-deficient leaves (Figure 4.9) show necrosis around the base of the leaves. The very low mobility of calcium is a major factor determining the expression of calcium deficiency symptoms in plants. Classic symptoms of calcium deficiency include blossom-end rot of tomato (burning of the end part of tomato fruits), tip burn of lettuce, blackheart of celery and death of the growing regions in many plants. All these symptoms show soft dead necrotic tissue at rapidly growing areas, which is generally related to poor translocation of calcium to the tissue rather than a low external supply of calcium. Very slow growing plants with a deficient supply of calcium may re-translocate sufficient calcium from older leaves to maintain growth with only a marginal chlorosis of the leaves. This ultimately results in the margins of the leaves growing more slowly than the rest of the leaf, causing the leaf to cup downward. This symptom often progresses to the point where the petioles develop but the leaves do not, leaving only a dark bit of necrotic tissue at the top of each petiole. Plants under chronic calcium deficiency have a much greater tendency to wilt than non-stressed plants.

Figure 4.9 Calcium deficiency symptoms in tomato. (Epstein and Bloom, 2004)

Chloride. These leaves (Figure 4.10) have abnormal shapes, with distinct interveinal chlorosis. Plants require relatively high chlorine concentration in their tissues. Chlorine is very abundant in soils, and reaches high concentrations in saline areas, but it can be deficient in highly leached inland areas. The most common symptoms of chlorine deficiency are chlorosis and wilting of the young leaves. The chlorosis occurs on smooth flat depressions in the interveinal area of

Figure 4.10 Chloride deficiency symptoms in tomato.(Epstein and Bloom, 2004)

the leaf blade. In more advanced cases there often appears a characteristic bronzing on the upper side of the mature leaves. Plants are generally tolerant of chloride, but some species such as avocados, stone fruits, and grapevines are sensitive to chlorine and can show toxicity even at low chloride concentrations in the soil.

Fgure 4.11 Copper deficiency symptoms in tomato. (Epstein and Bloom, 2004)

Copper. These copper-deficient leaves (Figure 4.11) are curled, and their petioles bend downward. Copper deficiency may be expressed as a light overall chlorosis along with the permanent loss of turgor in the young leaves. Recently matured leaves show netted, green veining with areas bleaching to a whitish gray. Some leaves develop sunken necrotic spots and have a tendency to bend downward. Trees under chronic copper deficiency develop a rosette form of growth. Leaves are small and chlorotic with spotty necrosis.

Figure 4.12 Iron deficiency symptoms in tomato.

Iron. These iron-deficient leaves (Figure 4.12) show strong chlorosis at the base of the leaves with some green netting. The most common symptom for iron deficiency starts out as an interveinal chlorosis of the youngest leaves, evolves into an overall chlorosis, and ends as a totally bleached leaf. The bleached areas often develop necrotic spots. Up until the time the leaves become almost completely white they will recover upon application of iron. In the recovery phase the veins are the first to recover as indicated by their bright green color. This distinct venial re-greening observed during iron recovery is probably the most recognizable symptom in all of classical plant nutrition. Because iron has a low mobility, iron deficiency symptoms appear first on the youngest leaves. Iron deficiency is strongly associated with calcareous soils and anaerobic conditions, and it is often induced by an excess of heavy metals.

Figure 4.13 Potassium deficiency symptoms in tomato. (Epstein and Bloom, 2004)

Potassium. Some of these leaves (Figure 4.13) show marginal necrosis (tip burn), others at a more advanced deficiency status show necrosis in the interveinal spaces between the main veins along with interveinal chlorosis. This group of symptoms is very characteristic of K deficiency symptoms.

The onset of potassium deficiency is generally characterized by a marginal chlorosis progressing into a dry leathery tan scorch on recently matured leaves. This is followed by increasing interveinal scorching and/or necrosis progressing from the leaf edge to the midrib as the stress increases. As the deficiency progresses, most of the interveinal area becomes necrotic, the veins remain green and the leaves tend to curl and crinkle. In some plant such as legumes and potato, the initial symptom of deficiency is white speckling or freckling of the leaf blades. In contrast to nitrogen deficiency, chlorosis is irreversible in potassium deficiency, even if potassium is given to the plants. Because potassium is very mobile within the plant, symptoms only develop on young leaves in the case of extreme deficiency. Potassium deficiency can be greatly alleviated in the presence of sodium but the resulting sodium-rich plants are much more succulent than a high potassium plant. In some plants over 90% of the required potassium can be replaced with sodium without any reduction in growth.

4.11 Observing Roots below Ground

Studying root growth below the soil surface requires a means of observing the root system directly. As early as 1873, the German botanist Julius von Sachs studied root systems by using simple soil-filled boxes with one glass wall. Since that time, facilities for studying root growth in the soil have become much more complex. Large laboratories with subterranean chambers for the observation of root growth have been constructed and allow the analysis of root growth while the aerial parts of the plant are exposed to natural field conditions. These laboratories are called **rhizotrons** (from the Greek *rhizos*, meaning "root," and *tron*, meaning "a device for studying").

In a rhizotron, roots grow in glass-walled chambers that line underground passageways. Grid lines on the glass walls indicate soil depth. Details of root morphology (root size and distribution) under natural growing conditions can be observed with specially designed microscopes, mounted adjacent to the glass walls of the root chambers. In addition, the growth of the roots over a period of time can be measured with time-lapse photography.

Because rhizotrons are expensive to construct and maintain, they have been largely supplanted by minirhizotrons or root periscopes. Minirhizotrons are transparent plastic tubes previously buried at an angle in the soil near the plants to be observed. An optical device such as a tilted mirror with a magnifying lens or a miniature videocamera is inserted into a tube to monitor roots growing along its surface (Figure 4.14). This device provides information on rooting density in the bulk soil, as well as on root growth and phenology. Plant physiologists use such information in conjunction with changes in nutrient and water levels in the soil to assess root activity through the soil profile.

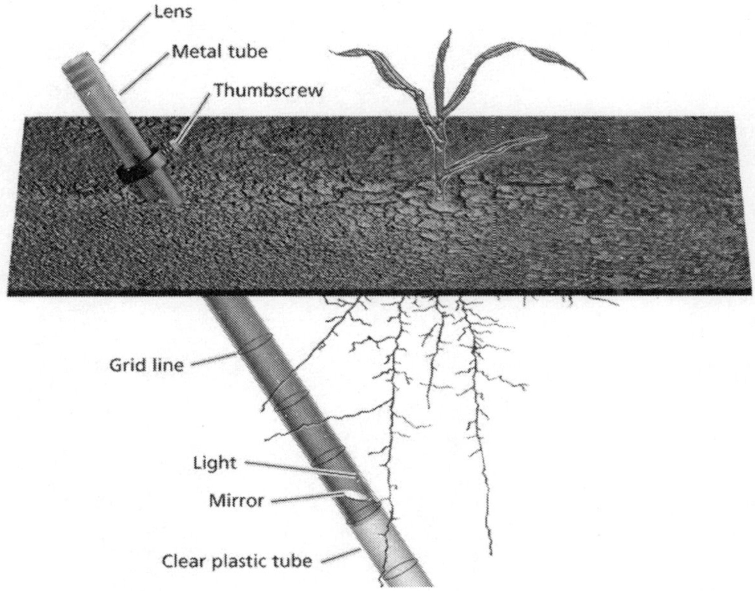

Figure 4.14 Drawing of a minirhizotron installed near an oat seedling. A metal tube that slides within a clear plastic tube is held in place by a thumbscrew. The metal tube contains the optics for viewing roots growing along the surface of the plastic tube. A dark grid scribed onto the surface of the plastic tube assists in positioning the metal tube. The optics include lenses, mirror, and lights. Many modern configurations use a miniature television camera instead of lenses to obtain the image.

Chapter **5**

Transport of Solute

5.1 Introduction

Plant cells are separated from their environment by a plasma mem brane that is only two lipid molecules thick. This thin sheet separates a relatively constant internal milieu from a highly variable external environment. It must accommodate and promote continual inward and outward traffic of molecules and ions, as the cell takes up nutrients and exports wastes. As the cell's only contact with the environment, the plasma membrane must also relay information about its physical environment, about molecular signals from other cells, and about the presence of invading pathogens. To accomplish these tasks, the plasma membrane must provide second-by-second control of the rate of movement of each of a multitude of solutes. The same is true of the internal membranes that separate the various compartments within each cell.

Molecular and ionic movement between different compartments in biological systems is known as **transport**. Transport can be considered at the level of individual cellular membranes, or between a plant and its environment, or from one part of the plant to another. However, such higher order transport is generally driven and controlled by transport at the cellular level. For example, the transport of sucrose from leaf to root through the phloem, referred to as *translocation,* may be driven and regulated by membrane transport into the phloem cells of the leaf, and from the phloem to the storage cells of the root.

When permeabilities and ion gradients are known, it is possible to calculate a diffusion potential for the membrane from the Goldman equation. In most plant cells, K^+ has both the greatest internal concentration and the highest membrane permeability, so the diffusion potential may approach the Nernst potential for K^+. Under some conditions, this phenomenon is observed. In plants and fungi, however, the experimentally measured membrane potentials are often much more negative than those calculated from the Goldman equation. For example, cells in stem and roots of young seedlings generally have membrane potentials of -130 to -110 mV, whereas their calculated diffusion potentials are usually only -80 to -50 mV. Thus, in addition to the diffusion potential, the membrane potential has second component. The excess voltage is provided by the plasma membrane electrogenic H^+ ATPase.

Whenever an ion moves into or out of a cell without being balanced by countermovement of an ion of opposite charges, a voltage is created across the membrane. Any active transport mechanism that results in the movement of a net electric charge will tend to move the membrane potential away from the value predicted by the Goldman equation. Such a transport mechanism is called an electrogenic pump and is common in living cells.

The energy required for active transport is often provided by the hydrolysis of ATP. In plants we can demonstrate the dependence of the membrane potential on ATP by observing the effect of cyanide (CN^-) on the membrane potential. Cyanide rapidly poisons the mitochondria, and the cell's ATP synthesis is inhibited, the membrane potential falls to the level of the Goldman diffusion potential, which is due to the passive movements of K^+, Cl^-, and Na^+. Thus, the membrane potentials of plant cell have two components: a diffusion potential and a component resulting from electrogenic ion transport (transport that results in the generation of a membrane potential). When cyanide inhibits electrogenic ion transport, the pH of the external medium increases while the cytosol becomes acidic because H^+ remains inside the cell. This is one piece of evidence that the active transport of H^+ out of the cell electrogenic.

As discussed earlier, a change in the membrane potential caused by an electrogenic pump will change the driving forces for diffusion of all ions that cross the membrane. For example, the outward transport of H^+ can create a driving force for the passive diffusion of K^+ into the cell. H^+ is transported electrogenically across the cell membrane not only in plants but also in bacteria, algae, fungi, and some animal cells, such as those of the kidney epithelia. ATP synthesis in mitochondria and chloroplasts also depends on an H^+-ATPase. In mitochondria and chloroplasts, this transport protein is sometimes called on ATP synthase because it forms ATP rather than hydrolyzing it. The mitochondrial ATP synthase is structurally different from the H^+-ATPases located in plasma membranes. The structure and function of membrane proteins involved in active and passive transport in plant cells is discussed as under.

5.2 Membrane Transport Proteins

Artificial membranes have been used extensively to study the permeability of pure phospholipid membranes. Because of the nonpolar nature of the membrane's interior, pure phospholipid bilayers are highly impermeable to ions or polar molecules. Nonpolar molecules such as O_2 pass through these membranes very rapidly, and the membrane permetabilities for small polar molecules (for example, water, CO_2 and glycerol) are slightly lower but still significant. Among polar molecules, water has an unusually high membrane permeability, probably because of its strong interactions with the polar head groups of the lipid and because of its small size. In general, as molecules increase in size and polarity, their membrane permeabilities decrease. Ions penetrate pure phospholipid bilayers so slowly that these artificial membranes may be considered impermeable to them.

When the permeabilities of artificial bilayers for ioins and molecules are compared with those of biological membranes, important similarities and differences emerge. For nonpolar molecules and many small polar molecules and many small polar molecules, both types of membranes have similar permeabilities. On the other hand, for ions and large polar molecules such as sugars, biological membranes are much more permeable than artificial bilayers. The reason is that, unlike artificial bilayers, biological membranes contain **transport proteins** that facilitate the passage of ions and other polar molecules. The molecules for which biological and artificial membranes exhibit similar permeabilities apparently diffuse directly through the lapid phase of the membrane. However, most of the substances that are important to cell nutrition and metabolism cannot diffuse across the lipid bilayer directly must be transported by membrane proteins. These transport proteins can be grouped into three main categories; *channels, carriers, and pumps.*

Transport proteins exhibit specificity in the solutes they transport, hence their great diversity in cells. The simple prokaryote *Haemophilus influenzae,* the first organism for which the complete genome has been sequenced, has only 1743 genes, yet more than 200 of these genes (greater than 10% of the genome) encode various proteins involved in membrane transport. Although a particular transport protein in usually highly specific for the kinds of substances it will transport, it generally also transports a family of related substances. For example, in plants a K^+ transporter on the plasma membrane may transport Rb^+ and Na^+ in addition to K^+, but K^+ is usually preferred. At the same time, the K^+ transporter is completely ineffective in transporting anions such as Cl^- or uncharged solutes such as sucrose. Similarly, a protein involved in the transport of natural amino acids may move glycine, alanine, and valine with equal base but not accept aspartic acid or lysine.

In this section we will consider the structures, functions, and physiological roles of the various channels, carriers, and pumps found in plant cells, especially on the plasma membrane and tonoplast. We begin with a discussion of the role of channels and carriers in promoting the diffusion of solutes across membranes. We then distinguish between primary and secondary active transport, and discuss the roles of the electrogenic H^+-ATPase and various symporters (proteins that transport two substances in the same direction simultaneously) in driving proton-coupled secondary active transport.

Membrane Transporters

Two types of membrane transporters enhance the movement of solutes across membranes: *channels and carriers.* In general, channels are transmembrane proteins that function as selective pores in the membrane. The size of a pore and the density of surface charges on its interior lining determine its transport specificity. Molecules that can move through a channel may or may not involve transient binding of the solute to the channel protein. In any case, as long as the channel pore is open, solutes that can penetrate the pore diffuse through it extremely rapidly: about 10^8 ions s^{-1} through each channel protein. However, channels do not usually stay open for long periods: They have "gates" that open and close the pore in response to external signals. Transport through channels is always passive, and because the specificity of transport depends on pore size and electric charge more than on selective binding, channel transport is limited mainly to ions or water. The region of the channel that determines specificity is called the **selective filter**.

Primary Active Transport

To carry out active transport, a carrier must couple the uphill transport of the solute with another, energy-releasing, event so that the overall free-energy change is negative. **Primary active transport** is coupled directly to a metabolic source of energy, such as ATP hydrolysis, an oxidation-reduction reaction (the electron transport chain of mitochondria and chloroplasts), or the absorption of light by the carrier protein (in halobacteria, bacteriorhodopsin). The membrane protein that carry out primary active transport are called pumps. Most pumps transport ions, such as H^+ or Ca^{2+}. However, pumps belonging to the "ATP-binding cassette" family of transporters can carry large organic molecules into the vacuole.

Ion pumps can be further characterized as either electrogenic or electroneutral. In general, electrogenic transport refers to ion transport involving the *net movement of charge* across the membrane. In contrast, electroneutral transport, as the name implies, involves no net movement of charge.

In the plasma membranes of plants, fungi, and bacteria, as well as in plant tonoplasts and other plant and animal endomembranes, H^+ is the principal ion that is electrogenically pumped across the membrane. The **plasma membrane H^+-ATPase** creates the gradient of electrochemical potentials of H^+ in the plasma membranes, while the **vacuolar H^+-ATPase** (V-ATPase) and the **H^+-pyrophosphatase** (H^+PPase) electrogenically pump protons into the lumen of the vacuole and the Golgi cisternae.

In plant plasma membranes, only H^+ and Ca^{2+} appear to be transported by pumps, and the direction of pumping is outward, not inward. Therefore another mechanism is needed to drive the active uptake of most mineral nutrients. The other important way that solutes can be actively transported across a membrane against their gradient of electrochemical potential is by coupling of the uphill transport of the solute to the downhill transport of another. This type of carrier-mediated cotransport is termed *secondary active transport,* and it is driven indirectly by pumps.

Secondary Active Transport

When protons are extruded from the cytosol by electrogenic H^+-ATPases, both at the plasma membrane and at the vacuole membrane, a membrane potential and pH gradient are created at the expense of ATP hydrolysis. This gradient of electrochemical potential for H^+, called the **proton motive force** (PMF), or ?*p*, represents stored free energy in the form of the H^+ gradient.

The proton motive force generated by electrogenic H^+ transport is used to drive the transport of many other substances against their gradient of electrochemical potentials in secondary active transport. The carrier is a transmembrane protein with a site on the outside of the membrane that can bind a proton. Proton binding causes a second site to be exposed. This site binds the ion or solute that is being actively transported. With both molecules bound, the transporter undergoes a conformational change that exposes the binding sites to the opposite side of the membrane. The cycle is completed by diffusion of the proton and the substrate molecule away from their binding sites, causing the transporter to regain its original, or "relaxed," conformation.

The example shown in secondary active transport is called a **symport** (and the protein involved is called a **symporter**) because the two substances are moving in the same direction through the membrane. **Antiport** (facilitated by a protein called an **antiporter**) refers to coupled transport in which the downhill movement of protons drives the active (unphill) transport of a solute in the opposite direction. In both types of secondary transport, the ion or solute being transported simultaneously with the protons is moving against its gradient of electrochemical potential, so its transport is active. However, the energy driving this transport is provided by the proton motive force rather than by ATP hydrolysis, and it is captured by the cell by the inward movement of H^+ through the transporter, which dissipates the proton motive force.

Typically, a cell depends on one primary active transport system coupled to ATP hydrolysis to set up a gradient of electrochemical potential of one ion-for example, H^+. Many other ions or organic substrates can then be transported by a variety of secondary active transport proteins, which energize the transport of their respective substrates by simultaneously carrying one or two H^+ ions down their energy gradient. Thus H^+ ions circulate across the membrane, outward through the primary active transport proteins, and back into the cell through the secondary transport proteins. In plants and fungi, sugars and amino acids are taken up by symport with protons.

This membrane depolarization is transitory, however because reduced membrane voltage allows the H^+ pump to work faster and thereby restore the membrane voltage and pH gradient in the presence of continuing glucose uptake.

Research has led to the view that most of the ionic gradients across membranes of higher plants are generated and maintained by electrochemical-potential gradients of H^+. In turn, these H^+ gradients are generated by the electrogenic proton pumps. Evidence suggests that Na^+ is transported out of the cell by a Na^+- H^+ antiporter and that Cl^-, NO_3^-, $H_2PO_4^-$, sucrose, amino acids, and other substances enter the cell via specific proton symporters. Potassium at very low external concentrations can be taken up by active symport proteins, but at higher concentrations K^+ can enter the cell by diffusion through specific K^+ channels. However, even transport through channels is driven by H^+-ATPase, in the sense that K^+ diffusion is driven by the membrane potential, which is maintained at a value more negative than the K^+ equilibrium potential by the action of the electrogenic H^+ pump.

5.3 The Goldman Equation

Many different ions permeate the membranes of living cells simultaneously; therefore, the Nernst potential for any single ionic species seldom describes the membrane's diffusion potential. Instead, the contributions of all the ions to the diffusion potential across the membrane are described by the Goldman equation:

$$\Delta E = \frac{RT}{F}\left(\ln \frac{P_{K+}\, C_{K+}{}^{o} + P_{Na+}\, C_{Na+}{}^{o} + P_{Cl-}\, C_{Cl-}{}^{i}}{P_{K+}\, C_{K+}{}^{i} + P_{Na+}\, C_{Na+}{}^{i} + P_{Cl-}\, C_{Cl-}{}^{o}} \right)$$

This equation relates the standing ion gradients across a membrane to the diffusion potential that develops. $P_K{}^+$, $P_{Na}{}^+$, and $P_{Cl}{}^-$ represent the membrane permeabilities for K^+, Na^+, and Cl^-, respectively. Although the equation should include terms for all ions passing through the membrane, K^+, Na^+, and Cl^- have the largest membrane permeabilities and highest concentrations in plant cells and therefore dominate the equation.

The relationship between the Goldman equation and the Nernst equation can be visualized if one imagines a membrane that is permeable to only one ion, say K^+, so that both $P_{Cl}{}^-$ and $P_{Na}{}^+$ equal zero. Under these conditions, the Goldman equation reduces to the Nernst equation for K^+. Although biological membranes are never permeable to only a single ionic species, artificial membranes can approximate this situation and are used in the manufacture of pH electrodes and other ion-selective electrodes.

5.4 ABC Transporters in Plants

The ATP-binding cassette (ABC) transporters are members of a large family of active transport proteins energized directly by ATP hydrolysis. Originally identified in microbial and animal cells, ABC transporters are able to use the energy of ATP hydrolysis directly to pump organic molecules (especially large anionic molecules) across a membrane. Like the P-type H^+-ATP synthases, ABC transporters form a phosphorylated intermediate during catalysis, and they are therefore inhibited by vanadate.

A number of genes for ABC transporters have been cloned from various plants. Genes for ABC proteins are recognizable by a predicted protein structure consisting of one or two copies each of two structural units, a transmembrane domain and a nucleotide-binding fold, which contains a characteristic signature sequence. The transmembrane domain provides the pathway for solute transport and determines the molecular specificity of the transporter.

An ABC transporter has been found to secrete anti-fungal terpenes across the plasma membrane of tobacco cells. More commonly, ABC transporters are found at the tonoplast, where they are sometimes referred to as glutathione conjugate pumps, or GS-X pumps, since they often transport molecules that have been covalently attached to glutathione. Glutathione (GS) is a tripeptide (Glu-Cys-Gly) that functions as an important cellular antioxidant molecule (see textbook p. 274 for molecular structure). The vacuolar GS-X pumps of plant cells function in herbicide detoxification, protection against oxidative damage, pigment accumulation, and the storage of antimicrobial compounds .

A family of enzymes called glutathione transferases (GSTs) is responsible for attaching glutathione to the molecule to be transported. The common chemical determinant for GS conjugation is a carbon–carbon double bond adjacent to an electron-withdrawing group (CH_2 = CH-Z) . Compounds that contain this determinant include the anthocyanins , IAA (indoleacetic acid, or auxin), and various phenolic compounds. Metal-binding polypeptides called phytochelatins, which are synthesized from glutathione, are also transported into plant vacuoles along with their chelated heavy metals via GS-X pumps.

Chapter **6**

Photosynthesis: The Light Reactions of Photosynthesis

6.1 Quantum Yield

The **quantum yield** (ϕ) of a process in which molecules give up their excitation energy (or "decay") is the fraction of excited molecules that decay via that pathway. Mathematically, the quantum yield of a process, such as photochemistry, is defined as follows:

$$\Phi = \frac{\text{yield of photochemical products}}{\text{total number of quanta absorbed}}$$

The value of Ö for a particular process can range from 0 (if that process is never involved in the decay of the excited state) to 1.0 (if that process always deactivates the excited state). The sum of the quantum yields of all possible processes is 1.0.

The quantum yield of formation of the products of photosynthesis, such as O_2, can be measured quite accurately. In this case the quantum yield is substantially lower than the value for photochemistry, because several photochemical events must take place before any O_2 molecules form. For O_2 production the measured maximum quantum yield is approximately 0.1, meaning that 10 quanta are absorbed for each O_2 molecule released. The reciprocal of the quantum yield is called the **quantum requirement**. The minimum quantum requirement for O_2 evolution is therefore about 10. Quantitative measurements of the absorption of light and the fate of the energy contained in the light are essential to an understanding of photosynthesis.

6.2 Photo System I&II

The puzzling red drop and enhancement effects were explained by experiments performed by Louis Duysens of the Netherlands . He suggested that two phtotchemical events were responsible for enhancement effects. One photochemical event produced an oxidation while the other produced a reduction. Chloroplasts contain cytochromes, iron-containing proteins that function as intermediate electron carriers in photosynthesis. Duysens found that when a sample of a red alga was illuminated with long-wavelength light, the cytochrome became mostly oxidized. If light of a shorter wavelength was also present, the effect was partly reversed (Figure 6.1). These **antagonistic effects** can be explained by a mechanism involving two photochemical events: one that tended to oxidize the cytochrome and one that tended to reduce it.

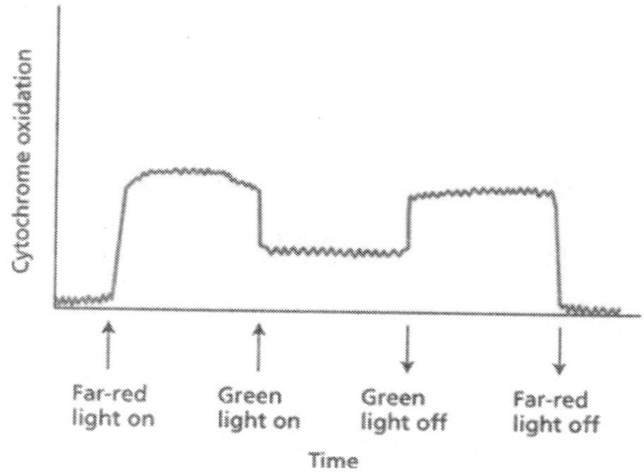

Figure 6.1 Antagonistic effects of light on cytochrome oxidation. Far-red light is very effective in oxidizing the cytochrome *f* in the chloroplast. If green light is also present, some of the cytochrome becomes reduced. The two wavelengths have opposite effects—hence the term "antagonistic." This experiment is one of the clearest demonstrations of the existence of two photochemical systems in photosynthesis: one that reduces cytochrome and one that oxidizes it. This particular experiment was done with a red alga, in which photosystem II is driven best by green light, and photosystem I is driven best by far-red light. (Duysens, *et al* ,*Nature* 190,190:510-511)

We know now that in the red region of the spectrum, one of the photoreactions, known as **photosystem I (PS-I)**, absorbs preferentially far-red light of wavelengths greater than 680 nm, while the second, known as **photosystem II (PS-II)**, absorbs red light of 680 nm well and is driven very poorly by far-red light. This wavelength dependence explains the enhancement effect and the red drop effect. Another difference between the photosystems is that Photosystem I produces a strong reductant, capable of reducing $NADP^+$, and a weak oxidant. Photosystem II produces a very strong oxidant, capable of oxidizing water, and a weaker reductant than the one produced by photosystem I. This reductant of photosystem II rereduces the oxidant produced by photosystem I, which explains the antagonistic effect. The PS-II reaction center is located in the grana lamellae, wherease PS-I reaction center (both with their associated antenna pigments and electro transport pigments) is found almost exclusively in the stroma lamellae and at the edge of the grana lamellae. Under many conditions the ratios of PS-II to PS-I is about 1.5 :1, but it can change when plants are grown in different conditions of light.

6.3 Bacterial Reaction Centers

In 1984, Hartmut Michel, Johann Deisenhofer, Robert Huber, and coworkers in Munich solved the three-dimensional structure of the reaction center from the purple photosynthetic bacterium *Rhodopseudomonas viridis*. This landmark achievement, for which a Nobel prize was awarded in 1988, was the first high-resolution X-ray structure determination for an integral membrane protein and the first structure determination for a reaction center complex. The structure is shown in Figure 6.2.

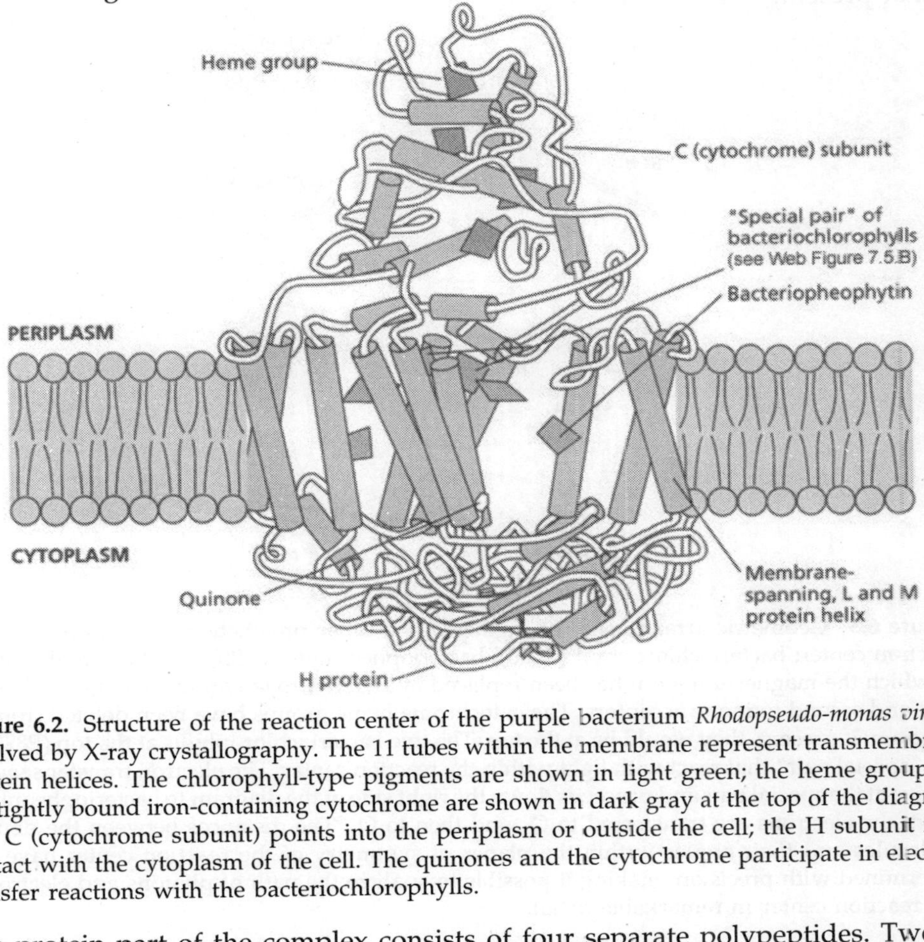

Figure 6.2. Structure of the reaction center of the purple bacterium *Rhodopseudo-monas viridis*, resolved by X-ray crystallography. The 11 tubes within the membrane represent transmembrane protein helices. The chlorophyll-type pigments are shown in light green; the heme groups of the tightly bound iron-containing cytochrome are shown in dark gray at the top of the diagram. The C (cytochrome subunit) points into the periplasm or outside the cell; the H subunit is in contact with the cytoplasm of the cell. The quinones and the cytochrome participate in electron transfer reactions with the bacteriochlorophylls.

The protein part of the complex consists of four separate polypeptides. Two of them, called L and M (for *l*ight and *m*edium mass) bind all of the bacteriochlorophyll, quinone, and carotenoid cofactors of the complex. The structure has a twofold symmetry about an axis perpendicular to the plane of the membrane, hinting at a dimeric nature of the reaction center. The ten transmembrane portions of the L and M peptides (five from each) are arranged in α helices, and there are almost no charged amino acid residues in the interior of the membrane. The H (*heavy*) protein has a single transmembrane helix and is localized mostly on the cytoplasmic side of the membrane. The C (*c*ytochrome) subunit is located in the periplasmic region (the region between the bacterial plasma membrane and the outer membrane). The geometric arrangement of the pigments and the quinones (electron acceptors) is shown in Figure 6.2. with the protein removed. A similar arrangement is found in the reaction center of another purple photosynthetic bacterium, *Rhodobacter sphaeroides*, except that the C subunit is not present.

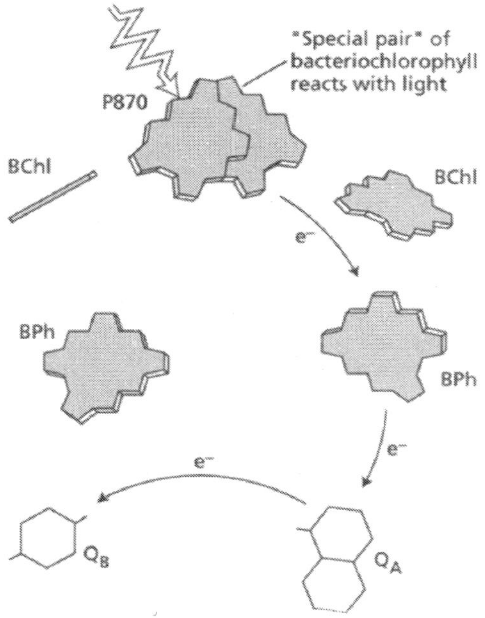

Figure 6.3 Geometric arrangement of pigments and other prosthetic groups in the bacterial reaction center: bacteriochlorophyll (BChl); bacteriopheophytin (BPh), a chlorophyll molecule in which the magnesium atom has been replaced by two hydrogen atoms; and Q_A and Q_B, the first and second quinone acceptors. The cytochrome heme groups have been deleted from this diagram; if present, they would be at the top. The two bacteriochlorophylls at the top (P870) are the "special pair" that react with light within the reaction center. The electron transfer sequence begins at the special pair and proceeds down the right side of the diagram to bacteriopheophytin. Next, the electrons are transferred to Q_A and then to Q_B. The distances between the depicted molecules and their angles within the planes of symmetry of the reaction center have been determined with precision, making it possible to analyze the path of photons and electrons in the reaction center in remarkable detail.

Two of the bacteriochlorophyll molecules are in intimate contact with each other and are known as the special pair (Figure 6.3). This dimer, whose existence was predicted from magnetic-resonance studies, is the photoactive portion of the complex. An electron is transferred from this dimer along the sequence of electron carriers on the right side of the complex. Detailed analysis of these structures, along with analysis of numerous mutants, has revealed many of the principles involved in the energy storage processes that are carried out by all reaction centers.

The bacterial reaction center structure is thought to be similar in many ways to that found in photosystem II from oxygen-evolving organisms, especially in the electron acceptor portion of the chain . The proteins that make up the core of the bacterial reaction center are relatively similar in sequence to their photosystem II counterparts, implying an evolutionary relatedness.

6.4 Evolution of Oxygen

The chemical mechanism of photosynthetic water oxidation is not yet known, although there is a great deal of indirect evidence about the process. If a sample of dark-adapted photosynthetic membrane is exposed to a sequence of very brief, intense flashes, a characteristic pattern of oxygen production is observed. Little or no oxygen is produced on the first two flashes, and maximal oxygen is released on the third flash and every fourth flash thereafter, until eventually the yield per flash damps to a constant value. This remarkable result was first observed by Pierre Joliot in the 1960s.

A schematic model explaining these observations, proposed by Kok and coworkers, has been widely accepted . This model for the photooxidation of water, called the **S state mechanism,** consists of a series of five states, known as S_0 to S_4, which represent successively more oxidized forms of the water-oxidizing enzyme system, or **oxygen-evolving complex** (Figure 6.4). The light flashes advance the system from one S state to the next, until state S_4 is reached. State S_4 produces O_2

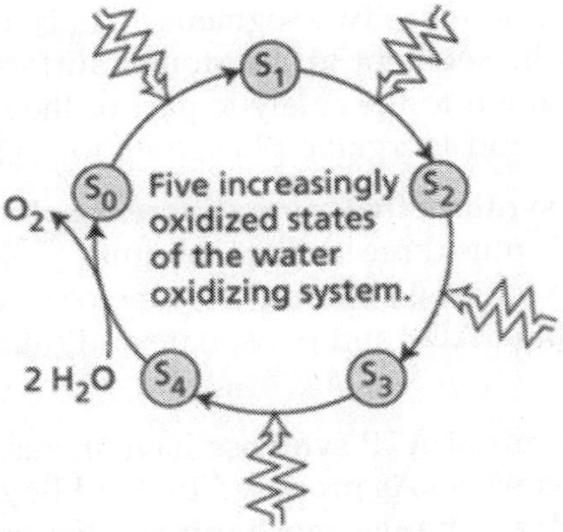

Figure 6.4.

without further light input and returns the system to S_0. Occasionally, a center does not advance to the next S state upon flash excitation, and less frequently, a center is activated twice by a single flash. These misses and double hits cause the synchrony achieved by dark adaptation to be lost and the oxygen yield eventually to damp to a constant value. After this steady state has been reached, a complex has the same probability of being in any of the states S_0 to S_3 (S_4 is unstable and occurs only transiently), and the yield of O_2 becomes constant. States S_2 and S_3 decay in the dark, but only as far back as S_1, which is stable in the dark. Therefore, after adaptation to the dark, approximately three-fourths of the oxygen-evolving complexes appear to be in state S_1 and one-fourth in state S_0. This distribution of states explains why the maximum yield of O_2 is observed after the third of a series of flashes given to dark-adapted chloroplasts.

This S state mechanism explains the observed pattern of O_2 release, but not the chemical nature of the S states or whether any partly oxidized intermediates, such as H_2O_2, are formed. Additional information has been obtained by measurement of the pattern of proton release as the S states are advanced with flashes. In addition to oxygen, hydrogen ions are a product of water oxidation. The release of protons is not strictly coupled to O_2 release.

6.5 Synthesis of ATP

The chemical mechanism by which ATP is synthesized is not yet understood in detail. However, considerable evidence now supports a mechanism, first proposed by Paul Boyer, in which the principal energy-requiring step is the release of bound ATP from the enzyme. It has also been proposed that during catalysis a large portion of the CF_1 complex rotates about a bearing consisting of the γ subunit . The γ subunit may act as a camshaft does, rotating alternately against the α and α subunits . The energy of the conformational movements is then translated into phosphoanhydride bond energy.

The thylakoid ATP synthase has two segments: CF_0 is a transmembrane segment, and CF_1 is a hydrophilic segment at the stomal surface. CF_0 traslocates protons from the chloroplast lumen to the catalytic part of the enzyme and CF_1 cataylses the conversion of ADP and inorganic phosphate to ATP.

The chloroplast ATP synthase has nine different subunits. CF_1 is made of two large subunits, α and β, plus three smaller subunits, γ, δ, and ε. Each CF_1 molecule has three copies of the α and β subunits and one copy of the γ, δ and ε subunits. The α and β subunits bind ADP and phosphate and catalyize the phosphorylation of ADP into ATP.

Studies on the mitochondrial ATP synthase have increased our understanding of how the ATP synthases work. As proposed by Paul Boyer in 1997, ATP synthesis takes place by a **binding change mechanism** (Figure 6.5). According to this model, the energy of the proton gradient is used to release a tightly bound form of ATP from a catalytic binding side of the enzyme. As shown in Web Figure 7.9.A, CF_1 has three nucleotide binding sites, L, T and O. ADP and inorganic phosphate are postulated to first bind to the O site, and then move to the L site. The T site binds ATP. As protons move from the lumen to the stromal region through the CF_0 channel, energy is released and the γ subunit of CF_1 rotates. The rotation causes conformation changes in the three nucleotide binding sites, which interconvert, thus changing the afiity of the sites for the nucleotides. As the T site converts into a O site, ATP is released. and another cycle starts. Direct observation of the rotation of the γ subunit of CF_1 has provided strong experimental support for the binding change mechanism.

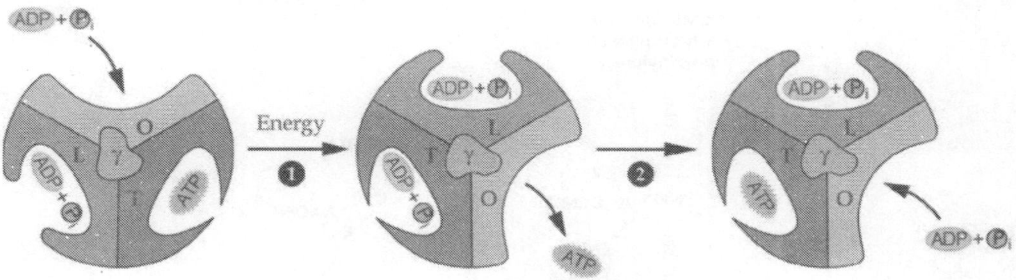

Figure 6.5 The model of Boyer's binding change mechanism. The α and β subunits configure three nucleotide binding sites: O, which provides the early binding site for ADP and inorganic phosphate, L, to which the ADP and inorganic phosphate bind after migrating from O, and T, which tightly binds ATP. Energy ensuing from the movement of protons from the chloroplast lumen to the stroma drives the rotation of the g subunit of CF_1, and the interconversion of the binding sites and the release of an ATP molecule.

6.6 Interaction of Photosynthetic Electron Flow by Herbicides

Herbicides of one major class—about half of the commercially important compounds—act by interrupting photosynthetic electron flow. Figure 6.6 shows the chemical structure of two of these compounds. The precise sites of action of many of these agents have been found to lie either at the reducing side of photosystem I (for example, in paraquat) or in the quinone acceptor complex in the electron transport chain between the two photosystems (for example, in diuron) (Figure 6.7.).

The use of herbicides to kill unwanted plants is widespread in modern agriculture. Many different classes of herbicides have been developed, and they act by blocking amino acid, carotenoid, or lipid biosynthesis or by disrupting cell division. Understanding the mode of action of herbicides has been an important tool in research on plant metabolism and has facilitated their application under different agricultural practices. Here, the chemical structure of two herbicides that block photosynthetic electron flow are shown. DCMU is also known as diuron. Paraquat has acquired public notoriety because of its use on marijuana crops.

Paraquat acts by intercepting electrons between the bound ferredoxin acceptors and NADP and then reducing oxygen to **superoxide** (O_2^-). Superoxide is a free radical that reacts nonspecifically with a wide range of molecules in the chloroplast, leading to the rapid loss of chloroplast activity. Lipid molecules in cell membranes are especially sensitive.

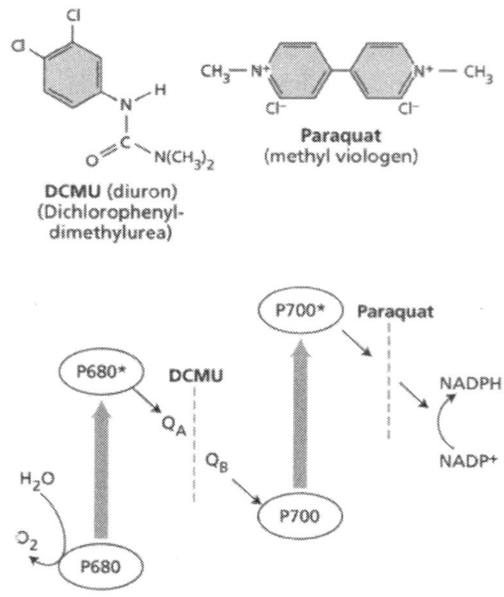

Figure 6.7. Sites of action of herbicides. Many herbicides, such as DCMU, act by blocking electron flow at the quinone acceptors of photosystem II, by competing for the binding site of plastoquinone that is normally occupied by Q_B. Other herbicides, such as paraquat, act by accepting electrons from the early acceptors of photosystem I and then reacting with oxygen to form superoxide, O_2^-, a species that is very damaging to chloroplast components, especially lipids.

The herbicides that act on the quinone acceptor complex compete with plastoquinone for the $\mathbf{Q_B}$ binding site. If herbicide is present, it displaces the oxidized form of plastoquinone and occupies the specific binding site for the quinone acceptor, which is thought to lie on the D1 herbicide-binding protein. The herbicide is not able to accept electrons, so the electron is unable to leave $\mathbf{Q_A}$, the first quinone acceptor. Thus, the binding of herbicide effectively blocks electron flow and inhibits photosynthesis. Many herbicides that act in this manner also inhibit electron flow in photosynthetic bacteria that have quinone-type electron acceptor complexes.

In recent years, herbicide-resistant biotypes of common weeds have appeared in areas where a single type of herbicide has been used continuously for several years. These biotypes can be orders of magnitude more resistant to certain classes of herbicides than the nonresistant plant is. In several cases the resistance factor has been traced to a single amino acid substitution in the D1 protein. This change, presumably in the quinone (and herbicide) binding region of the peptide, lowers the binding affinity of the herbicide, making it much less effective.

The possibility of fine-tuning the herbicide sensitivity of crop plants by making subtle changes in the proteins of the PS-II reaction center has created a great deal of interest in the agricultural chemical industry. Through biotechnology, it is now possible to make a crop plant that is resistant to a particular herbicide, which can then be applied to control undesirable plants that are not resistant. The success of this approach will depend on whether undesirable side effects of the herbicide-resistant mutations can be controlled and on how rapidly the weeds acquire resistance through natural selection or gene transfer. Thus, it is a race between science and evolutionary change.

A complementary approach is to insert resistance factors to certain pests, such as insects, into the plant via genetic engineering. The plants then produce the toxic species themselves, eliminating the need for insecticides.

Chapter 7

The Dark Reactions of Photosynthesis

7.1 Rubisco: A Key Enzyme

With the exception of the carbon fixed by a some prokaryotic organisms, most of the carbon fixed on Earth is processed by the Calvin cycle. The concentration of rubisco, the carboxylating enzyme of the Calvin cycle, is generally high. For example, it accounts for 50% or more of the total protein in plant leaves, and its concentration within chloroplasts is extremely high (*ca.* 0.2 g/ml).

Rubisco occurs in two functionally analogous forms, having different structure, distribution, and O_2 sensitivity. Oxygenic phototroph (cyanobacteria, chloroplasts from high plants) and many photosynthetic bacteria contain a form of the enzyme, (form I), made up of eight large (L) catalytic subunits (about 55 kDa each) and eight small (S) subunits (about 14 kDa each), giving a molecular mass of about 560 kDa for the complete protein (L8S8). The 8 L subunits of form I are arranged as an octameric core surrounded by two layers of four S subunits, with each layer located on opposite sides of the molecule (Figure 7.1).

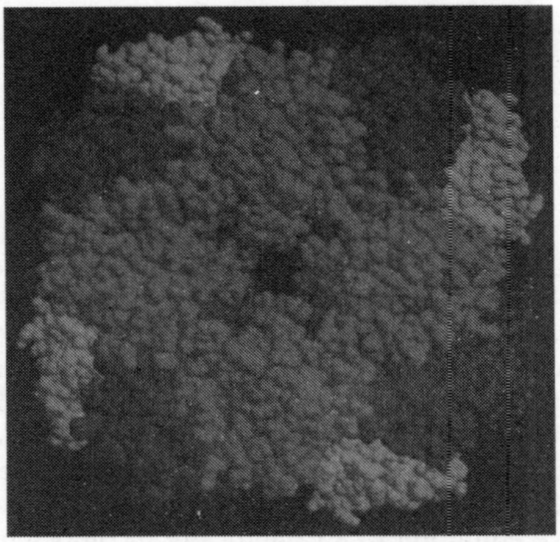

Figure 7.1. A model for the structure of rubisco in chloroplasts from higher plants. Rubisco consists of 8 large (L) and 8 small (S) subunits arranged as 4 dimers. Small subunits are shown in red (only four of the small subunits are seen), large subunits are shown in blue and green, in order to show the boundaries of the dimers.

Form II of rubisco is found in some photosynthetic purple nonsulfur bacteria, and it is composed of two L subunits, each 50 kDa. Eukaryotic algal dinoflagellates also use a dimeric enzyme of proteobacterial ancestry. A rubisco-like enzyme lacking catalytic activity has recently been identified in photosynthetic green sulfur bacteria which lack the Calvin cycle. The evidence suggests that the enzyme may function in sulfur oxidation to provide electrons for the photosynthetic electron transport chain. Genome sequences show that the rubisco-like protein also occurs in certain heterotrophic bacteria.

7.2 Rubisco Activase

Rubisco catalyzes the incorporation of CO2 to the C-2 of the enediol form of ribulose-1,5-bisphosphate and the subsequent cleavage of the unstable intermediate to yield two molecules of 3-phosphoglycerate The conversion of the inactive form of rubisco to an active state, which is required for both the carboxylation and oxygenation reactions, requires the carbamylation of a specific lysine residue (in spinach, Lys-201) (Figure 7.2.). As a consequence of the binding of the CO2–

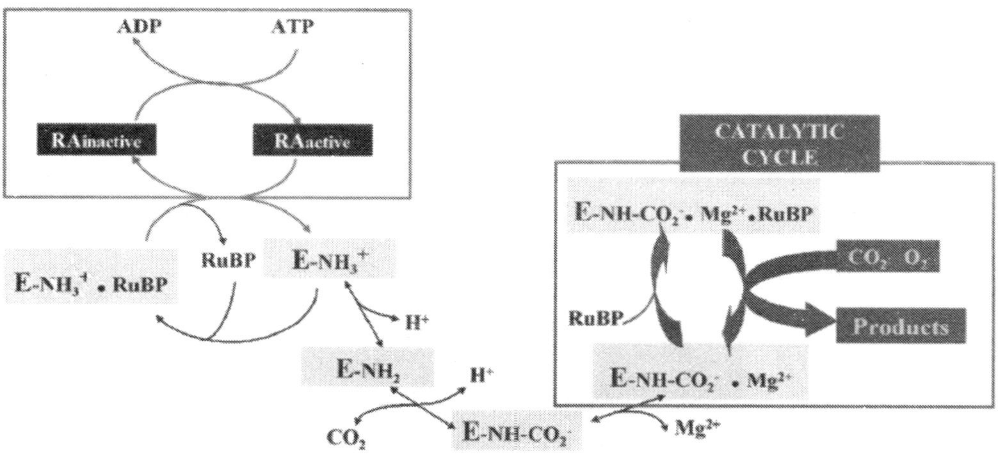

Figure 7.2. Modulation of rubisco by rubisco-activase. The combination or the interaction of rubisco (E) with modulators distributes the enzyme among different states (yellow boxes). The unprotonated form of a specific lysine residue (E-NH_2) reacts with CO_2 releasing a proton and forming the carbamate (E-NH-CO_2^-) that is stabilized by the cofactor Mg^{2+} (E-NH-CO_2^-. Mg^{2+}). At this stage (blue box), rubisco is catalytically competent and thereby binds ribulose-1,5-bisphosphate (RuBP) yielding the ternary complex (E-NH-CO_2^- . Mg^{2+} . RuBP) in which the latter is converted to the enediol form. This entails the incorporation of CO_2 or O_2 in the active site to generate the products of carboxylation or oxygenation, respectively. The high affinity of ribulose-1,5-bisphosphate for free rubisco (E-NH^{3+}) generates the dead-end complex (E-NH^{3+} . RuBP) thereby displacing the above equilibria to a catalytically incompetent state. Rubisco-activase (RA) interacts noncovalently with the binary complex and facilitates the release of ribulose-1,5-bisphosphate (red box) freeing the enzyme for the next carbamylation cycle. This process requires the concurrent hydrolysis of ATP.

molecule to the lysine residue, the enzyme acquires a triad of anionic residues that provide the binding site for the cofactor Mg2+ (in spinach, the other two residues are Asp-203 and Glu-204). At this stage, the incorporation of ribulose-1,5-bisphosphate causes conformational changes in rubisco that allows the enzyme to sequester the sugar bisphosphate from the solvent and, upon the binding of a CO2 or O2 molecule, to start the catalytic cycle. However, the tight binding of ribulose-1,5-bisphosphate to the uncarbamylated enzyme molecules displaces the equilibrium to the dead end complex, rubisco-ribulose-1,5-bisphosphate, so that the rate of reaction declines rapidly until it finally stops. How is this problem solved?

A combination of genetic, physiological and biochemical studies have provided the solution to the long-term riddle of rubisco activation. In vivo, some mutants of *Arabidopsis thaliana* lack light-mediated activation of rubisco and can only grow at high concentrations of CO_2. Surprisingly, the mutation does not impair the response of the enzyme to modulation by CO_2 and Mg^{2+} in vitro. These contradictory results led to the discovery of a protein, rubisco-activase, that uncouples ribulose-1,5-bisphosphate from decarbamylated active sites and, in so doing, promotes the access of CO_2 and Mg^{2+} for the carbamylation of the enzyme.The importance of this process is that the rate at which rubisco-activase enhances the catalytic capacity of rubisco is linked to the hydrolysis of ATP (50 molecules of ATP hydrolyzed / 1 rubisco site activated). Two isoforms of rubisco activase, (42 and 46 kDa) found in several species differ in the C-terminal region as result of the differential splicing of a common pre-mRNA. Both isoforms catalyze the activation of rubisco and the hydrolysis of ATP but the smaller isoform has higher capacity than the larger isoform for both activities. Evidence for the fact that the two processes are independent are provided by studies showing that the digestion of native rubisco-activase with trypsin or the modification of its primary structure by site-directed mutagenesis do not elicit the same responses in the stimulation of rubisco and the hydrolysis of ATP. More importantly, the deletion of some residues at the C-region of rubisco-activase doubles the rate of rubisco activation with only a minor effect on the hydrolysis of ATP.

Rubisco activation by the activase requires a mutual recognition of both proteins. Rubisco-activase from tobacco (Solanaceae) fails to activate rubisco from spinach or Chlamydomonas (non-Solanaceae) and, similarly, the rubisco-activase from spinach or Chlamydomonas fails to stimulate rubisco from tobacco. However, the substitution of some residues on the surface of the large subunit of Chlamydomonas rubisco shifts the specificity from non-Solanaceae to Solanaceae rubisco-activase. These experiments underscore the contribution of specific amino acids to protein-protein interaction and provide experimental evidence for a recognition step separated from the subsequent activation process.

7.3 Thioredoxins

Thioredoxins are 12 kDa proteins that have a catalytically active disulfide group and occur in most if not all organisms. Thioredoxins have an active site with two redox-active cysteine residues (-Trp-Cys-Gly-Pro-Cys-). The active site is located on the first α helix of a motif consisting of three flanking α helices and a four-stranded β sheet that has been identified in several redox-active proteins (Figure 7.3).

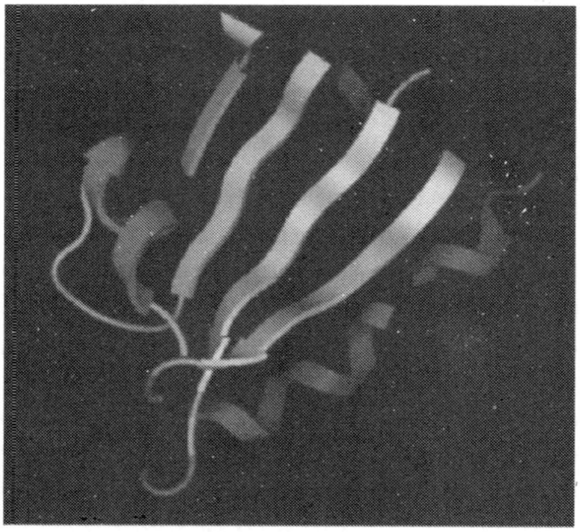

Figure 7.3. Thioredoxin-2, from *Anabaena*. The alpha-beta-alpha topology of the protein is shown.

The oxidized form of each thioredoxin contains a disulfide (—S—S—) bridge that is reduced to the sulfhydryl (—SH) level by either reduced ferredoxin or NADPH. The reduced form of thioredoxin is an excellent catalyst for the reduction of intramolecular (intrachain) disulfide bonds of proteins that are only slowly reduced by glutathione or glutaredoxin, the other major cellular sulfhydryl reductants. Upon reduction, the disulfide bond of enzymes targeted specifically by a thioredoxin causes a dramatic shift (usually increase) in catalytic activity. First observed to activate enzymes of photosynthesis, thioredoxins are now known to play a broad regulatory role in many cell types.

The enzymes regulated by thioredoxin show differing degrees of specificity. Enzymes associated with the Calvin cycle, fructose-1,6-bisphosphate phosphatase, sedoheptulose-1,7-bisphosphate phosphatase, ribulose-5-phosphate kinase, NADP:glyceraldehyde-3-phosphate dehydrogenase and rubisco activase, are preferentially activated by thioredoxin f. One enzyme, ATP synthase, which catalyzes the synthesis of ATP in photophosphorylation, is activated by both chloroplast thioredoxins.

Thioredoxin m effectively activates NADP:malate dehydrogenase, an enzyme of the C4 photosynthetic carbon cycle and C3 redox shuttle, and deactivates glucose-6-phosphate dehydrogenase. Thioredoxin of the "m" or bacterial, type has been found to regulate other light-dependent chloroplast processes, such as the translation of mRNA.

Plants have a third class of thioredoxin, the h type—located in the cytosol, endoplasmic reticulum, and mitochondrion that is reduced by NADPH in a reaction catalyzed by a flavoprotein enzyme, NADP:thioredoxin reductase (NTR):

$$\text{NADPH} + \text{H}^+ + \text{thioredoxin}_{\text{ox}} \xrightarrow{\text{NTR}} \text{NADPH}^+ + \text{thioredoxin}_{\text{red}}$$

The function of thioredoxin *h* is currently being investigated in several plant systems, including seed, phloem and pollen. In cereals, it has been found to promote the mobilization of carbon and nitrogen of the endosperm early in grain germination. Thioredoxin *h* appears to act (1) by reducing the major seed storage proteins, thereby enhancing their susceptibility to proteases, (2) by activating a newly discovered type of calcium-linked protease (thiocalsin), and (3) by neutralizing (via reduction of their intrachain disulfide bonds) low-molecular-weight proteins that inhibit enzymes of starch degradation.

Through studies of gene inactivation, thioredoxin *h* has been identified as a requirement for the development of pollen self-incompatibility and to be a prominent component of phloem. Thioredoxin, again linked to NADP, has also been found to function widely in animal cells in processes ranging from transcription to development of the immune response to embryogenesis.

These biochemical and molecular studies of thioredoxin have led to possible applications in technology and medicine. Thioredoxin may be used to improve foods by lowering allergenicity and increasing digestibility. Ongoing studies have uncovered the potential use of thioredoxin as a marker in diagnosing and monitoring certain human diseases, in neutralizing venom neurotoxins and factors involved in the inflammatory response, and in improving the success of organ transplants. Thioredoxin is an example in which principles learned in basic research are being applied to contemporary societal problems.

7.4 C_2 Oxidative Photosynthetic Carbon Cycle

Evidence in support of the C_2 oxidative photosynthetic carbon cycle stems from several types of experiments. Using mutant *Arabidopsis* plants lacking phosphoglycolate phosphatase, C. Somerville and W. L. Ogren demonstrated that radioactivity from $^{14}CO_2$ accumulates in 2-phosphoglycolate under conditions favoring photorespiration (high concentrations of O_2, low concentrations of CO_2), but not under nonphotorespiratory conditions (low O_2, high CO_2). In addition, glycolate synthesis in vivo is dependent on O_2 and is inhibited by CO_2 in a competitive manner, as expected from the primary role of rubisco in the cycle.

Several studies have shown that, as with the Calvin cycle, the enzymes of the C_2 oxidative photosynthetic carbon cycle catalyze all the proposed reactions in vitro at rates higher than the in vivo rate of photorespiration. In addition, the participating enzymes are localized in the choroplast (rubisco, phosphoglycolate phosphatase, glycerate kinase), the peroxisome (glycolate oxidase, catalase, hydroxypyruvate reductase, and the two aminotransferases), and the mitochondrion (glycine decarboxylase and serine hydroxymethyltransferase).

7.5 Three Kinds of C_4 Metabolism

There are three variations of the basic C_4 pathway (Figure 7.4). The variations differ principally in the C_4 acid transported into the bundle sheath cells (malate or aspartate) and in the manner of decarboxylation.

The variations are named after the enzymes that catalyze the respective decarboxylation reactions: NADP-dependent malic enzyme (NADP-ME) found in chloroplasts; NAD-dependent malic enzyme (NAD-ME) in mitochondria; and phosphoenolpyruvate carboxykinase (PEP-CK) in the cytosol. The discussion in the text traces the route of carbon for the NADP-ME type, such as sugarcane or corn.

The Carbon Path

The primary carboxylation reaction, catalyzed by PEP carboxylase, which is common to all three variants, occurs in the cytosol of the mesophyll cells and uses HCO_3^- rather than CO_2 as a substrate . In the NADP-ME variant, oxaloacetate is rapidly reduced to malate in the mesophyll chloroplasts by NADPH using NADP:malate dehydrogenase . The malate formed enters the chloroplast of the bundle sheath cell and there undergoes oxidative decarboxylation, yielding pyruvate. The CO_2 released within the bundle sheath cells is converted to carbohydrate by the Calvin cycle.

The residual C_3 acid is transported back to the mesophyll as pyruvate and converted to phosphoenolpyruvate in the mesophyll chloroplast . This reaction takes place in the mesophyll chloroplast of all three C_4 variants. The cycle then starts again with carboxylation of the phosphoenolpyruvate to produce oxaloacetate.

Another process localized in the mesophyll chloroplast of all three C_4 variants is the reduction of 3-phosphoglycerate to triose phosphates, which occurs within the Calvin cycle proper in C_3 chloroplasts. In contrast, all C_4 variants export 3-phosphoglycerate from the bundle sheath for reduction in the mesophyll chloroplast. After reduction, the triose phosphates are transported back to the bundle sheath chloroplasts, for use in the operation of the Calvin cycle there.

One possible reason for this remarkable split of the Calvin cycle between the bundle sheath and the mesophyll chloroplast is to restrict electron transport activity and oxygen evolution in the bundle sheath, and thus avoid oxygen levels that would favor photorespiration. The transport of metabolites of the C_4 pathway is accompanied by movements of protons and other ionic species in order to maintain pH and charge balance.

7.6 Fructans

Although starch is the dominant carbohydrate reserve in many species, it may be almost absent in the storage organs of some plants. Fructans are less familiar reserve carbohydrates but occur in 15% of flowering species that mainly belong

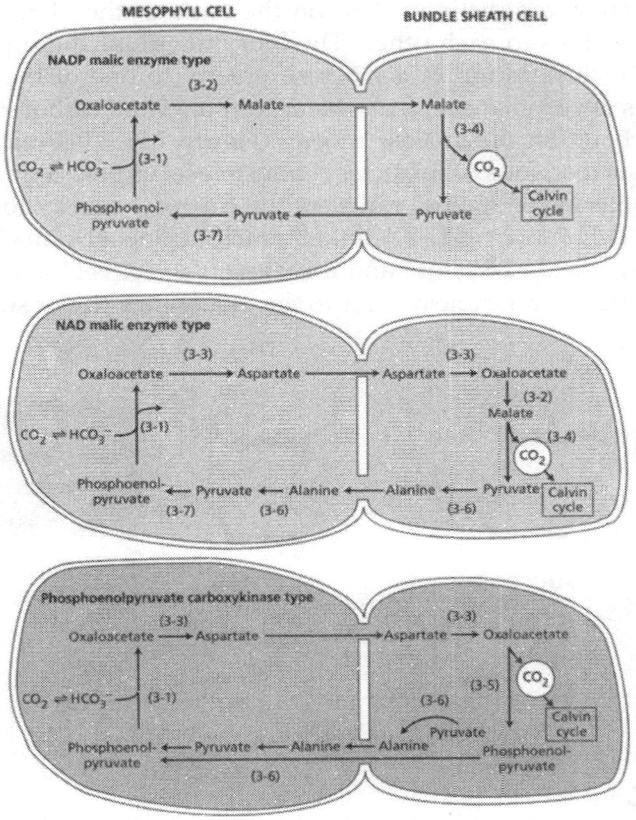

Figure 7.4. The three variants of the C_4 photosynthetic carbon cycle. The variants differ principally in (1) the nature of the four-carbon acid (malate or aspartate) transported into the bundle sheath cell and of the three-carbon acid (pyruvate or alanine) returned to the mesophyll cell and (2) the nature of the enzyme that catalyzes the decarboxylation step in the bundle sheath cell. The three variants are named after the enzymes that catalyze the decarboxylation reactions. Representatives of each variant include maize, crabgrass, sugarcane, sorghum (NADP malic enzyme); pigweed, millet (NAD malic enzyme); guinea grass (phosphoenolpyruvate carboxykinase) (Taiz and Zeiger, 1998 Sinaur Associates Inc,MA).

to the Asteraceae, Campanulaceae and Boraginaceae (dicots), and Poaceae and Liliaceae (monocots). This type of polysaccharide, whose basic unit is fructose, was found in 1804 in a hot water extract from *Inula helenium* and later named "inulin." Fructans can replace starch as a reserve carbohydrate in some plants, though in many species both polysaccharides coexist. Moreover, the carbohydrate reserve substances vary among plant species and even within a single plant species throughout the seasons and under various climatic conditions . In contrast to starch—which is stored in plastids—fructans appear in vacuoles of different organs: for example, in bulbs (onion), tubers (Jerusalem artichoke, dahlia), taproots (chicory), leaves, and in stems (wheat, barley).

Chemically, fructans are distinguished on the basis of the glycosidic bond that links fructose residues to each other. The biosynthesis of these polysaccharides starts with the incorporation of a fructose moeity to one of the three primary hydroxyl groups of vacuolar sucrose—that is, carbon-1 and carbon-6 in the fructose moeity, and carbon-6 in the glucose moeity (Figure 7.5). This reaction, catalyzed by the enzyme sucrose:sucrose 1-fructosyl transferase, implies a transglycosylation in which one molecule of sucrose transfers the fructosyl moeity to another. Thus, the addition of β-D-2,1- or β-D-2,6-linked fructofuranosyl units to the fructose moeity of sucrose yields 1-kestose and 6-kestose, respectively, while the addition of a β-D-2,6-linked fructofuranosyl unit to the glucose moeity of sucrose produces neo-kestose.

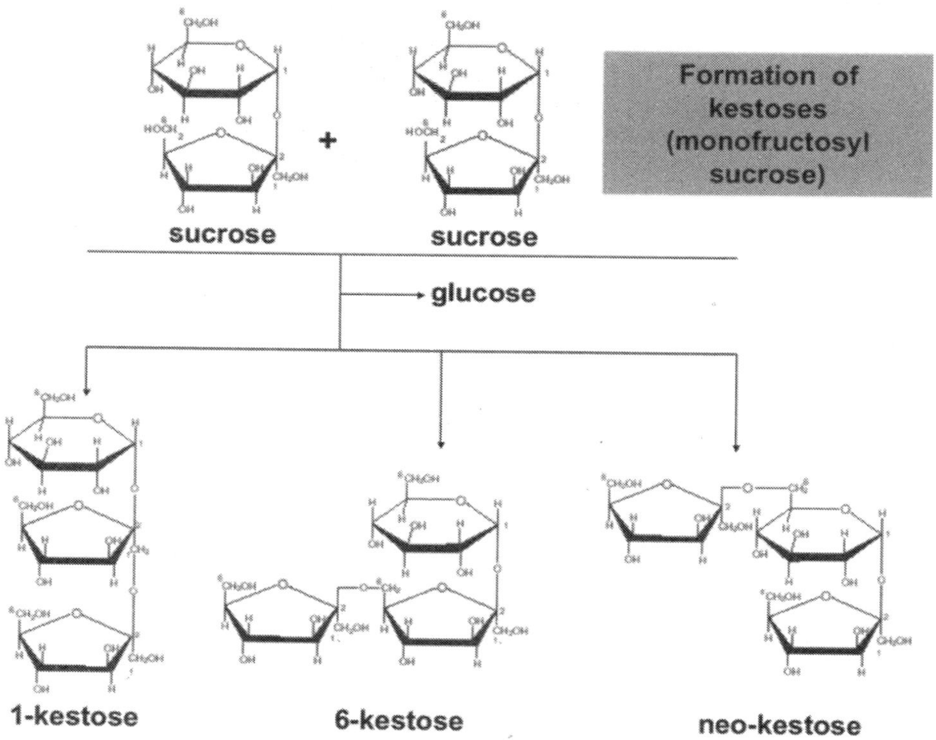

Figure 7.5 GF (sucrose) + GF (sucrose) → GFF (kestose) + G (glucose)

7.7 Plastidic Phosphate Translocators

Chloroplasts are metabolic powerhouses within leaf cells. They are not only the main site of carbon, nitrogen, and sulfur assimilation but also share metabolic pathways—such as the photorespiratory carbon cycle—with other plant cell compartments. A continuous exchange of metabolites and ions with the cytosol is necessary for playing these roles. In all vascular plants, plastids are surrounded by an envelope that restricts the nonspecific diffusion of polar molecules. Thus, a pair of concentric membranes modulate the exchange of metabolites between the stromal interior and the cytosol. Although many pore-forming proteins that are substrate-specific were characterized in the outer membrane, the inner membrane is considered the main permeability barrier between the cytosol and the chloroplast. All transporters of the inner membrane are nuclear-encoded membrane proteins whose precursors are synthesized in the cytosol and post-translationally imported into the inner membrane.

Chloroplast transporters collectively grouped as phosphate translocators catalyze a strict 1:1 counter-exchange (antiport) of orthophosphate with triose phosphates, pentose phosphates, and hexose phosphates. The Arabidopsis genome contains six genes that encode four classes of functional plastidial phosphate translocators (Table 7.1): single copies of the triose phosphate/phosphate translocator (TPT) and the phosphoenolpyruvate/phosphate translocators (PPT); and two copies each of the glucose 6-phosphate/phosphate translocators (GPT) and the xylulose 5-phosphate/phosphate translocator (XPT).

Table 7.1 : Plastidic Phosphate Translocators

Translocator	MATDB entry	Compounds transported
TPT	TPT (At5g46110)	triose phosphates, 3-phosphoglycerate
PPT	PPT1 (At5g54800), PPT2 (At3g01550)	phosphoenolpyruvate, 2-phosphoglycerate
GPT	GPT1 (At5g33320), GPT2 (At1g61800)	glucose 6-phosphate, triose phosphates
XPT	(At5g17630)	xylulose 5-phosphate, triose phosphates

1. The Triose Phosphate/Phosphate Translocator (TPT)

Triose phosphates, generated by the Calvin cycle at the expense of photosynthetically generated ATP and NADPH, flow to the cytosol across the chloroplast envelope. In turn, the orthophosphate released in the cytosol from biosynthetic processes is transported back into chloroplasts to replenish the ATP necessary for sustaining the assimilation of CO_2 and photosynthetic electron transport. Both the efflux of triose phosphates and the influx of orthophosphate are driven by a dimer composed of two identical subunits—the chloroplastic TPT. This membrane protein functions as an antiport system that exchanges phosphorylated—mainly three-, but also admits five- or six-carbon compounds—for orthophosphate. Hence, the chloroplast TPT exports the fixed carbon for the cytosolic synthesis of sucrose, the main product allocated to heterotrophic plant organs for further metabolism or conversion into storage products (e.g., starch, fructans). Therefore, the chloroplast TPT represents *the diurnal path for carbon export.*

Mutants defective in chloroplast TPT rely on different paths to transfer the photoassimilate from the chloroplast to the cytosol and, in so doing, compensate for the deficiency in TPT activity. Therefore, knockout mutations of this translocator increase the allocation of recently fixed carbon into transitory starch, followed by the degradation of the starch in the light and the subsequent export of neutral sugars to the cytosol. On the other hand, the simultaneous knockout of chloroplast TPT and starch synthesis drastically impairs plant growth.

2. Phosphoenolpyruvate Translocator (PPT)

PPT in the inner chloroplast membrane also links the stromal metabolism of all plants with the surrounding cytosol. C4-plants of the NADP-malic enzyme type (e.g., maize) under active photosynthesis produce pyruvate in bundle sheath cells. This three-carbon compound returns to mesophyll cells for the conversion to phosphoenolpyruvate by the chloroplast-localized pyruvate-phosphate dikinase (Figure 7.6). Subsequently, the phosphoenolpyruvate flows from the stroma to the cytosol for the primary carbon fixation via PEPCase. To ensure high rates of phosphoenolpyruvate efflux for the proper functioning of this cycle, the envelope of mesophyll chloroplasts bears the PPT, which concurrently drives the chloroplast uptake of orthophosphate to restore the lost phosphorus.

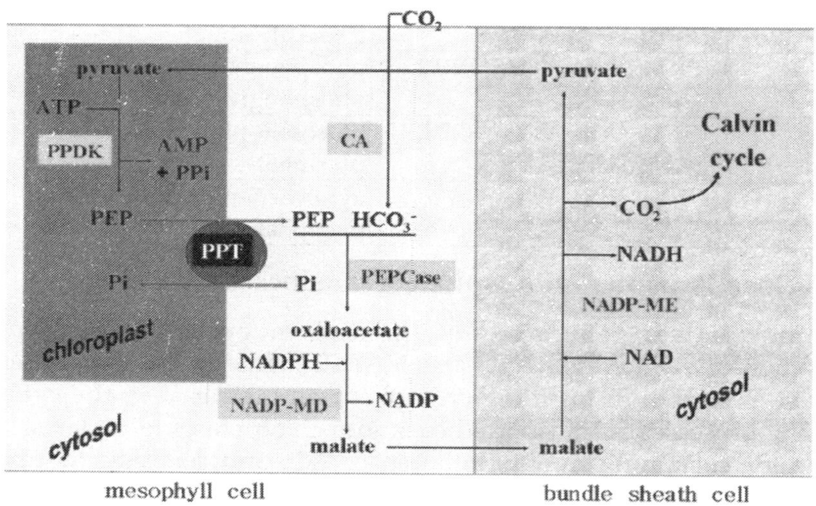

Figure 7.6 Mobilization of phosphoenolpyruvate in mesophyll chloroplasts of C4 plants of the NAD-malic enzyme type. The pyruvate produced by the malic enzyme (ME) in bundle sheath cells flows to mesophyll cells for conversion to phosphoenolpyruvate (PEP) by the pyruvate-phosphate dikinase (PPDK) localized at the chloroplast stroma. The efflux of PEP to the cytosol through the phosphoenolpyruvate/phosphate translocator (PPT) restores the acceptor of HCO3- required by the phosphoenolpyruvate carboxylase (PEPCase) for the primary carbon fixation. CA: carbonic anhydrase; NADP-MD: NADP-dependent malate dehydrogenase.

On the other hand, CAM plants of the malic enzyme-type decarboxylate during the day the malate accumulated at night. Again, two successive processes account for the production of phosphoenolpyruvate, the primary acceptor of CO_2. First, the pyruvate formed in decarboxylation of the malate enters into the chloroplast stroma for its transformation to phosphoenolpyruvate catalyzed by the pyruvate-phosphate dikinase. Second, the phosphoenolpyruvate leaves the chloroplast in exchange for orthophosphate via the PPT.

Knowledge on the function of the PPT in C_3 plants or nongreen tissues of C_4 plants is scarce. Apparently, the uptake—rather than the export—of phosphoenolpyruvate is the primary function of PPT in these tissues. Chloroplasts and most nongreen plastids are devoid of phosphoglyceromutase and enolase and, consequently, are unable to convert 3-phosphoglycerate to phosphoenolpyruvate. However, phosphoenolpyruvate combines with erythrose 4-phosphate, an intermediate of the pentose-phosphate pathway, for the synthesis of aromatic compounds through the shikimic acid pathway (Figure 7.7. Thus, the provision of cytosolic phosphoenolpyruvate to the stroma via the PPT would be essential for the biosynthesis of flavonoids, lignins, alkaloids.

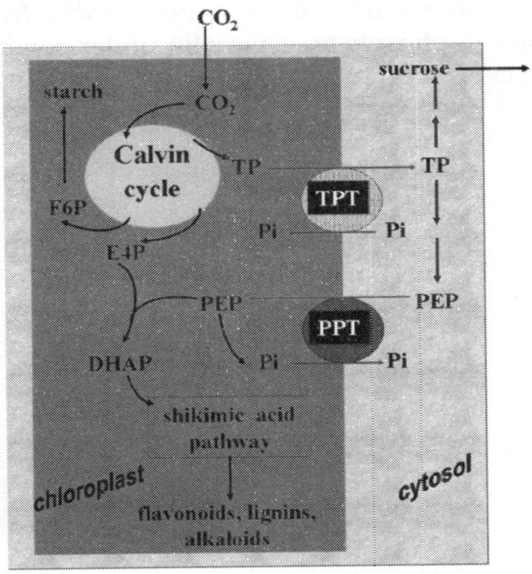

Figure 7.7 Mobilization of phosphoenolpyruvate in chloroplasts of C_3 plants. Under active photosynthesis, TPT furnish the cytosol with carbon skeletons for the biosynthesis of phosphoenolpyruvate. Upon the release of orthophosphate, the uptake of the cytosolic phosphoenolpyruvate, which combines in the stroma with erythrose 4-phosphate, triggers the formation of 3-deoxy-D-arabino-heptulosonic acid-7-phosphate (DHAP). DHAP is a scaffold for the biosynthesis of shikimic acid-5-phosphate, which in turn starts the pathway leading to the aromatic amino acids tyrosine and phenylalanine, the precursors of flavonoids, lignins, and alkaloids.

3. The Glucose 6-Phosphate/Phosphate Translocator (GPT)

The genome of Arabidopsis harbors two paralogous GPT genes, *AtGPT1* and *AtGPT2*, that, apparently, function in plastids of nongreen tissues for the import of glucose 6-phosphate. Two transgenic lines, bearing T-DNA insertions in the *GPT1* gene, exhibited not only impaired pollen and ovule development but also were lethal in the homozygous state. At variance, the disruption of the *GPT2* genes have no effect on growth and development. On the basis of drastic gametogenesis defects, it appears that the import of glucose 6-phosphate into nongreen plastids by the translocator AtGPT1 is crucial for the maturation of the pollen and the development of the female gametophyte.

4. The Xylulose 5-Phosphate/Phosphate Translocator (XPT)

The XPT constitutes the fourth subfamily of phosphate translocators that exhibits substrate specificity similar to the GPT but lacks the capacity to transport glucose 6-phosphate. In Arabidopsis, the lack of cytosolic transketolase and transaldolase halts the cytosolic oxidative pentose phosphate pathway at the stage of pentose phosphates. In this context, the putative function of the XPT would be to furnish the stromal pentose phosphate pathway with cytosolic five-carbon skeletons in the form of xylulose 5-phosphate, especially under conditions that require its removal from the cytosol.

Chapter **8**

Electron Transport Chain and Oxidative Phosphorylation during Respiration

8.1 The Electron Transport Chain

The electron transport chain (ETC) in the mitochondrial inner membrane of most eukaryotes consists of four large protein complexes, complex I to complex IV. The electron transport activity of complexes I, III and IV (the latter two, often jointly called the cytochrome pathway) is coupled to extrusion of protons across the inner mitochondria. The electrochemical proton gradient that is formed and maintained in the process is used by the ATP synthase (also called complex V) to make ATP (Figure 8.1). The plant ETC, however, is highly branched. As a consequence, the plant oxidative phosphorylation system has several alternative pathways of electron transport via type II NAD(P)H dehydrogenases and the alternative oxidase (AOX), and additionally, an alternative proton transporter called the uncoupling protein (UCP). These proteins mediate bypasses around the proton-translocating multi-protein complexes (I–V): type II NAD(P)H dehydrogenases bypass complex I, AOX bypasses complexes III and IV, and UCP bypasses the ATP synthase. The type II NAD(P)H dehydrogenases and the AOX transport electrons without pumping protons, whereas the UCP allows proton flow from the intermembrane space to the matrix without ATP synthesis.

8.2 The Alternative Oxidase

The presence of the AOX is one of the features that sets plant (and fungal) mitochondria apart from mammalian mitochondria. This enzyme has therefore received a lot of attention from plant scientists and we now have a fairly good understanding of the regulation of its structure and activity.

The AOX is a quinol-oxygen oxidoreductase and it does not pump protons. The oxygenase transfers electrons from ubiquinol (the reduced form of ubiquinone) to oxygen and generates water as the end-product of the reaction (Figure 8.1), so four electrons must be transferred to oxygen. The precise mechanism is not known, but the protein has been shown to contain a catalytical di-iron centre.

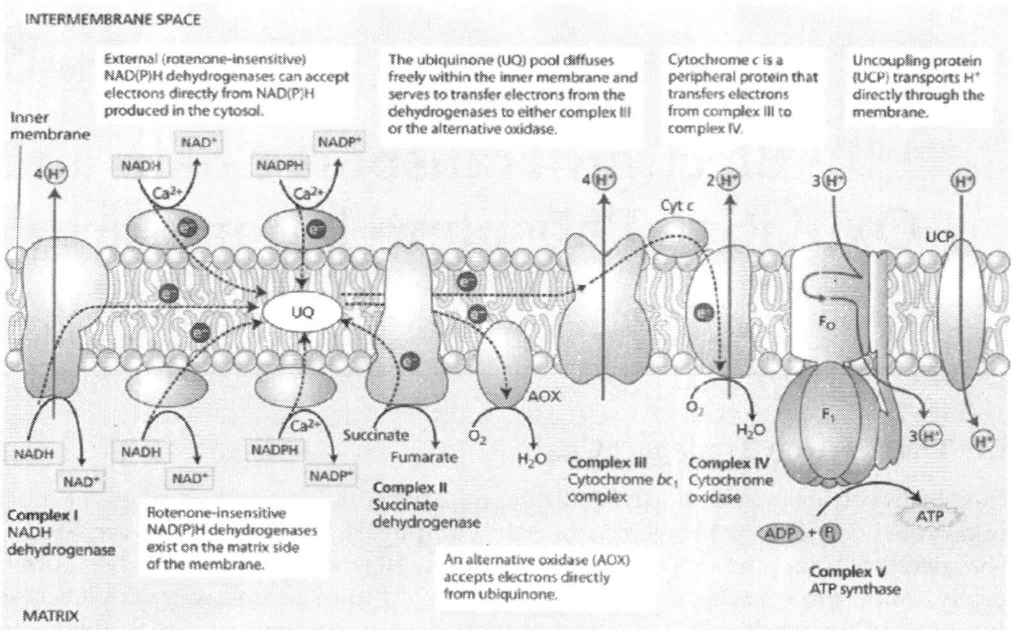

Figure 8.1 Organization of the electron transport chain and ATP synthesis in the inner membrane of plant mitochondria. In addition to the four standard protein complexes found in nearly all other mitochondria, the electron transport chain of plant mitochondria contains five additional enzymes (depicted in green). None of these additional enzymes pumps protons. Additionally, the uncoupling protein directly bypasses the ATP synthase by allowing passive proton influx. This multiplicity of bypasses in plants, where animals have only the uncoupling protein, gives a greater flexibility to plant energy coupling. Specific inhibitors—rotenone for complex I, antimycin for complex III, cyanide for complex IV, and salicylhyroxamic acid (SHAM) for the alternative oxidase—are important tools used to investigate mitochonondrial electron transport. Some also have commercial uses, such as rotenone—which is used as an insecticide and to remove unwanted fish from lakes. Because plants have the alternative pathways, they can survive exposure to inhibitors of the respiratory complexes.

The functional form of the enzyme is a dimer with the two polypeptides either covalently or non-covalently bound to each other (Figure 8.2). The predicted structure places the protein embedded into the inner leaflet of the inner mitochondrial membrane, with the active site and the regulatory cysteines exposed to the matrix

When electrons from NADH oxidation flow through the AOX, at least two of the three sites of proton translocation (complexes III and IV, see Figure8.1) are bypassed, so energy conservation in the form of ATP is much smaller when the AOX is active. The activity of the AOX must therefore be carefully regulated. There are often several AOX isoforms in each plant species—for example, five genes encoding AOX proteins are present in Arabidopsis. The relative amounts of three isoforms of AOX vary with development, and cellular redox status, as well as external factors.

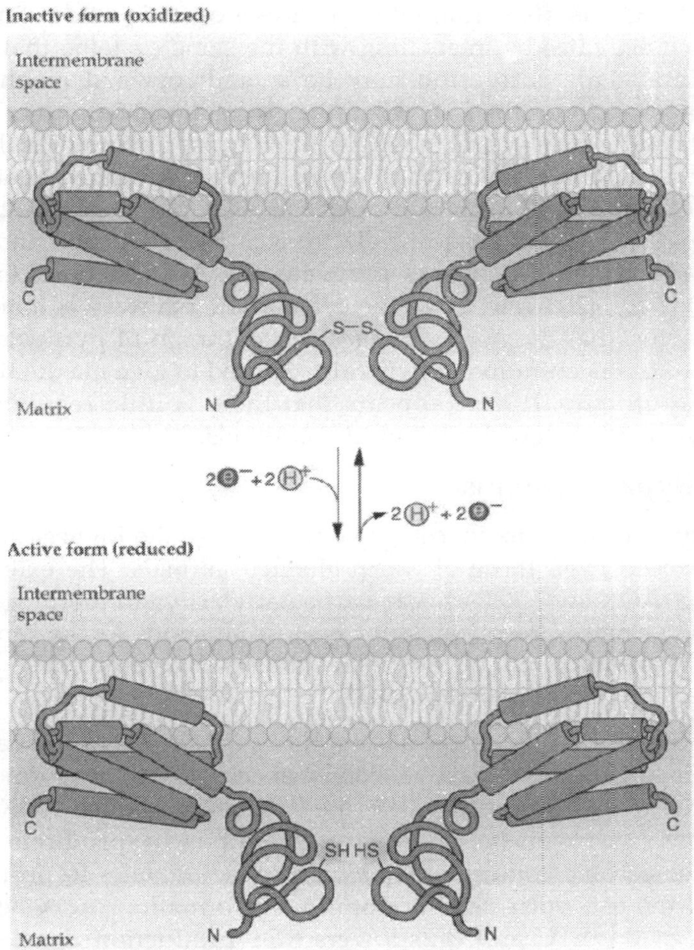

Figure 8.2 The structure and regulatory features of the alternative oxidase in plant mitochondria. The C and N indicate the C- and N-terminal ends of the protein, respectively (Siedow and Umbach,2000,:*Biochem. Biophys. Acta* 1459:432-439).

The AOX is present in the inner membrane as a dimer in which the two subunits can be covalently linked through a disulfide bridge between cysteine residues. The reduced non-covalently linked dimer is the active form, and this form appears to be the predominant one *in vivo*. It was recently shown that a mitochondrial thioredoxin system regulates AOX. Thioredoxins regulate enzymes by reducing disulfide bridges, and it is likely that the regulation involves the redox-active cysteines. The presence of thioredoxin regulation connects the AOX activation level to the mitochondrial NAD(P)H pool and the oxidation of citric acid cycle metabolites by NADP-utilizing enzymes.

The activity of AOX is also stimulated by 2-oxo organic acids—in particular, pyruvate, which activates by interacting with the same cysteine that is involved in dimerization . This activation may be a feed-forward mechanism that upregulates the capacity of the ETC when the glycolytic end-product is available. Both the cytochrome pathway and the AOX pathway use ubiquinol as the substrate. In the absence of pyruvate, the cytochrome pathway is saturated at 40% Q_{red}/Q_{tot} (40% of the UQ is reduced), at which point the AOX pathway is not engaged at all. The AOX pathway starts accepting electrons at 50% reduction and is not fully saturated at 80%. In the presence of pyruvate, the AOX becomes engaged at a much lower Q_{red}/Q_{tot} (20%), whereas the cytochrome pathway is unaffected. It is puzzling, however, that the normal *in vivo* concentration of pyruvate appears to be higher than the concentration of pyruvate required to give maximal stimulation of AOX activity *in vitro*. It thus appears that there is little regulation of AOX activity via pyruvate under physiological conditions.

8.3 Energy Bypass Proteins

The respiratory rates of some thermogenic flowers are the highest among plants and, in fact, exceed even those of warm-blooded animals. The external NADH dehydrogenase, AOX and UCP appear particularly active in thermogenic flowers where they are primarily responsible for heat production. At the same time it is clear that the respiratory rate of most plant tissues is insufficient to generate an appreciable increase in tissue temperature.

AOX is up-regulated by cold stress, although this is clearly not to heat the tissue. However, during cold stress, reactive oxygen species (ROS) are formed, and also other stress treatments that lead to elevated ROS will induce AOX. ROS (e.g., superoxide and hydrogen peroxide) are formed as by-products of electron transport under aerobic conditions. ROS can cause damage to proteins, lipids, and DNA and the cell must therefore limit their formation. AOX helps prevent overreduction of the ETC and thus lowers ROS production demonstrated the importance of the AOX for limiting ROS formation in living plant cells. Overexpression of AOX lowered the steady state level of ROS, whereas suppression increased it. This result also points to the mitochondrion as an important site of ROS generation in plant cells. With the recent discovery of ROS stimulation of plant UCP and the increased ROS resistance in plants overexpressing UCP , the picture has become more complex. By decreasing the resistance for electron transport through the proton-pumping complexes, UCP activity may, like AOX, lead to decreased overreduction and ROS formation. Another factor speaking in favour of UCP having a role in ROS protection is that the AOX is inhibited by an ROS product. For further information on mitochondrial involvement in ROS metabolism.

Both type II NAD(P)H dehydrogenases and AOX may act as reductant overflow enzymes by reoxidizing surplus NADH to oxygen. This can prevent inhibition of the citric acid cycle under conditions in which the ADP concentration is low, and thus the proton-pumping complexes are restricted by a high membrane potential. In this way, the citric acid cycle can supply carbon skeletons for biosynthesis (e.g., 2-oxoglutarate for nitrogen assimilation), even if the ATP level in the cell is high. The concept of overflow is also relevant for photorespiratory conditions, under which there is a massive flow of glycine that has to be oxidized by the mitochondria. As much as half of the NADH produced in this process is probably exported, but the other half has to be oxidized promptly. There are several reasons to believe that a reductant overflow via energy bypass proteins is involved. The ND_{in}(NADH) and AOX capacity in leaf mitochondria is higher than in mitochondria from heterotrophic tissue, and both proteins are down-regulated by dark treatment. Consistent with these observations, AOX activity *in vivo* is up-regulated by light, AOX capacity is induced by greening of etiolated tissue, and expression of the *nda*1 gene [encoding an internal NAD(P)H dehydrogenase] in potato and Arabidopsis is light-dependent.

Mutant tobacco plants, lacking the proton-pumping complex I can still survive, apparently aided by the action of ND_{in}(NADH), which oxidizes matrix NADH with a lower ATP yield. These plants, however, show decreased photosynthetic rates, indicating that the ND_{in}(NADH) cannot fully compensate for the missing complex I. Also, external NAD(P)H dehydrogenases have been suggested to support photosynthesis. These enzymes may reoxidise NAD(P)H that is shuttled out of the chloroplast in order to balance the NAD(P)H to ATP ratio in the stroma. In summary, there are many indications that mitochondria—especially the energy bypass proteins—have roles in supporting photosynthesis, for example by oxidizing excess NAD(P)H.

The AOX pathway is involved in the acidification/deacidification cycle of CAM plants. CAM plants fix CO_2 in the form of malate during the night. This is done by cytosolic PEP carboxylase, and the malate produced is stored in the vacuole. The CO_2 fixed is released during the daytime by malic enzymes, located in cytosol and mitochondria). The oxygen discrimination technique was used to measure the contribution of the AOX pathway to whole-leaf respiration in the CAM-plant, *Kalanchoë daigremontiana.* They found that the respiratory rate of the leaves increased by 35% in the early hours of the day. This increase was due to increased AOX activity (Table 8.1). A more detailed time course showed that the AOX activity increased two- to threefold during the early day at the same time as the malate was metabolized. As soon as the malate had disappeared by mid-day, the AOX activity returned to the low level observed during the night (Figure 8.3). A reasonable conclusion is that a substantial amount of the malate from the vacuole is oxidized via malic enzyme in the mitochondria. This generates NADH in the matrix, which is oxidized to a large extent via the AOX. The released CO_2 can then be refixed by the Calvin cycle in the chloroplast.

Table 8.1 Respiration, oxygen isotope discrimination and partitioning of respiratroy pathways in *Kalanchoe* leaf discs during acidification and deacidification (in the first 6h of the light period)

Parameter	Acidification	Deacicification
Measured		
Unhibited respiration	1.81	2.56
Rate (? mol O_2 uptake m^{-2} s^{-1})	22.38	25.00
Discirmination (0/00)		
Caluclated*		
Cytochrome pathway flux (? mol O_2 uptake $m^{-2}s^{-1}$)	1.33	1.40
Alternative pathway flux (? mol O_2 uptake $m^{-2}s^{-1}$)	0.52	1.16

Source : Robinson *et al*. (1992) Plant Physiol. 95:157-163

*Based on the following measurement. Cytochrome oxidase and alternative oxidase discrimination was 19.8 and 31.2 o/oo, respectively

Respiration, oxygen isotope discrimination and partitioning of respiratory pathways in *Kalanchoë* leaf discs during acidification and deacidification (in the first 6 h of the light period).

8.4 Relationship Between Respiration and Crop Yields

Plant respiration can consume an appreciable amount of the carbon fixed each day during photosynthesis over and above the losses due to photorespiration. To what extent can changes in a plant's respiratory metabolism affect crop yields?

Attempts to establish a quantitative relationship between respiratory energy metabolism and the various processes going on in the cell have led to a break-up of respiration into two components. In growth respiration, reduced carbon is processed to bring about the addition of new biomass. The other component, maintenance respiration, is needed to keep existing, mature cells in a viable state. Utilization of energy by maintenance respiration is not well understood, but estimates indicate that it can represent more than 50% of the total respiratory flux.

Although numerous questions remain regarding these issues, there are several empirical examples of relations between plant respiration rates and crop yield. In the forage crop, perennial ryegrass (*Lolium perenne*), yield increases of 10% to 20% were correlated with a 20% decrease in the leaf respiration rate . Similar correlations have been found for other plants, including corn (*Zea mays*) and tall fescue (*Festaca arundinacea*). However, later investigations have shown that "selection for low rates of mature leaf respiration is not an appropriate method to select for high-yielding cultivars in perennial grasses".

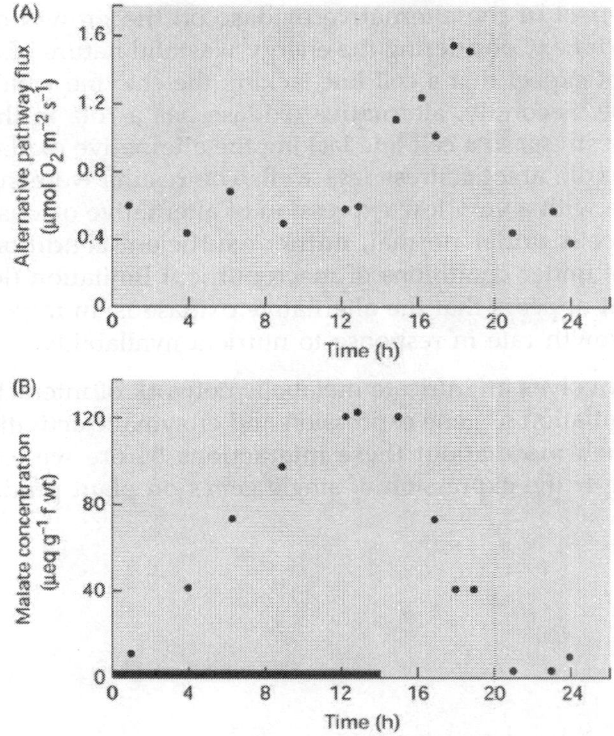

Figure 8.3 The participation of the AOX in malate oxidation in *Kalanchoë diagremontiana* during a night-day cycle. (A) Changes in flux through the AOX pathway in leaf discs. (B) Changes in malate concentration of leaf discs. The solid bar at the bottom of the figure represents the dark period. (Robinson *et al.* 1992.*Plant Physiol.* 95:157-163)

Although there is a potential for increasing crop yield through reduction of respiration rates, a better understanding of the sites and mechanisms that control plant respiration is needed before such changes can be exploited commercially in a systematic fashion by plant physiologists, geneticists, and molecular biologists. Furthermore, much remains to be learned about the general applicability of such observations and the conditions under which slower-respiring lines could be at a disadvantage, causing a reduction in crop yields rather than an increase.

Two recent studies illustrate the present difficulty of predicting the effect of directed changes at the molecular level on plant productivity:

The productivity of transgenic tomato plants with a reduced activity of mitochondrial malate dehydrogenase was increased as compared to the wild type. Although the respiratory activity of mitochondria isolated from the transgenic plants was unchanged or higher than in the wild type, the rate of leaf respiration was reduced in the transgenic plants. Photosynthesis was markedly increased in the transgenic lines, possibly linked to increased levels of ascorbate.

Studieds on the effect of the alternative oxidase on the growth of tobacco cell cultures has been done.tConsidering the energy-wasteful nature of the alternative oxidase one might expect that a cell line lacking the enzyme would grow faster than the wild type. Secondly, alternative oxidase has a role in the response of plants to oxidative stress, so a cell line lacking the alternative oxidase would also be expected to handle abiotic stress less well. The results were surprising—the transgenic cell line with a very low expression of alternative oxidase grew as fast as the wild-type cells under normal, nutrient-sufficient conditions, and faster than the wild type under conditions of macronutrient limitation (low phosphate or low nitrogen). It appears that the alternative oxidase is an important factor in modulating the growth rate in response to nutrient availability.

Plant respiration involves an intricate metabolic network of interacting pathways and a complex regulation of gene expression and enzymatic activities. We clearly need to know much more about these interactions before we can predict the effects of changing fr the expression of single genes on plant productivity

Chapter 9

Translocation

9.1 Studies on Phloem Transport

In classic experiments on the translocation of organic solutes performed by the Italian anatomist Marcello Malpighi in 1686, the bark of a tree was removed in a ring around the trunk . In 1928, T. G. Mason and E. J. Maskell observed that this treatment, called **girdling**, has no immediate effect on transpiration, since water moves in the xylem, interior to the bark However, sugar transport in the trunk is blocked at the site where the bark has been removed. Sugars accumulate above the girdle—that is, on the side toward the leaves—and are depleted below the treated region. (Figure 9.1)

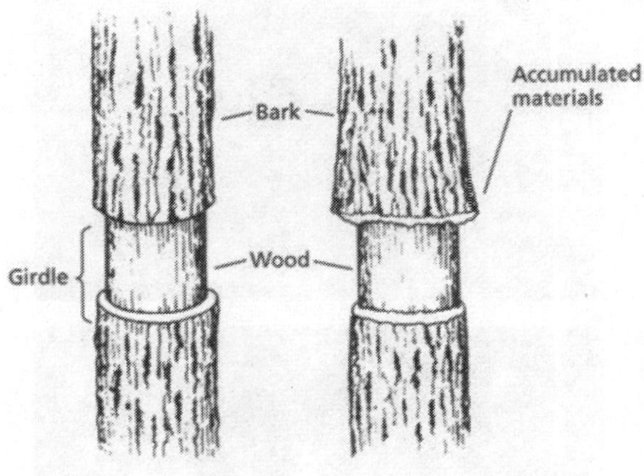

Figure 9.1 Tree trunk immediately after girdling (left) and later (right). Girdling is the removal of the bark of a tree in a ring around the trunk. At right, materials translocated from the leaves have accumulated in the region above the girdle and caused it to swell.

Eventually the bark below the girdle dies, while the bark above swells and remains healthy. Mason and Maskell concluded that sugar is transported in the bark of the tree and that the sieve elements are the cellular channels of sugar transport. The latter conclusion was based on an observed high correlation between leaf and bark sucrose contents and on the high sucrose concentration calculated to be present in the sieve elements.

More sophisticated experiments on phloem translocation became possible in the

1940s, when radioactive isotopes became available for scientific research. Labeled organic compounds can be introduced into the plant in various ways. For example, carbon dioxide can be labeled with ^{14}C or ^{11}C and supplied to an intact mature leaf enclosed in a sealed chamber. These two isotopes of carbon, ^{11}C and ^{14}C, have very different half-lives (the half-life is the time required for the radioactivity to decrease by one-half). The half-life for ^{11}C is 20 minutes; that for ^{14}C is 5,600 years. For this reason, the isotopes are used in different types of experiments. Carbon-14 is useful for long term experiments analyzing the compounds in which the radioactivity is present. Carbon-11 would disappear before such an analysis could be made. Carbon-11 is most useful when the researcher wishes to repeat measurements on the same plant. Since only a few hours are needed for the isotope in the plant to disappear, the radioactivity used in one experiment does not interfere with next.

The labeled carbon dioxide is incorporated into organic compounds via the CO_2-fixing reactions of photosynthesis and eventually into transport sugars. In another type of experiment, researchers bypass the photosynthetic reactions by supplying labeled transport sugar, usually sucrose, directly to the leaf.

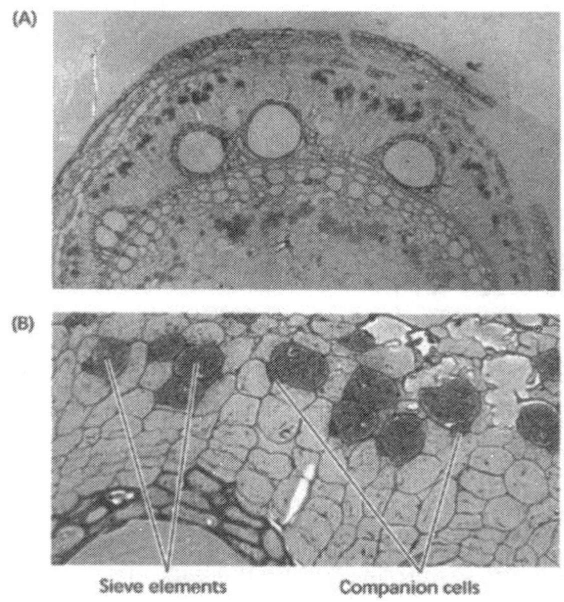

Figure 9.2 Photosynthate from the leaves appears in the sieve elements of phloem in the stem. $^{14}CO_2$ was supplied to a source leaf of morning glory (*Ipomea nil*). ^{14}C was incorporated into sugars synthesized in the photosynthetic process, which were then transported to other parts of the plant. The location of the label is revealed in the tissue cross sections by the presence of dark grains on the film. (A) shows a low magnification of the cross section of the stem (50×), revealing dark spots resulting from the silver grains in the film, shown in higher magnification (325×) in (B). The label is confined almost entirely to the sieve elements of the phloem.

To identify the pathway of sugar translocation, we must determine the tissue or cellular location of the isotope, usually by means of a technique called **autoradiography**. In tissue autoradiography, tissue of labeled plants is rapidly frozen, freeze-dried, embedded in paraffin or resin, and sliced into thin sections. The sections are then coated with a film of photographic emulsion.

During a period of exposure, radiation from the radioactive compound fed to the plant exposes the film, and when the film is developed, silver grains appear wherever the label was present in the tissue. A comparison of the tissue section and the pattern of silver grains shows the location of the label in the tissue. In terms of the transport pathway for sugars, the label initially appears in the sieve elements of the phloem, confirming the results of the earlier experiments (Figure 9.2.).

9.2 Identification of matirial of Phloem Sap

The ideal way to collect phloem sap would be to tap into a single sieve element by using a tiny syringe. Fortunately, nature has provided just such a probe: the aphid stylet. Aphids are small insects that feed by inserting their mouthparts, consisting of four tubular stylets, into a single sieve element of a leaf or stem (Figure 9.2). The high turgor pressure in the sieve element forces the cell contents through the insect's food canal and into its gut, where amino acids are selectively removed and some sugars are metabolized. The excess sap, still rich in carbohydrates, is then excreted as "honeydew" (the droplet seen exuding from the body of the aphid which can be collected for chemical analysis. Sap can be collected from aphid stylets cut from the body of the insect, usually with a laser, after the aphid has been anesthetized with CO_2. Exudate from severed stylets provides a more accurate picture of the substances present in the sieve elements than does honeydew (aphid excreta), the composition of which has been altered by the insect.

Experimental techniques using aphids have substantial advantages in the study of phloem physiology. Of particular importance is the fact that aphids tap a single sieve element, so there is no problem of contamination from other cell types. Exudation from severed stylets can continue for hours, suggesting that the aphid prevents normal sealing mechanisms from operating or that the drop in pressure that occurs when the sieve element is penetrated is not sufficient to trigger sealing mechanisms.

The disadvantages of using aphids include difficulties in placing the insects at a desired location and severing the stylets without disrupting them. In addition, aphids may induce reactions in the host plant by secreting saliva into the plant tissues. However, the magnitude of these reactions over the period of time necessary to collect sap for analysis is probably negligible.

Figure 9.3 Using aphids to tap a single sieve element and collect phloem sap. (A) An aphid (*Longistigma caryae*) feeding on a branch of linden (*Tilia americana*). The high turgor pressure in the sieve element forces the cell contents through the food canal of the aphid and into its gut. The honeydew excreted by the aphid consists of sieve element sap from which selected solutes, especially amino acids, have been removed in the gut of the insect. The high turgor pressure in the sieve element forces the cell contents through the food canal of the aphid and into its gut. The insect in the photo is releasing a droplet of honeydew, which it does about once every 30 minutes. (B) Transverse section through the bark of linden, showing the tips of aphid stylets that had been exuding before they were sectioned. The tips of the stylets are inside a single sieve element; the sheath of saliva secreted by the aphid ends outside the cell (Zimmerman and Brown, 1971. *Trees: Structure and Function*. Springer-Verlag Berlin)

9.3 Loading of Sieve Elements

Early research on phloem loading focused on the apoplastic pathway . Many studies devoted to testing these predictions have provided solid evidence for apoplastic loading in several species. Some of the more important evidence is:

1. Transport sugars are found in the apoplast. Sucrose is the predominant apoplastic sugar in species that transport mainly sucrose in the phloem, for example, sugar beet (*Beta vulgaris*) and broad bean (*Vicia faba*). Treatments and events that alter the rate of translocation from the source leaf also change the flux of sucrose through the apoplast or the apoplastic sucrose level.

2. Suçrose supplied exogenously to a source leaf of sugar beet accumulates in the sieve elements and companion cells of the minor veins, as does sucrose derived from photosynthetic CO_2 fixation. Similar observations have been made in broad bean, pea (*Pisum sativum*), castor bean (*Ricinus communis*), and other species.

3. PCMBS (*p*-chloromercuribenzenesulfonic acid) is a reagent that inhibits the transport of sucrose across plasma membranes but does not enter the symplast.

PCMBS inhibits the uptake of sucrose from the apoplast when the sugar is supplied exogenously to sugar beet (Figure 9.4). Of greater importance, PCMBS also inhibits the export of sucrose synthesized from CO_2 in the mesophyll, implying that short-distance transport in sugar beet normally includes an apoplastic step. Assimilate loading is *also* inhibited by PCMBS in broad bean, pea, and other species.

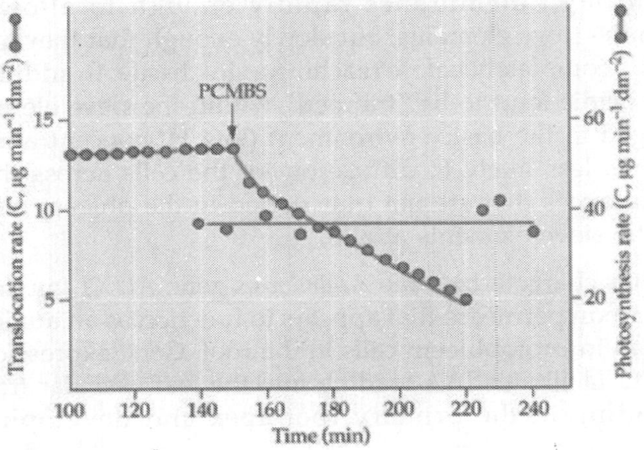

Figure 9.4 Effects of PCMBS on phloem transport from a sugar beet leaf. Export pf labeled photosynthate from a sugar beet leaf decreases upon addition of PCMBS. Photosynthesis was not affected, indicating that PCMBS did not alter the photosynthetic properties of the leaf.(Fisher,2000,*In Biochemistry and Molicular Biology of Plants*,.American Society of Pl Physiologist, Rockvill,Md.)

4. Experiments with transgenic plants further support the interpretation that sucrose moves through the apoplast. (Transgenic organisms carry cloned genes integrated into their DNA by a variety of recombinant-DNA techniques.) Transgenic tomato plants that have a sucrose-cleaving enzyme (invertase) in their apoplast show a very slow growth rate and fail to mobilize the starch in their source leaves during a prolonged dark period . Apparently, hydrolyzing sucrose in the apoplast inhibits phloem loading in source leaves of this species. This result is expected only if the phloem-loading pathway includes an apoplastic step.

9.4 Diffusion of Organic Acids in Phloem

Many substances, such as organic acids and plant hormones, are found in the phloem sap at lower concentrations than carbohydrates. These substances are probably not actively loaded into the sieve element–companion cell complex but enter the sieve elements via other pathways and mechanisms. They may be taken up directly by diffusion across the phospholipid bilayer of the plasma membrane of the sieve element–companion cell complex or by a passive transporter in the plasma membrane of those cells, or they may diffuse into the sieve elements via the symplast.

Once in the sieve elements, these substances are swept along in the translocation stream by bulk flow, the motive force being generated by the active loading of only certain sugars or amino acids. Many substances not normally found in plants, such as herbicides and fungicides, can be transported in the phloem because of their ability to diffuse through membranes at an intermediate rate. In other words, they diffuse through membranes rapidly enough to allow considerable accumulation in the sieve elements, but slowly enough that they are not lost from the sieve elements completely before reaching a sink tissue. In addition, compounds that are weakly acidic tend to be "trapped" within the sieve elements, becoming negatively charged in the basic environment (low H^+ concentration) of the sieve elements and thus less likely to diffuse out of the cells across the hydrophobic membrane . Substances that are not transported in the phloem (such as calcium) may not enter the sieve elements at all.

A recent study has characterized the *Arabidopsis* gene *AUX1* (auxin influx carrier) that encodes an auxin permease that appears to function as an auxin carrier. *AUX1* has been localized in protophloem cells in the root. Gene expression studies have shown that *AUX1* facilitates IAA loading into the leaf vascular transport system, and IAA unloading in the primary root apex and developing lateral root primordium.

9.5 Sugars in the Phloem

In symplastic phloem unloading, transport sugars such as sucrose move through the plasmodesmata to the sink cells. In the sink cells, sucrose can be metabolized in the cytosol or the vacuole before being stored or entering metabolic pathways

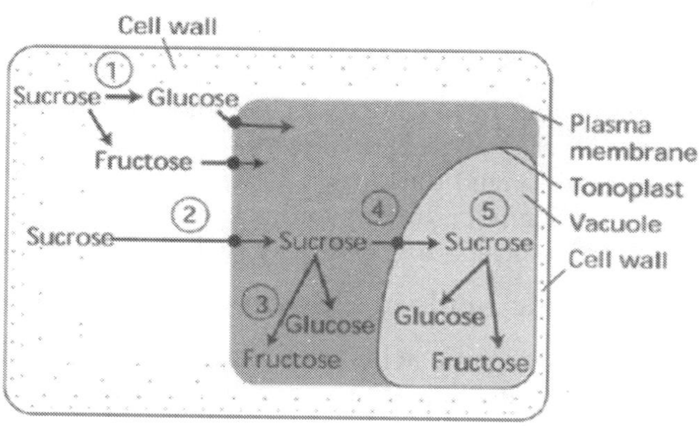

Figure 9.5 The possible fates of sucrose unloaded apoplastically in sink tissues. (1) Sucrose that enters the apoplast can be split into glucose and fructose by a wall invertase before entering a cell from a sink tissue, or (2) sucrose can be taken up into the cell unaltered. (3) Once in the symplast of the cell from the sink tissue, sucrose can be split into glucose and fructose by a cytoplasmic invertase, or (4) sucrose can enter the vacuole unaltered. (5) Once in the vacuole, sucrose can be split into glucose and fructose by a vacuolar invertase, or it can remain unaltered.

associated with growth of the tissue. When phloem unloading is apoplastic, however, there is an additional opportunity for metabolic change. The transport sugar can be partly metabolized in the apoplast, or it can cross the apoplast unchanged (Figure 9.5). For example, sucrose can be hydrolyzed into glucose and fructose in the apoplast by invertase, and glucose and/or fructose would then enter the sink cells. The fact that many monosaccharide transporters have been localized mainly in sink tissues supports the possible existence of this pathway.

9.6 Unloading Process in Developing Seeds and Organs

Developing seeds have proven to be a most interesting system in which to study unloading processes. In legumes such as soybean, the embryo can be removed from the seed coat. In this way, unloading from the seed coat into the apoplast can be studied without the influence of the embryo, and uptake into the embryo can also be investigated separately. Studies with legumes have shown that both entry of sucrose into the apoplast and uptake into the embryo are mediated by transporters and are active. In cereals like wheat, only uptake into the embryo is active; the loss of sucrose (sucrose efflux) from the maternal tissues is passive (down the concentration gradient), because the subsequent active step keeps the sucrose concentration in the apoplast low. In corn, the cell wall invertase helps maintain a low apoplastic sucrose concentration by splitting the disaccharide into monosaccharides. In general, sugar–proton symport mechanisms appear to function in the uptake of sugars from the apoplast, as in sucrose uptake into the soybean embryo.

Storage organs often accumulate sugars to high concentrations, for example, in sugar beet taproot and sugarcane stem. This sugar accumulation requires active membrane transport, since energy is required to move sugars into storage compartments against a concentration gradient.

Sugar transport into the vacuoles of storage cells such as those of sugar beet is thought to be accomplished by a **sucrose–proton antiport** . In this case, a vacuolar H^+-ATPase pumps protons into the vacuole; the antiport carrier then moves sucrose into the vacuole in exchange for protons, which exit the vacuole down their electrochemical-potential gradient.

9.7 Assimilate Allocation and Partitioning

The photosynthetic rate determines the total amount of fixed carbon available to the leaf. However, the amount of fixed carbon available for translocation depends on subsequent metabolic events. The regulation of the diversion of fixed carbon into the various metabolic pathways is termed **allocation**.

The vascular bundles in a plant form a system of pipes that can direct the flow of photoassimilates to various sinks: young leaves, stems, roots, fruits, or seeds. However, the vascular system is often highly interconnected, forming an open network that allows source leaves to communicate with multiple sinks. Under these conditions, what determines the volume of flow to any given sink? The differential distribution of photoassimilates within the plant is termed **partitioning**.

Photosynthetic rate and sink demand are linked processes, particularly in species which store starch. Increased sink demand is correlated with increased photosynthesis, and vice versa. An accumulation of photosynthate (starch, sucrose, or hexoses), in the source leaf could be one of the signals coupling photosynthetic rate with sink demand. Possible mechanisms include:

1. **Inhibition by starch**. When sink demand is low, high starch levels in the source could physically disrupt the chloroplasts, interfere with CO_2 diffusion, or block light absorption. Little evidence supports this hypothesis.

2. **Phosphate availability**. When sink demand is low, photosynthesis could be restricted by a lack of free orthophosphate in the chloroplast (Du *et al.* 2000). Under conditions of low demand by the sink, sucrose synthesis is usually reduced, and less phosphate is thus available for exchange with triose phosphate from the chloroplast (via the phosphate translocator). If starch synthesis, which releases orthophosphate in the chloroplast, could not release phosphate fast enough, a deficiency in phosphate would ensue. ATP synthesis and thus CO_2 fixation would decline. A study with potato plants transformed with antisense DNA to the phosphate translocator provides support for this hypothesis. The transformed plants, which displayed reduced phosphate translocator activity, allocated proportionately more carbon into starch and less into sucrose. These effects were accompanied by a reduction in the light- and CO_2-saturated rates of photosynthesis in young plants.

3. **Regulation by sugars**. High sugar levels decrease the transcription rate and expression of genes for many photosynthetic enzymes. The changes in gene expression occur over the same time frame as the source adjustments already described. For example, in source leaves of spinach (*Spinacia oleracea*), mRNA for several photosynthetic enzymes decreased when soluble carbohydrates accumulated as a result of inhibition of export from the leaf. Although transcript levels began to decline almost immediately, changes in photosynthetic enzyme activity were apparent only after several days. In this species at least, photosynthesis appeared to be inhibited because of changes in gene expression, not because of phosphate limitation, as discussed earlier.

After giving an overview of allocation and partitioning, we will examine the coordination of starch and sucrose synthesis in some detail. We will conclude by discussing how sink demand might regulate photosynthetic rate in the source leaf, and how sources and sinks communicate with each other.

Allocation Includes the Storage, Utilization, and Transport of Fixed Carbon

1. *Synthesis of storage compounds.* Starch is synthesized and stored within chloroplasts and, in most species, is the primary storage form that is mobilized for translocation during the night. Such plants are called starch storers. In some organs of certain grasses, fructans (polymerized fructose molecules) are the storage compounds, rather than starch. (Seed also item 3 below)
2. *Metabolic utilization.* Fixed carbon can be utilized within various compartments of the photosynthesizing cell to meet the energy needs of the cell or to provide carbon skeletons for the synthesis of other compounds required by the cell.
3. *Synthesis of transport compounds.* Fixed carbon can be incorporated into transport sugars for export to various sink tissues. A portion of the transport sugar can also be stored temporarily in the vacuole. In most species studied, this stored sucrose is highly transitory, providing a buffer against short-term changes in sucrose synthesis. But in some species, such as barely (*Hordeum vulgare*), fixed carbon is stored primarily as sucrose for use during the night, and very little is stored as starch. Such plants are called sucrose storers.

Allocation is a key process in sink tissues as well. Once the transport sugars have been unloaded and enter the receiver cells, they can remain as such or can be transformed into various other compounds. In storage sinks, fixed carbon can be accumulated as sucrose or hexose in vacuoles or as starch in amyloplasts. In growing sinks, sugars can be utilized for respiration and for the synthesis of other molecules required for growth.

Transport Sugars Are Partitioned among the Various Sink Tissues

The greater the ability of a sink to store or metabolize imported sugars (the process of allocation), the greater is its ability to compete for assimilates being exported by the sources. Such competition determines the distribution of transport sugars among the various sink tissues of the plant (assimilate partitioning), at least in the short term.

Of course, events in sources and sinks must be synchronized. Partitioning determines the patterns of growth, and that growth must be balanced between shoot growth (photosynthetic productivity) and root growth (water and mineral uptake). So an additional level of control lies in the interaction between areas of supply and demand. Turgor pressure in the sieve elements could be an important means of communication between sources and sinks, acting to coordinate rates of loading and unloading. Chemical messengers are also important in signaling to one organ the status of the other. Such chemical messengers include plant growth regulators (hormones) and nutrients, such as sucrose, potassium, and phosphate.

There has been a surge in research on assimilate allocation and partitioning, with the goal of improving yields of crop plants. In general, efforts by plant breeders to increase yield by increasing net photosynthetic rates have been unsuccessful. However, significant improvements in yield have resulted from increase in shoot harvest index, the ratio of commercial or edible yield such as grain to total shoot yield (the latter including inedible portions of the shoot). An understanding of partitioning should enable plant breeders to select and develop varieties with improved transport to edible portions of the plant.

However, allocation and partitioning in the whole plant must be coordinated such that increased transport to edible tissue does not occur at the expense of other essential processes and structures. Crop yield will also be improved if photoassimilates that are normally "lost" by the plant are retained. For example, losses due to nonessential respiration or exudation from roots could be reduced. In the latter case, care must be taken not to disrupt essential processes outside the plant-for example, the growth of beneficial microbial species in the vicinity of the root.

Enzymes Involved in Allocation

The quantity of sucrose available for export during the day is influenced by various biochemical reactions and carrier-mediated events, including the rate of CO_2 fixation. In fact, increasing the rate of photosynthesis in a source leaf generally results in an increase in the rate of translocation from the source. The fraction of fixed carbon that is channeled to the so-called transport pool is determined by other processes. Control points include the allocation of triose phosphates: (1) to the regeneration of intermediates in the C_3 photosynthetic carbon reduction (PCR) cycle; (2) to starch synthesis; or (3) to sucrose synthesis, and the distribution of sucrose between transport and temporary storage pools. Various enzymes operate in the pathways that process the triose phosphates, and the control of these steps is complex.

One area of importance is the coordination of starch and sucrose synthesis during the day. The rate of starch synthesis in the chloroplast must be coordinated with sucrose synthesis in the cytosol. Triose phosphate (glyceraldehydes-3-phosphate and dihydroxyacetone phosphate) produced in the chloroplast by the C_3 PCR cycle can be used for either starch or sucrose synthesis (disregarding for the moment the utilization of fixed carbon for regeneration of ribulose bisphosphate in the PCR cycle and for processes serving the photosynthesizing cell itself).

One mechanism for coordinating starch and sucrose synthesis involves a membrane carrier called the **phosphate translocator**, located in the inner chlorophast membrane. Since sucrose is synthesized in the cytosol, the triose phosphate destined for sucrose synthesis must leave the chloroplast. At the same time, ATP synthesis in the chloroplast requires a supply of phosphate from the cytosol. One carrier, the phosphate translocator, accomplishes the exchange of triose phosphate from the chloroplast for phosphate from the cytosol. A balance between sucrose and starch synthesis can be achieved because phosphate is released into the cytosol whenever sucrose is synthesized.

The regulatory events occur in the following sequence: (1) sucrose synthesis in the cytosol, (2) release of phosphate, (3) exchange of phosphate from the cytosol for triose phosphate from the chloroplast, and (4) reduction in the availability of triose phosphate for starch synthesis in the chloroplast. Thus, sucrose synthesis diverts triose phosphate away from starch synthesis and storage. For example, it has been shown that when the demand for sucrose by other parts of a soybean plant is high, less carbon is stored as starch by the source leaves.

The key enzymes in the cytoplasm that are involved in the regulation of sucrose synthesis are **sucrose phosphate synthase** (SPS) and fructose-1, 6-bisphosphatase; the key chlorophast enzyme involved in the regulation of starch synthesis in ADP -glucose pyrophosphorylase. The activities of these enzymesare activities are controlled by the concentrations of key metabolites.

There is often a correlation between SPS activity and sucrose export in the light. SPS catalyzes the formation of sucrose phosphate, which is subsequently hydrolyzed to sucrose and inorganic phosphate. High sink demand removes sucrose, the end product of these reactions, from the source, "pulling" the reaction in the forward direction. The resulting increase in phosphate release in the source cytoplasm leads to increased efflux of triose phosphate from the chloroplast via the phosphate translocator. Glucose-6-phosphate (which is form triose phosphate) stimulates the activity of SPS, thus providing the sucrose to meet the increased sink demand.

There is, however, a limit to the amount of carbon that normally can be diverted from starch synthesis in species that store carbon primarily as starch. Studies of allocation between starch and sucrose under different conditions suggest that a fairly steady rate to translocation throughout the 24-hour period is a priority for most plants. Such control is evident in studies of starch synthesis in soybean. In soybean, short day lengths (7 hours lights, 17 hours dark) led to increased rates to starch synthesis at the expense of translocation from the source. On long days (14 hours light, 10 hours dark), rates of starch accumulation were lower. By synthesizing more starch and translocating less sucrose during the short day period, the plants has more stored material available for translocation during the long dark period. The biochemical basis for such "anticipatory" controls is not known.

The examples discussed so far have shown a delicate balance between the synthesis of sucrose for translocation and the synthesis of starch, at least in species that store carbon in the form of starch. The controls we have discussed operate during daylight hours. Starch must then be degraded during periods of low to zero photosynthesis to provide fixed carbon for export. Thus, the form in which fixed carbon is supplied by the chloroplast for sucrose synthesis and export differs between the day and the night, allowing differential regulation of the two pathways.

The use of mutants and transgenic plants enables use to ask a whole new set of questions; for example, what happens when one of the competing processes, such as starch synthesis, is inhibited or even eliminated? The results have revealed the amazing flexibility of plants. For example, starch-deficient tobacco mutants synthesize only trace amounts of starch but are able to compensate for a lack of stored carbon by doubling the rate of sucrose synthesis and export during the day and by switching most of their growth to the day. On the other hand, plants with enhanced starch synthesis during the day often export more of their fixed carbon during the night.

Sink Tissues Compete for Available Translocated Assimilate

As discussed earlier, translocation to sink tissues depends on the position of the sink in relation to the source and on the vascular connections between source and sink. Another factor determining the pattern of transport in whole plants in competition between sinks. Fox example, reproductive tissue (seeds) can compete with growing vegetative tissue (young leaves and roots) for assimilates in the translocation stream. Competition is indicated by numerous experiments in which removal of a sink tissue from a plant generally results in increased translocation to alternative, presumably competing sinks.

Sink Strength is a Function of Sink Size and Sink Activity

Various experiments indicate that the ability of a sink to mobilize assimilates toward itself, the **sink strength**, depends on two factors: **sink size** and **sink activity**.

$$\text{Sink strength} = \text{sink size} \times \text{sink activity}$$

Sink activity is the rate of uptake of assimilates per unit weight of sink tissue, and sink size is the total weight of the sink tissue. Altering either the size or the activity of the sink results in changes in transport patterns. For example, the ability of a pea pod to import ^{14}C depends on the dry weight of that pod as a proportion of the total reproductive dry weight.

Changes in sink activity can be more complex, since various activities in sink tissues can potentially limit the rate of uptake by the sink. These activities include unloading from the sieve elements, metabolism in the cell wall, retrieval from the apoplast in some cases, and metabolic processes that use the assimilate in either growth or storage. Cooling a sink tissue inhibits activities that require metabolic energy and results in a decrease in the speed of transport toward the sink. Because cooling is so general, however, it does not permit discrimination among processes.

Sink activity and thus sink strength are also thought to be related to the presence and activity of the sucrose splitting enzymes **acid invertase** and **sucrose synthase,** since they catalyze the first step in sucrose utilization. Acid invertase displays optimum activity at a low pH (3.5 to 5.0), catalyzes the irreversible breakdown of sucrose to glucose and fructose, and can be bound to the cell wall or found in a soluble from, probably in the vacuole. Sucrose synthase catalyzes a reversible reaction, but it is usually involved in sucrose breakdown, and generates UDP glucose and fructose in the presence of UDP.

All there enzymes (sucrose synthase and bound and soluble acid invertase) are thought to be involved in the partitioning, utilization, and storage of sucrose-their specific roles depending on the plant organ and its developmental stage. Invertase in sink cells is thought to be involved in early sink growth and expansion; sucrose synthase is associated with polysaccharide synthesis during the storage phase of sink development.

How might these enzymes in involved in partitioning between sinks? In sinks that have an apoplastic stem in phloem unloading, cell wall-bound invertase could enhance transport by maintaining a large sucrose concentration gradient between the cell releasing the sucrose and the apoplast. In sinks that have entirely symplastic unloading, invertase or sucrose synthase in receiver cells could maintain a low sucrose concentration there, with the resulting steep gradient driving diffusion from the sieve elements to the sink cells. As already noted, a stronger sink can draw phloem transport toward itself by affecting the sieve element turgor gradient.

Plasmodesmata May Control Sites for Whole-Plant Translocation

Plasmodesmata are clearly important in the short-distance pathways of phloem loading and unloading. Although originally considered to be simple pores connecting the cytoplasma of neighboring cells, plasmodesmata have a complex structure and can exercise dynamic control of the intercellular diffusion of small molecules. For example, large pressure differences between cells close plasmodesmata, and the degree of closure depends on the pressure difference. Callose deposition, regulated by cytoplasmic calcium levels, is one established mechanism for sealing plasmodesmata, and some plant hormones, such as IAA and ABA, alter intracellular calcium levels via the inositol trisphosphate-diacylglycerol second-messenger system.

Information macromolecules, such as RNA and protein, also move from cell to cell in plants via plasmodesmata. Virally encoded "movement proteins" interact directly with plasmodesmata to allow the passage of viral nucleic acids. Tobacco and potato plants transformed with the tobacco-mosaic-virus movement protein show altered allocation patterns in source leaves and, at least in tobacco, modified whole-plant partitioning patterns. The source leaves of the transformed potato plants retained less sugar and starch than normal plants retain. It remains to be seen whether the changes in allocation and partitioning observed in these transformed plants were due to interactions of their altered phenotypes or to a direct effect on the coordination of growth and development throughout the plant.

Plasmodemata have been implicated in nearly every aspect of phloem translocation, from loading to long-distance transport (remember that pores in sieve areas and sieve plates are modified plasmodesmata) to allocation and partitioning. They will certainly constitute an important avenue of research in the exciting days ahead for transport in the phloem.

Chapter **10**

Growth and Development

10.1 Growth Kinematics

As we know, new cells form continually in the apical meristems. Cells enlarge slowly in the apical meristem and more rapidly in the subapical regions. The resulting increase in cell volume can range from severalfold to one hundredfold, depending on the species and environmental conditions. Classically, plant growth has been analyzed in terms of cell number or overall size (i.e., mass). However, these measures tell only part of the story.

Tissue growth is neither uniform nor random. The derivatives of the apical meristems expand in predictable and site-specific ways, and the expansion patterns in these subapical regions largely determine the size and shape of the primary plant body. The total growth of the plant can be thought of as the sum of the local patterns of cell expansion.

The analysis of the motions of cells or "tissue elements" (and the related problem of cell expansion) is called *kinematics*. The advantage of the kinematic approach is that it allows one to describe the growth patterns of organs mathematically in terms of the expansion patterns of their component cells.

10.2 Measurment of Plant Growth

Growth in plants is defined as an irreversible increase in volume. The largest component of plant growth is cell expansion driven by turgor pressure. During this process, cells increase in volume manyfold and become highly vacuolate. However, size is only one criterion that may be used to measure growth.

Growth also can be measured in terms of change in fresh weight—that is, the weight of the living tissue—over a particular period of time. However, the fresh weight of plants growing in soil fluctuates in response to changes in water status, so this criterion may be a poor indicator of actual growth. In these situations, measurements of dry weight are often more appropriate.

Cell number is a common and convenient parameter by which to measure the growth of unicellular organisms, such as the green alga *Chlamydomonas* (Figure 10.1). In multicellular plants, however, cell number can be a misleading growth measurement because cells can divide without increasing in volume.

For example, during the early stages of embryogenesis, the zygote subdivides into progressively smaller cells with no net increase in the size of the embryo. Only after it reaches the eight-cell stage does the increase in volume begin to mirror the increase in cell number. Because the zygote is an especially large cell, this lack of correspondence between an increase in cell number and growth may be unusual, but it points out the potential problem in equating an increase in cell number with growth.

Although cell number may not always be a reliable measure of plant growth, under most circumstances dividing cells, particularly in meristems, double in volume during their cell cycle. Therefore, an increase in cell number, such as the increase brought about by the activity of the apical meristems, does contribute to plant growth. However, the largest component of plant growth is the rapid cell expansion that occurs in the subapical region after cell division ceases.

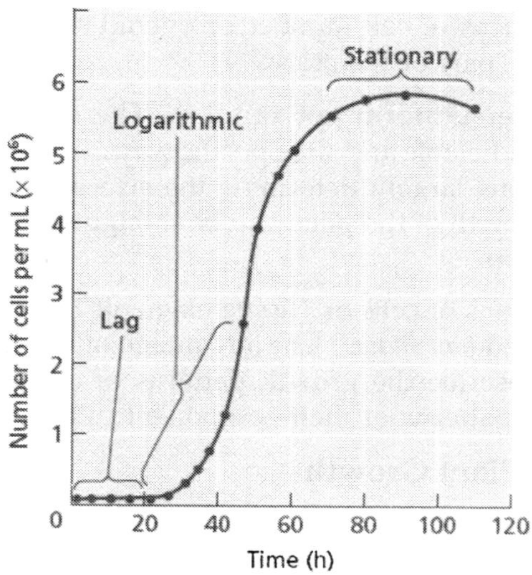

Figure 10.1 Growth of the unicellular green alga *Chlamydomonas*. Growth is assessed by a count of the number of cells per milliliter at increasing times after the cells are placed in fresh growth medium. Temperature, light, and nutrients provided are optimal for growth. An initial lag period during which cells may synthesize enzymes required for rapid growth is followed by a period in which cell number increases exponentially. This period of rapid growth is followed by a period of slowing growth in which the cell number increases linearly. Then comes the stationary phase, in which the cell number remains constant or even declines as nutrients are exhausted from the medium.

Because all the cells of the plant axis elongate under normal conditions, the greater the number of cells produced by the apical meristem, the longer the axis will be. For example, when Arabidopsis plants are transformed with a gene that encodes cyclin, a key component of the cell cycle regulatory machinery, the cells of the apical meristem progress through their cell cycles more rapidly, so more cells form per unit time. As a result, the roots of these transgenic plants have more cells and are substantially longer than the roots of wild-type plants grown under similar conditions.

New cells form continually in the apical meristems. With each new round of cell division and associated cell expansion, the older derivatives are displaced a small distance from the apex. As the cells recede farther from the apex, the rate of displacement is greatly accelerated. By viewing plant growth as a process of cell displacement from the apex, we can apply the principles of kinematics.

10.3 Activity of Meristematic Cells

Moving fluids such as waterfalls, fountains, and the wakes of boats can generate specific forms. The study of the motion of fluid particles and the shape changes that the fluids undergo is called **kinematics**. The ideas and numerical methods used to study these fluid forms are useful for characterizing meristematic growth. In both cases, an unchanging form is produced even though it is composed of moving and changing elements.

An example of an unchanging form composed of changing and displaced elements in plants is the hypocotyl hook of a dicot, such as the common bean . As the bean seedling emerges from the seed coat, the apical end of the hypocotyl bends back on itself to form a hook. The hook is thought to protect the seedling apex from damage during growth through the soil. During seedling growth (in soil or dim light) the hook migrates up the stem, from the hypocotyl into the epicotyl and then to the first and second internodes, but the form of the hook remains constant.

If we mark a specific epidermal cell on the seedling stem located close to the seedling apex, we can watch it as it flows into the hook summit, then down into the straight region below the hook . The mark is not crawling over the plant surface, of course; plant cells are cemented together and do not experience much relative motion during development. The change in position of the mark relative to the hook implies that the hook is composed of a procession of tissue elements, each of which first curves and then straightens as it is displaced from the plant apex during growth. The steady form is produced by a parade of changing cells.

A root tip is another example of a steady form composed of changing tissue elements. Here, too, the form is observed to be steady only when distance is measured from the root tip. A region of cell division occupies perhaps 2 mm of the root tip. The elongation zone extends for about 10 mm behind the root tip. Phloem differentiation is first observed beginning at 3 mm from the tip, and functional xylem elements may be seen at about 12 mm from the tip. A marked cell near the tip will seem to flow first through the region of cell division, then through the elongation zone and into the region of xylem differentiation, and so on. This shifting implies that developing tissue elements first divide and elongate, and then differentiate.

In an analogous fashion, the shoot bears a succession of leaves of different developmental stages. During a period of 24 hours, a leaf may grow to the same size, shape, and biochemical composition that its neighbor had a day earlier. Thus, shoot form is also produced by a parade of changing elements that can be analyzed with kinematics. Such an analysis is not merely descriptive; it permits calculations of the growth and biosynthetic rates of individual tissue elements (cells) within a dynamic structure.

Growth in shoots and roots is localized in regions at the tips of these organs. Regions with expanding tissue are called **growth zones**. With time, meristems move away from the plant base by the growth of the cells in the growth zone.

If successive marks are placed on the stem or root, the distance between the marks will change, depending on where they are within the growth zone. In addition, all of these marks will move away from the tip of the root or shoot, but their rate of movement will differ depending on their distance from the tip.

From another perspective, if you were to stand at the tip of a root that had marks placed at intervals along the axis, you would see that all marks would move farther away from you with time. The reason is that discrete regions on the plant axis experience displacement as well as expansion during growth and development.

As a given region of the plant axis moves away from the apex, its growth velocity increases (the rate of elongation accelerates) until a constant limiting velocity is reached equal to the overall organ extension rate. The reason for this increase in growth velocity is that with time, progressively more tissue is located between the moving particle and the apex, and progressively more cells are expanding, so the particle is displaced more and more rapidly. In a rapidly growing maize root, a tissue element takes about 8 hours to move from 2 mm (the end of the meristematic zone) to 12 mm (the end of the elongation zone).

Beyond the growth zone, elements do not separate; neighboring elements have the same velocity (expressed as the change in distance from the tip per unit of time), and the rate at which particles are displaced from the tip is the same as the rate at which the tip moves through the soil. The root tip of maize is pushed through the soil at 3 mm h^{-1}. This is also the rate at which the nongrowing region recedes from the apex, and it is equal to the final slope of the growth trajectory.

10.4 The Growth Velocity Profile

The velocities of different tissue elements are plotted against their distance from the apex to give the spatial pattern of growth velocity, or **growth velocity profile** (Figure 10.2(A). Growth velocity increases with position in the growth zone. A constant value is obtained at the base of the growth zone. The final growth velocity is the final, constant slope of the growth trajectory equal to the elongation rate of the organ, as discussed in the previous section. In the rapidly growing maize root, the growth velocity is 1 mm h^{-1} at 4 mm, and it reaches its final value of nearly 3 mm h^{-1} at 12 mm.

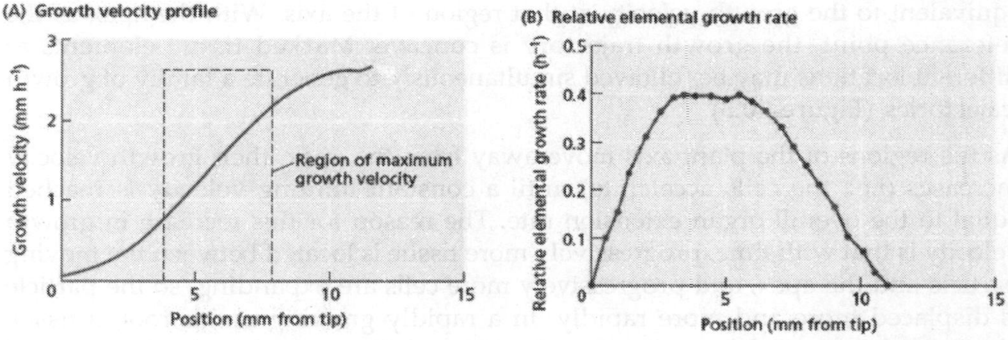

Figure 10.2 The growth of the primary root of Zea mays (maize) can be represented kinematically by two related growth curves. (A) The growth velocity profile plots the velocity of movement away from the tip of points at different distances from the tip. This tells us that growth velocity increases with distance from the tip until it reaches a uniform velocity equal to the rate of elongation of the root. (B) The relative elemental growth rate tells us the rate of expansion of any particular point on the root. It is the most useful measure for the physiologist because it tells us where the most rapidly expanding regions are located.(Silk, 1994, *Biomimetics* 2:199-213)

If the growth velocity is known, the **relative elemental growth rate**, which represents the fractional change in length per unit time, can be calculated (see below). The relative elemental growth rate shows the location and magnitude of the extension rate and can be used to quantify the effects of environmental variation on the growth pattern (Figure 10.2(B).

10.5 The Relative Elemental Growth Rate

One way to characterize the growth pattern along an axis mathematically is to plot the distance of a tissue element from the apex versus time. The resulting curve is called the growth trajectory. The slope of the curve at any given point is

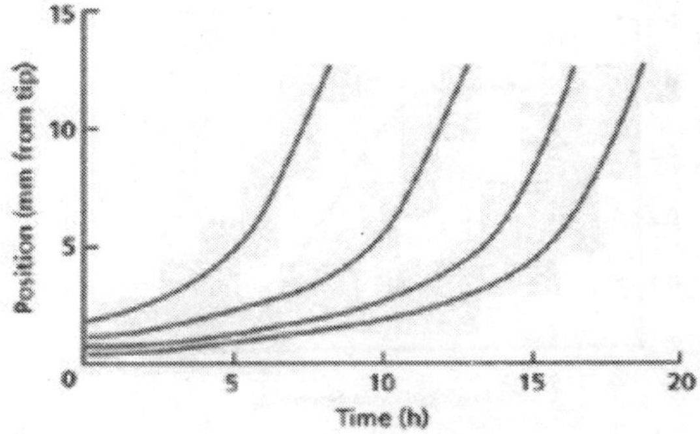

Figure 10.3 Growth trajectories of *Zea mays* (maize):Growth trajectories plot the positions of points spaced along the root axis as a function of time. Note that distance is measured from the root tip. This is termed a material specification of growth because real or material particles are followed.

equivalent to the growth velocity at that region of the axis. With the apex as the reference point, the growth trajectory is concave. Marked tissue elements at different locations may be followed simultaneously to generate a family of growth trajectories (Figure 10.3).

As the regions of the plant axis move away from the apex, their growth velocity increases (i.e., the cells accelerate) until a constant limiting velocity is reached equal to the overall organ extension rate. The reason for this increase in growth velocity is that with time, progressively more tissue is located between the moving particle and the apex, and progressively more cells are expanding, so the particle is displaced more and more rapidly. In a rapidly growing maize root, a tissue element takes about 8 hours to move from 2 mm (the end of the meristem) to 12 mm (the end of the elongation zone). The root tissue element accelerates through this region (Figure 10.3).

Beyond the growth zone, elements do not separate; neighboring elements have the same velocity, and the rate at which particles are displaced from the tip is the same as the rate at which the tip is pushed away from the soil surface. The root tip of maize is pushed through the soil at 3 mm h^{-1}. This is also the rate at which the nongrowing region recedes from the apex, and it is equal to the final slope of the growth trajectory.

The growth trajectory is a material description of growth because an individual element is followed through time. The growth trajectory provides a way to convert spatial information into a developmental time course. For example, if the amount of suberin in the root increases with position, the growth trajectory will give information on the time required for the observed accumulation of suberin in cells moving between any two spatial points.

To characterize the local expansion rate, we must consider the growth velocities at the apical and basal ends of a small segment of tissue. If the apical end of the

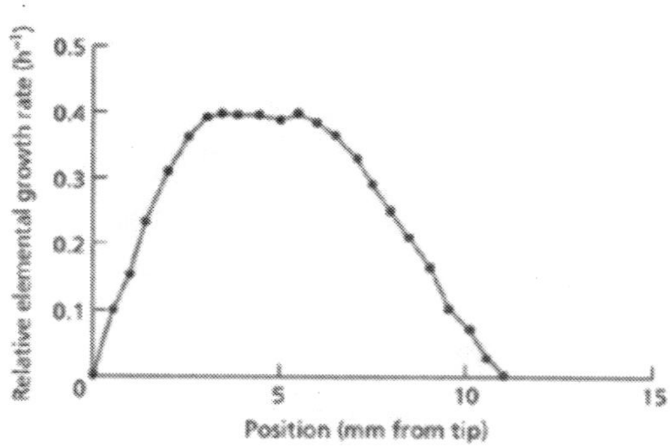

Figure 10.4 Relative elemental growth rate of *Zea mays* (maize): The relative elemental growth rate tells us the rate of expansion of any particular point on the root. It is the most useful measure of root growth rate for the physiologist because it tells where the most rapidly expanding regions are located.

segment is moving faster than the basal end, the segment is elongating. The velocity difference, divided by the segment length, gives the relative growth rate of the segment. Now, imagine that the segment shrinks to a point x. To find the relative growth rate, r, at point x, we can use calculus: The value of r is given by the velocity gradient, dv/dx, where v is the growth velocity. This measure of the local growth rate is called the relative elemental growth rate . It represents the fractional change in length per unit time and has units of h^{-1}. If the growth velocity is known, the relative elemental growth rate can be evaluated by differentiation of the growth velocity with respect to position (Figure 10.4). Or we can approximate r by measuring the relative growth rates of closely spaced segments of initial length L. For each segment, $r = (1/L)(dL/dt)$.

The relative elemental growth rate profile, $r(x)$, is a spatial description of the expansion pattern in the growth zone. In the maize root, the relative elemental growth rate has a maximum of 0.40 h^{-1} at 4 mm and drops to zero by 12 mm. That is, the tissue element at 4 mm is expanding 40% h^{-1}, while tissue beyond 12 mm has ceased expansion. Again, to understand the profile in physiological terms, we must consider that individual cells expand more and more rapidly as they are displaced to 4 mm and then less and less rapidly as they are displaced through the basal part of the growth zone.

The relative elemental growth rate profile shows the location and magnitude of the extension rate and can be used to quantify the effects of environmental variation on the growth pattern. It is also the basis for studying the physiology of growth. The graph of $r(x)$ can be compared to the distribution pattern of any substance thought to be regulating the growth rate. A close correlation between the two curves would provide circumstantial evidence for the role of the substance as a limiting factor for growth.

10.6 Cellular Polarity

Plant cells perceive their position within a tissue and assume the polarity appropriate for that tissue. In addition, cells can perceive and act on information coming from their environment to alter their polarity and redirect growth when necessary. Cellular polarity must be established to orient subsequent expansion. Cells that are initially apolar, such as zygotes, spores, and pollen grains, later become polarized and initiate tip growth and have been especially valuable in the analysis of the induction and establishment of polarity.

A wide variety of external gradients, including temperature, light intensity, pH, ions, and electric-potential gradients, can polarize and orient the axis of growth of these cells. Thus, polarity can be experimentally manipulated without the complexities associated with multicellular tissues. Much of our understanding of the establishment of polarity in tip-growing cells is derived from studies of rhizoid emergence in zygotes of the brown alga *Fucus*.

Unlike higher plants, and even unlike most other algae, certain brown algae, particularly *Fucus* and *Pelvetia*, shed their eggs and sperm into the sea, where fertilization occurs. The zygote initially lacks a cell wall and is completely apolar. However, a polar axis is established within a few hours of fertilization in response to unilateral light. The axis is not stable at first and can be changed many times simply by a change in the direction of the light.

The polar axis becomes fixed within 10 to 14 hours after fertilization, after which it cannot be altered by a change in the direction of the light, and a structure that will become the rhizoid begins to protrude from the darkened side.

Even though the zygote is still a single cell, it is polarized, with two structurally different poles, the thallus and rhizoid poles, which have radically different fates. The first zygotic division is asymmetric and occurs at right angles to the polar axis. It establishes two cells, each having the identity of the pole from which it emerges. The smaller *rhizoid cell* produces a filament of cells and the holdfast that attaches the plant to rocks. The larger *thallus cell* divides to produce a globular mass of cells, from which the main body of the plant, the thallus, forms

A dye known as fluorescent dihydropyridine (FL-DHP), which binds to calcium channels in mammalian cells and can interfere with calcium uptake, has been used as a probe to examine the polarization of the *Fucus* zygote cell membrane. During the first 6 to 8 hours after fertilization, and in response to unilateral light, FL-DHP receptors, presumably calcium channels, are redistributed and become localized on the dark side of the *Fucus* zygote at what will become the rhizoid pole. If the direction of the light is changed, the location of the FL-DHP–binding proteins also changes, again becoming concentrated on the dark side of the zygote.

One of the earliest events in the establishment of polarity after exposure of the zygotes to unilateral light is the induction of a positive current that flows into the cell on the shaded side and loops back through the external medium (Figure 10.5). This **transcellular current**, which is due largely to calcium ions, is on the order of 100 pA (1 picoampere = 10^{-12} ampere) and develops in advance of any morphological sign of polarity.

The relevance of this phenomenon to the establishment of cell polarity is indicated by the fact that the site at which positive charge flows into the cell is also the site of subsequent rhizoid outgrowth. Transcellular ionic currents also have been observed in some tip-growing cells of higher plants, including root hairs and germinating pollen grains, but they have not yet been detected in plant cells that expand by diffuse growth.

The ionic current in polarizing *Fucus* zygotes likely arises from the redistribution of ion channels in the plasma membrane. This current generates an electric-potential difference across the membrane that could alter other membrane properties, such as the distribution of membrane proteins. The current also alters the ionic composition of the cytosol; the magnitude of the effect depends on the ions that form the current, which vary in different cell types. Because the Ca^{2+} concentration in the cytosol is usually very low (between 10^{-6} and 10^{-8} µM), even a small increase in intracellular Ca^{2+} concentration could have a marked effect.

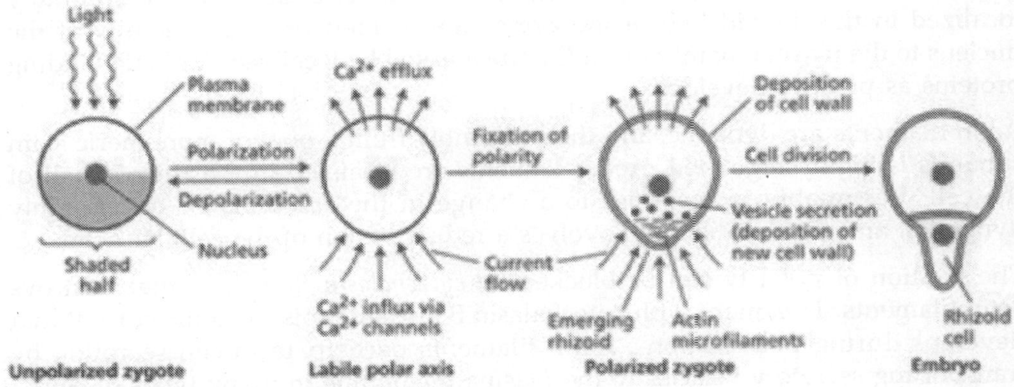

Figure 10.5 A summary of the events involved in the establishment of polarity in zygotes of the brown alga *Fucus*. (A) The zygote is polarized by an asymmetric stimulus from the environment, such as unilateral light. (B) A current flows (charged calcium ions move) through the polarized but still spherical zygote at the site at which the rhizoid will emerge, driven by Ca^{2+} uptake in the shaded half of the cell, from which the rhizoid will emerge. The current (Ca^{2+}) flows out on the opposite side. (C) Cell polarity becomes fixed when actin microfilaments assemble at the site of rhizoid emergence and a cell wall is assembled around the zygote. The cell wall completely surrounds the zygote, but its composition differs in the rhizoid and thallus halves. Vesicles containing sulfated polysaccharides are transported by actin filaments to the plasma membrane and deposited in the cell wall only at the site of rhizoid emergence. (D) Finally, the zygote divides, and the rhizoid cell grows at its tip.

The fluorescent calcium-indicator dye *fura*-2 has been used to detect intracellular cytosolic calcium gradients in both plant and animal cells. Since the excitation spectrum of fura-2 shifts when it binds calcium, we can determine the intracellular concentrations of both forms of the dye (with and without bound calcium) by exciting them with the appropriate two wavelengths. The ratio of the two emissions provides a measure of the calcium concentration that is independent of the dye concentration. Using this technique to measure the cytosolic Ca^{2+} gradient in growing *Fucus* rhizoids, researchers observed a gradient from 450 nM at the tip to 100 nM at the base.

This calcium ion gradient may be important for the polarized assembly of actin filaments and vectoral discharge of Golgi-derived vesicles at the rhizoid pole. Calcium ion gradients also have been shown to be established in other tip-growing cells, such as germinating pollen grains, where the direction of growth and the region in which Golgi vesicles carrying cell wall components are deposited are predicted by the calcium gradient .

Inhibition of protein synthesis by treatment with the drug *cycloheximide* delays rhizoid formation but does not stop the establishment of polarity. This result suggests that although rhizoid formation requires the synthesis of new proteins, the establishment of polarity does not, so polarity development must involve a redistribution of existing components. Actin filaments become preferentially localized in the rhizoid half of the zygote and extend from the vicinity of the nucleus to the plasma membrane at the site marked by localized FL-DHP–binding proteins as polarity develops .

Actin filaments are dynamic, and their assembly from a pool of monomeric actin subunits is highly regulated. Actin filaments are localized in the rhizoid half of the cell presumably in response to a change in the assembly or disassembly dynamics, and this localization involves a redistribution of the cellular actin.

The fixation of polarity can be blocked by *cytochalasin B*, a drug that destroys actin filaments. Treatment with cytochalasin B also disrupts the ionic current that develops during polarization . Actin filaments participate in cell secretion by transporting secretory vesicles to the plasma membrane in many types of cells.

Cytoplasmic organelles become asymmetrically distributed before the polarity of the cell is fixed (Figure 10.6.). Golgi-derived vesicles accumulate at the site of the future outgrowth in the rhizoid half of the *Fucus* zygote as polarization is induced by light . For several hours after the induction of polarity, the asymmetric distribution of vesicles can be changed by subsequent treatment with light from a different direction.

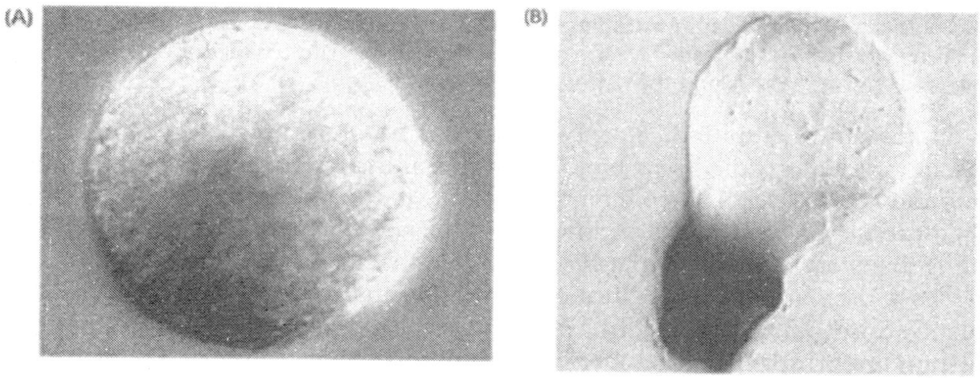

Figure 10.6 Cytoplasmic vesicles that contain specific cell wall components become asymmetrically distributed early in the development of *Fucus* zygote polarity. (A) A toluidine blue stained *Fucus* zygote before fixation of the polar axis. The toluidine blue dye stains a wall component in vesicles or in walls. (B) The embryo after establishment of rhizoid–thallus polarity, with toluidine blue staining confined to the cell wall in the rhizoid portion of the embryo. (Quatrano and Shaw 1997 *Trends Plant Sci* 2: 15-21)

During this phase when changes are possible, cytoplasmic vesicles accumulate at the prospective site of rhizoid formation on the shaded side of the cell. Subsequently, the vesicles fuse with the plasma membrane, secreting their contents into the cell wall, leading to rhizoid emergence. Evidence suggests that calcium gradients play a role in triggering vesicle discharge at the pole.

The fixation of light-induced polarity in *Fucus* zygotes also requires the presence of the cell wall. Removal of the wall with cell wall–digesting enzymes does not prevent the induction of polarity by unilateral light, but it does prevent fixation of the axis. The orientation of the embryonic axis remains unstable until a new cell wall is synthesized.

From these observations it is possible to envisage a chain of events that might occur during the establishment of cell polarity in tip-growing cells.

- The ionic currents that arise early in the process may initiate the formation of an intracellular Ca^{2+} gradient.
- The increase in intracellular calcium at the tip may help organize and stabilize the microfilament components of the cytoskeleton.
- The cytoskeleton directs vesicle traffic to the growing tip.
- A calcium gradient, perhaps immediately adjacent to the plasma membrane, may promote vesicle fusion with the plasma membrane, although this remains to be demonstrated.
- Finally, cell wall deposition fixes the polarity of the cell, possibly by interacting with transmembrane proteins.

10.7 Azolla Root Development

Our undersanding of the mechanics of meristem function has been enhanced by investigations of the apical meristems of ferns and other primitive vascular plants. The root and shoot apical meristems of these organisms have a single large apical cell, which resides at the center of the meristem. The apical cell is the ultimate source of all the cells in the meristem, as well as the root or shoot body. Brian Gunning at the Australian National University studied in detail the patterns of cell divisions in roots of the water fern *Azolla*.

Azolla roots are unusual in that they do not grow throughout the life of the plant. The *Azolla* root is a **determinate** organ, and the root meristem has a genetically determined limit to its growth. This condition is in contrast to most roots, in which the activity of the meristem is **indeterminate**—growth persisting as long as environmental conditions permit it.

The *Azolla* apical cell, which is shaped something like a pyramid with four more or less triangular sides, divides only 50 to 55 times, cutting off derivative cells in each of its four planes, before it stops dividing entirely. With the exception of the root cap, all the cells in the *Azolla* root can be traced to the apical cell, whose division planes are determined with great precision.

The cells that will become the root body are generated by divisions parallel to the three internal triangular faces of the apical cell, forming a wedge-shaped cell known as a **merophyte**. Each of the three derivative cells subsequently divides a set number of times to form a sector that makes up one-third of the root axis. The planes and numbers of divisions of the apical cell and each of its derivatives are very precise. We can recognize two different kinds of divisions—formative and proliferative—which differ in the fate of the daughter cells produced. Both daughter cells resulting from a **proliferative division** have the same fate.

Formative divisions may be either periclinal or anticlinal, but they differ from proliferative divisions in that the two daughter cells have different fates. That is, one of the daughters may remain undifferentiated while the other daughter cell will differentiate as a particular cell type. Formative divisions establish the files of cells that make up the various tissues of the root; proliferative divisions increase the number of cells in each file.

In the root, most formative divisions are periclinal (longitudinal), while proliferative divisions tend to be anticlinal (transverse). Longitudinal, formative divisions within the merophytes give rise to progenitor cells for the files of cells in the root. For example, the division that generates the endodermis and inner cortical cells occurs in the fourth merophyte on both sides of the root, as do the divisions that delineate the vessel element, pericycle, outer cortex, and root hair-epidermis cell files. Once formed, the progenitor cells usually complete several transverse, proliferative divisions before the progeny differentiate.

Shoot Branching

The arrangement of branches constitutes one of the most conspicuous aspects of shoot architecture. During vegetative growth, the formation of branches is guided by a complex interplay between genetically determined programs of development and the environment. As the plant develops, environmental cues are in some way perceived and interpreted to enable adaptive growth responses, ranging from complex shade avoidance responses, to replacement of shoots lost through herbivory. Root systems show a comparable, but usually less appreciated complexity, which, like shoots, reflects interactions between the environment's intrinsic programs. Although these programs have the potential to yield an infinite variety of forms, certain common themes suggest basic control mechanisms may be widely conserved among plants. This section examines the phenomenon of branching from several perspectives, beginning with a discussion of its relationship to shoot development and then shifting to a consideration of roots.

In the seed plants, branching is often constrained by phyllotaxy, since the predominant form of branching in most species is axillary (i.e., meristems form at the base of leaves). Superimposed on this basic pattern are further controls that regulate when and where such axillary meristems become active. In many species, axillary buds may lie dormant for long periods, becoming active only when the plant senses favorable growth conditions. Alternatively, buds can be activated by intrinsic factors, such as those associated with the onset of reproductive development.

Formation of the axillary meristem can be linked to the activity of several classes of transcription factors. Seed plants provide examples where secondary shoot systems can form in a variety positions on the plant, ranging from the small plantlets on leaf margins of Kalenchoe, to root-derived tuberous shoots. However, in most plants, such axillary shoots typically form in close association with the adaxial side of the leaf base. Because of this, the pattern of branching is often linked to some degree with the pattern of phyllotaxy. The earliest stages of axillary meristem development begin with the appearance of a bulge at the leaf primodia, soon after its initiation. *KNOTTED*-like genes are expressed from this early stage, in contrast to their down-regulation in adjacent leaf primordia. As the axillary develops further, it establishes continuity with the vascular network of the parent growth axis and begins to initiate lateral organs such as leaves or bracts. In many plants, this formative period is followed by a period of dormancy, which can be broken by either intrinsic or external cues.

Although the programs that underlie the formation of axillary buds are undoubtedly complex, some insight into their nature has been offered by the analysis of mutants in which the process is somehow altered. As previously discussed, mutations that affect the polarity of the leaf also influence the pattern of meristem development, suggesting that factors associated with adaxial patterning of the leaf somehow promote meristem formation. Mutations that activate such programs, such as those associated with the dominant alleles of *PHABULOSA* and *PHAVOLUTA*, lead to the **ectopic** formation of axillary meristems in both adaxial and abaxial positions. The related transcription factor, *REVOLUTA*, which functions in the normal establishment of a polarized vasculature, appears to also play a role. Loss of functions mutations that lead to vascular polarity defects can also block the formation of axillary meristems. The presence of START domains in these CLASS III HD-ZIP class transcription factors suggest that their activity may be sensitive to an unidentified sterol-like signalling molecule.

Expression analyses suggest that other genes, such as *CUC*, that are required for the formation of the primary SAM in the embryo, are likely to play a similar role for axillary meristems. However, certain types of mutants in which only axillary meristems are affected point towards a novel class of more specific genes. One of these, *lateral suppressor*, was first described in tomato as a mutant in which lateral branches are completely blocked . Cloning of this gene revealed a VHIID-type transcription factor, suggesting that axillary branching is normally dependent on regulation of transcription by the LAS protein. The mechanism that LAS functions in appears to be well-conserved, as LAS-like genes have been described in both Arabidopsis and rice, which have similar branch-deficient mutant phenotypes.

Some insights into the potential targets of LAS have been provided by analyses of mutants in Arabidopsis where loss of gene function is linked to a more diffuse pattern of STM expression. Thus, LAS may enable axillary meristem formation by activating genes normally involved in SAM formation. It remains unclear what signals are responsible for the localized expression of LAS.

Once formed, axillary meristems may lie dormant or become activated through a combination of environmental and intrinsic factors, and it appears that many of these responses are determined in part via hormone-signaling pathways. The plant growth hormone ABA, know for its role in promoting seed dormancy, has analogous activities in maintaining the dormancy of axillary buds. Factors that regulate synthesis, turnover, perception, and response of ABA are likely to play important roles in both processes.

Several lines of evidence implicate cytokinins as promoting axillary meristem growth. In classic regeneration experiments, a relatively high ratio of cytokinin to auxin was shown to be essential for regeneration of shoots from undifferentiated callus, whereas low ratios led to the formation of roots. Comparable effects are observed in *Agrobacterium*-infected plants. Plants infected with cytokinin-producing *Agrobacterium tumefaciens* strains develop ectopic shoots, whereas auxin-producing *Agrobacterium rhizogenes* develop ectopic roots. In addition to its role in initiating shoot meristems, cytokinin also appears capable of activating shoots when applied exogenously. Combining cytokinin's effects on both the initiation and activation of axillary meristems has the potential to cause massive increases in branching, as suggested by the highly branched phenotype of *supershoot,* which have been linked to increased levels of cytokinin.

The activity of axillary meristems can also be repressed by factors associated with the growth of the primary shoot, a phenomenon known as **apical dominance**. As previously discussed, this type of response can help ensure the survival of the plant where growing shoot tips have been damaged by herbivory or other environmental stresses. Apical dominance phenomena are also likely to be at least partly responsible for some of the complex branching patterns associated with flowering, including the elaboration of characteristic inflorescence architectures, or the release of more proximal vegetative buds to enable perennial growth.

Although the phenomenon of apical dominance is well known and has been exploited throughout the history of agriculture to yield more productive plant forms, its basis still remains incompletely understood. Apical dominance can be inferred when the decapitation of a plant leads to the activation and growth of axillary buds. As this activation can be suppressed by the application of auxin to the decapitated plant, and since growing shoots are a rich source of this hormone, it is reasonable to suppose that auxin plays some role in apical dominance. Indeed, physiological studies show that auxin is normally transported basipetally from the growing shoot of intact plants via auxin-efflux transporters. An association with auxin is reinforced by the observation that mutants such as *axr1*, which render the plant less-sensitive to auxin, lead to increased branching. However, auxin is normally not efficiently transported into buds, suggesting that its suppression of axillary bud-growth is likely to be mediated indirectly.

The possibility that cytokinin might have a role in mediating apical dominance is suggested by its pattern of synthesis in plants and its growth-promoting activities. Cytokinin is known to be synthesized in roots; from there it is transported acropetally in xylem sap, as well as in nodal regions associated with axillary buds. Furthermore, the inhibitory effects of auxin on bud-growth can be bypassed by exogenous cytokinin, suggesting cytokinin acts downstream of auxin.

The involvement of a novel hormone-signalling pathway in the regulation of bud-growth has been inferred by genetic analyses of branching in Arabidopsis, pea and petunia. In Arabidopsis, a suite of mutants with **m**ore **ax**illary branches encoded by the *MAX 1, 2, 3,* and *4* genes has been analyzed by a combination of surgical and molecular techniques. Grafting experiments suggest that *MAX 1, 3,* and *4* can act nonautonomously. For example, the excessive branching of a shoot, which is mutant for any one of these genes, can be rescued by grafting it onto a normal root stock. By contrast, *max2* mutant shoots remain highly branched when grafted. These studies suggested that *MAX1, 3* and *4* are involved in the synthesis of a mobile signal, whereas the *MAX2* gene product mediates perception and response to the signal (e.g., by acting as a receptor). When various mutant versions of *MAX* genes are combined, there is no significant increase in branching, suggesting that these genes act in a linear response pathway rather than acting separately on independent pathways.

Molecular cloning of the *MAX* genes have born out models suggested by grafting analyses. Both *MAX3* and *MAX 4* encode related gene products that are predicted to act as dioxygenases on carotenoid-type molecules. *MAX1*, by contrast, encodes

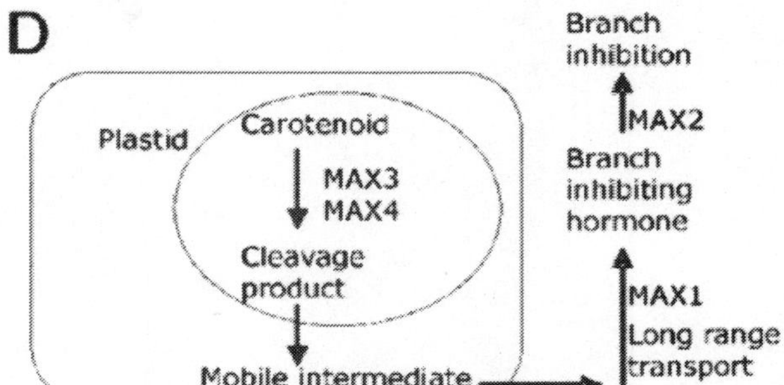

Figure 10.7 Model for proposed functional relationships between *MAX* gene activities in relationship to branching.

a cytochrome P450-type enzyme, which is proposed to act downstream of *MAX3* and *MAX 4* to convert the carotenoid product of these genes into a hormone with branch-inhibiting activity (Booker *et al.* 2005). The identification of this unknown hormone through a combination of biochemical and genetic methods should offer valuable insights into the regulation of bud-growth and may offer new tools for crop improvement.

In contrast to *MAX 1, 3,* and *4,* which encode proteins involved in the synthesis of the branch-inhibiting hormone, *MAX2* has been shown to encode an F-box protein. As described in more detail elsewhere, F-box proteins can, when activated, target specific proteins for degradation. In the case of *MAX2,* one can hypothesize one or more proteins that activate bud-growth are normally targeted for destruction by the activated *MAX2* gene product. By this model, the perception of the carotenoid branch-inhibiting hormone would, in some way, activate the *MAX2* gene product. By analogy with the auxin receptors, which have also been shown to encode F-box proteins, it is conceivable that the MAX2 protein itself acts as a receptor, with binding of the branch-inhibiting hormone activating the ability of MAX2 to target specific proteins for degradation. Presumably, these proteins, which would be stabilized in the absence of MAX2 activity, would in some way promote branching.

Chapter 11

Phytochrome

11.1 Chemical Structure

Phytochrome is a soluble protein with a molecular mass of about 250 kDa. It occurs as a dimer made up of two equivalent subunits. Each subunit consists of two components: a light-absorbing pigment molecule called the chromophore, and a polypeptide chain called the apoprotein. The apoprotein monomer has a molecular mass of about 125 kDa. Together, the apoprotein and its chromophore make up the holoprotein. The chromophore of phytochrome is a linear tetrapyrrole termed phytochromobilin. There is only one chromophore per monomer of apoprotein, and it is attached to the protein through a thioether linkage to a cysteine residue (Figure 11.1.).

Methods for phytochrome purification were developed through research pioneered by the agricultural biochemist H. William Siegelman, working at the USDA lab at Beltsville, Maryland. After surveying many different plants and plant tissues, Siegelman found that the young etiolated seedlings of cereals, such as oat and rye, were particularly rich sources of phytochrome. Some initial success in purifying phytochrome from etiolated oats (*Avena sativa*) was reported in 1964 , and many workers subsequently used dark-grown oat seedlings for phytochrome extractions. As a result,While on a fresh-weight basis there is 50 times more phytochrome in etiolated tissue than in equivalent green tissue of the same age, the overall content of phytochrome in etiolated plants is low—about 0.2% of the total extractable protein. Thus, to purify even milligram quantities usually requires more than a kilogram fresh weight of starting material. Etiolated seedlings are also rich in proteases, and much of the phytochrome is susceptible to proteolysis. To minimize the loss of phytochrome, researchers carry out these extractions rapidly, at a low temperature, and in the presence of protease inhibitors. The stability of phytochrome during purification is also favored by conversion to the Pfr form with red light. For these reasons, successful purification of fully intact, native phytochrome from etiolated oat seedlings was not reported until 1983. Hence, as on today much of the biochemical information on phytochrome is derived from studies on oat seedlings, although there is a considerable amount of data on phytochrome isolated from dark-grown rye (*Secale* spp.) and pea (*Pisum sativum*) seedlings as well.

Researchers have visualized the Pr form of phytochrome using electron microscopy by spraying a solution of the protein onto mica flakes, drying it under vacuum, and "shadowing" the dried molecules by using platinum atoms to enhance its three-dimensional structure. As Figure 11.2. shows, phytochrome molecules

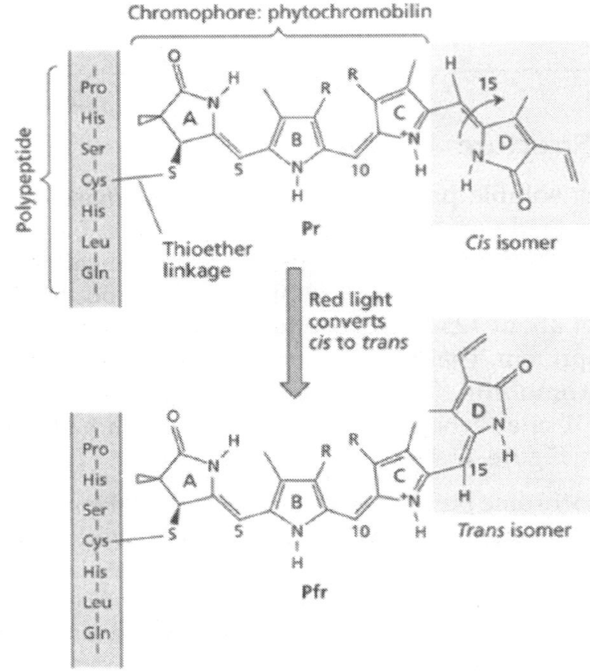

Figure 11.1. Structure of the Pr and Pfr forms of the chromophore (phytochromobilin) and the peptide region bound to the chromophore through a thioether linkage. The chromophore undergoes a *cis-trans* isomerization at carbon 15 in response to red and far-red light.)

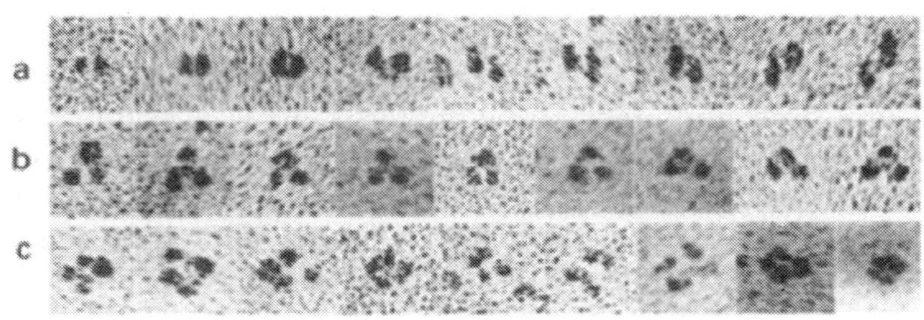

Figure 11.2 Phytochrome from pea, purified and then sprayed onto a mica flake. After being dried, the specimen was shadowed with platinum to enhance the three-dimensional structure. A carbon replica was prepared and viewed in the electron microscope. Various images were produced, including pairs of spots (a), three spots (b), and clusters of spots (c). The variation in the pattern in A is caused by different orientations of the protein.

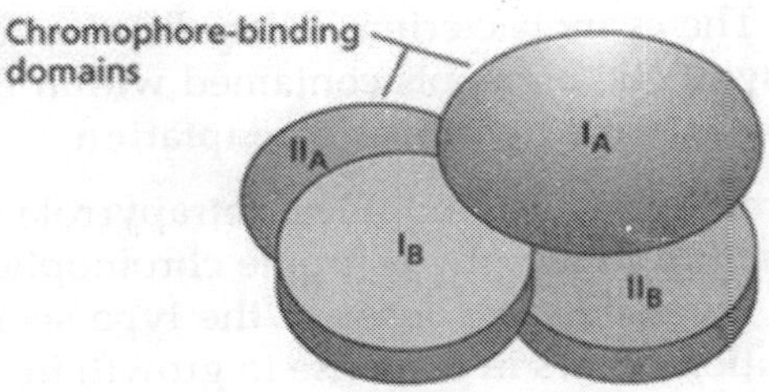

Figure 11.3 Structure of the phytochrome dimer, based on a type of X-ray scattering that does not require crystallization. The monomers are labeled I and II. Each monomer consists of a chromophore-binding domain (A) and a smaller non-chromophore domain (B). The molecule as a whole has an ellipsoidal rather than globular shape.

appear as a collection of spots, the number depending on the orientation of the protein. On the basis of this information and more accurate measurements using X-ray scattering, the model shown in Figure 11.3 has been proposed The polypeptide folds into two major domains separated by a "hinge" region. The larger N-terminal domain is approximately 70 kDa and bears the chromophore; the smaller C-terminal domain is approximately 55 kDa and contains the site where the two monomers associate with each other to form the dimer.

11.2 Evolution of Phytochrome from a Bacterial Two-Component Receptor

Bacteria often use **two-component signaling systems** to sense and respond to environmental signals such as nutrients or light. The first component is a **sensor protein,** consisting of an *input domain* that receives the input signal, linked to a *transmitter domain* that transmits the signal to the second component. The second component is the **response regulator protein,** composed of a *receiver domain* that accepts the signal from the transmitter of the sensor protein, linked to an *output domain* that initiates the cellular response .

These two components interact by a series of phosphorylation and dephosphorylation reactions. When it receives a signal, the sensor autophosphorylates a specific histidine residue in the transmitter domain. This phosphate is then transferred to a specific aspartate residue on the receiver domain of the response regulator But what has all this to do with phytochrome?

Although they lack the specific histidine residue that becomes autophosphorylated, the C termini of phytochromes have some sequence similarity to the transmitter domains of bacterial sensor proteins . This sharing of sequences raises the interesting possibility that phytochrome evolved from the sensor protein of a bacterial two-component system.

Provocative new evidence for the idea that phytochrome evolved from the histidine kinase of a bacterial two-component system has recently emerged from studies of cyanobacteria. The cyanobacterium *Fremyella* uses light to regulate the composition of its photosynthetic pigments contained within the light-harvesting phycobilisomes , a response termed **chromatic adaptation.**

Like phytochromes, phycobiliproteins bind linear tetrapyrrole chromophores that are closely related in structure to the phytochrome chromophore, but there is no evidence for red/far-red reversible responses of the type seen in higher plants. Instead, chromatic adaptation occurs in response to growth in red light and green light.

When the gene for the photoreceptor for this response (*RcaE*) was cloned, it was found to encode a 74 kDa polypeptide, the N-terminal end of which is related to the N terminus of higher-plant phytochromes, while the C-terminal end is related to histidine kinases . However, the lack of the conserved chromophore-binding site and of red/far-red reversibility suggests that this protein is a distant relative of higher-plant phytochromes.

Publication of the complete sequence of the genome of the cyanobacterium *Synechocystis* in 1996 led to great excitement because this organism contains a gene that has between 50 and 60% amino acid identity with *Arabidopsis* phyE along its length, including a bona fide chromophore attachment site. In addition, the amino acid sequence of the C terminus of the *Synechocystis* pigment protein is related to the KinA sensor protein of *Bacillus subtilis*. This *Synechocystis* gene has been named *cph1* (cyanobacterial *ph*ytochrome *1*)

Recently it was shown that the purified Cph1 apoprotein can assemble with phycocyanobilin, a cyanobacterial linear tetrapyrrole, in vitro, and the resultant holoprotein exhibits red/far-red reversibility and absorption spectra similar (but not identical) to higher-plant phytochrome assembled in vitro with the same chromophore (Figure 11.2.)

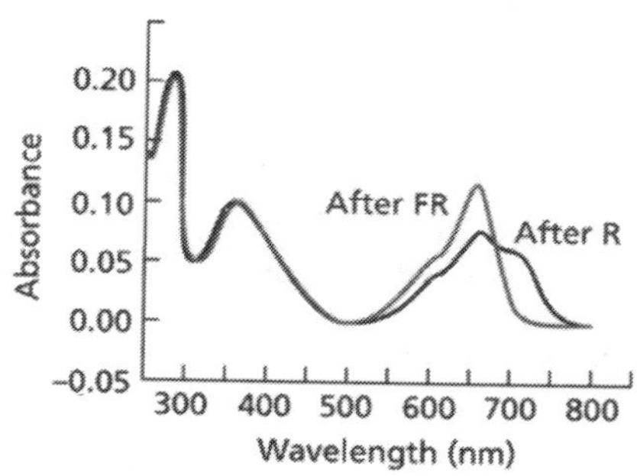

Figure 11.4. Absorption spectra of cyanobacterial phytochrome after irradiation with red (R) or far-red (FR) light. Note the similarities to and differences from the plant phytochrome spectra shown

Perhaps the most exciting finding of all was the identification, 10 base pairs downstream from the cyanobacterial phytochrome gene, of another gene having the features of a bacterial response regulator. Since the C-terminal end of the cyanobacterial phytochrome contains all the conserved features of a histidine kinase transmitter module, and the downstream response regulator resembles aspartate kinase receiver modules it was now possible to test whether this phytochrome acts as part of a light-regulated transmitter–receiver pair, as had been proposed earlier .

Using the purified proteins, the Cph1 protein was shown to phosphorylate itself in a manner consistent with its having a histidine kinase activity, and further, this phosphate could be transferred to the response regulator Surprisingly, Cph1 is far more active as a kinase in the Pr form than in the Pfr form, suggesting that the light signal is transduced through regulation of Pr abundance, rather than through Pfr abundance as in higher plants.

Taken together, the results from these studies suggest that plant phytochromes originally evolved from a class of photoreversible transmitter histidine kinases in cyanobacteria. Are plant phytochromes also kinases? In support of this idea, a phytochrome gene isolated from the moss *Ceratodon* contains an additional C-terminal sequence that has kinase homology .

In addition, highly purified preparations of higher-plant phytochrome have been found to contain a polycation-stimulated protein kinase activity that is also greater in the Pr form than in the Pfr form . However, the kinase activity associated with these phytochromes is of the serine–threonine type, and consistent with this observation, these proteins lack the conserved histidines of the bacterial sensor proteins. More recently, it was shown that recombinant oat phyA is a serine-threonine kinase with the same biochemical properties as the plant-derived photoreceptor . Moreover, R-light stimulated its autophosphorylation, supporting the hypothesis that Pfr is the active form. Polycations were shown to stimulate Pr phosphorylation more than Pfr, a result consistent with previous studies.

Thus, while higher-plant phytochromes share a homology with bacterial two-component signaling systems, the reaction mechanism and components may have diverged considerably. Nevertheless, the discovery of cyanobacterial phytochromes has opened up many exciting new areas of research that have already contributed much to our understanding of phytochrome evolution and function.

11.3 Involvement of G-Proteins and Calcium in Phytochrome Responses

Early experiments with intact seedlings and cultured cells indicated that the control of gene expression by phytochrome involves the action of a heterotrimeric G-protein and the calcium-binding regulatory protein **calmodulin**.

Subsequently, a single-cell system has been developed that can be microinjected with various proposed components of the pathway and monitored for changes in gene expression . Although these experiments have not yet led to the identification of specific pathway components, they have been informative about the involvement of G-proteins and calmodulin.

Microinjection of phytochrome (mainly phyA) purified from etiolated oat seedlings into single cells prepared from the tomato *aurea* mutant (a chromophore biosynthetic mutant that has reduced levels of all phytochromes) results in stimulation of chloroplast development, accumulation of anthocyanin pigment, and expression of a GUS-coding sequence fused to a phytochrome-regulated promoter sequence.

Etiolated seedlings of the *aurea* mutant show poor induction of these responses when transferred to light, probably because of the reduction in the levels of functional phyA. Inhibitors of G-proteins block the phytochrome-stimulated events in the single-cell system, while activators or intermediates stimulate the response. Increased levels of Ca^{2+} in the medium also stimulate *GUS* expression and chloroplast development when calmodulin is present, but they do not affect anthocyanin biosynthesis.

In the *Aurea* cell system, cGMP has been shown to act as a signaling intermediate in at least two signal transduction pathways . Specifically, the pathway that activates anthocyanin biosynthesis requires cGMP. Inhibitors of cGMP action inhibit this pathway; addition of cGMP activates it. In contrast, the pathway that requires calcium to activate chloroplast development and gene expression has a variable requirement for cGMP. Activation of genes that encode several chloroplast proteins, including RBCS and LHCB, does not require cGMP, while activation of genes that encode some photosystem I polypeptides and the cytochrome *b6/f* complex does.

The difference in signaling requirements is due in part to the different *cis*-acting regulatory elements found upstream of these genes. Specific DNA elements upstream of *RBCS* genes appear to be the targets for the calcium pathway, while other distinct elements located upstream of anthocyanin biosynthesis genes are the targets for cGMP. The differential regulation of the expression of phytochrome-regulated genes can now be explained in terms of their *cis*-acting regulatory elements, different combinations of which afford both tissue-specific and light-specific expression.

11.4 Effects of Phytochrome on Ion Fluxes

Phytochrome can rapidly alter the properties of membranes. We have already seen that low-fluence red light is required before the dark period to induce rapid leaflet closure during nyctinasty, and that fluxes of K^+ and Cl^- into and out of extensor and flexor cells mediate the response. However, the rapidity of leaf closure in the dark (lag time about 5 minutes) would seem to rule out mechanisms based on gene expression. Instead, rapid phytochrome-induced changes in membrane permeability and transport appear to be involved.

The role of the plasma membrane proton pump in *Samanea* pulvini (the structures at the base of the petiole that include the extensor and flexor cells) was studied by use of a proton-sensitive liquid membrane electrode . During phytochrome-mediated leaflet closure, the apoplastic pH of the flexor cells (the cells that swell during leaflet closure) decreased, while the apoplastic pH of the extensor cells (the cells that shrink during leaflet closure) increased. Thus, the plasma membrane H^+ pump of the flexor cells appears to be activated by darkness (provided that phytochrome is in the Pfr form), whereas the H^+ pump of the extensor cells appears to be deactivated under the same conditions.

Increased H^+ pump activity on the plasma membranes of the flexor cells should increase the membrane potential, thereby allowing more K^+ to enter the cells . Conversely, a reduction in the plasma membrane H^+ pump activity in the extensor cells would promote the loss of K^+ ions. Consistent with this model, the reverse pattern of apoplastic pH change was seen during leaflet opening.

Studies have also been carried out on phytochrome regulation of K^+ channels. Since it is difficult to study membrane channels in the intact pulvinus, Richard Crain and his colleagues at the University of Connecticut isolated protoplasts (cells without their cell walls) of both extensor and flexor cells from *Samanea* leaves . They determined the open or closed state of the K+ channels indirectly by measuring the membrane potential changes that occurred when the extracellular K^+ concentration was raised from 20 to 200 m*M*. When the extracellular K^+ concentration was raised, K^+ entered the protoplasts and depolarized the membrane potential only if the K^+ channels were open. When the extensor and flexor protoplasts were transferred to constant darkness, the state of the K^+ channels exhibited a circadian rhythmicity during a 21-hour incubation period and the two cell types varied reciprocally, just as they do in vivo. That is, when the extensor K^+ channels were open, the flexor K^+ channels were closed, and vice versa. Thus, the circadian rhythm of leaf movements has its origins in the circadian rhythm of K^+ channel opening.

To test the effects of phytochrome on K^+ channels, researchers first treated protoplasts with white light, then exposed them to a brief pulse of either red light alone or red light followed by far-red light and transferred them to continuous darkness. Red light caused the K^+ channels of the flexor protoplasts to open, and far-red light reversed the effect of red light. The extensor protoplasts were closed by both treatments. In contrast to red light, blue light caused the K^+ channels of the extensors to open, and those of the flexors to close. Thus, phytochrome seems to regulate specifically the K^+ channels of the flexor cells. Blue light regulates primarily the extensor cells, although it can also affect the flexors at specific times of day .

On the basis of the evidence thus far, we can conclude that phytochrome brings about leaflet closure by regulating the activities of the primary proton pumps and/or the K+ channels of the flexor and extensor cells. Although the effect is rapid, it is not instantaneous, and it is therefore unlikely to be due to a direct effect of phytochrome on the membrane. Instead, phytochrome acts indirectly via one or more signal transduction pathways, as in the case of the regulation of gene expression by phytochrome.

However, some effects of red and far-red light on the membrane potential are so rapid that phytochrome may also interact directly with the membrane. Such rapid modulation has been measured in individual cells and has been inferred from the effects of red and far-red light on the surface potential of roots and oat (*Avena*) coleoptiles. The lag between the production of Pfr and the onset of measurable potential changes varies from organism to organism; for example, it is about 1.7 s for depolarization (inside becomes less negative relative to outside the plasma membrane) in the giant alga *Nitella* and 4.5 s for hyperpolarization in *Avena.*

Changes in the bioelectric potential of cells imply changes in the flux of ions across the plasma membrane. The *Mougeotia* microbeam irradiation experiments of Haupt seemed to support the idea that phytochrome is localized on the plasma membrane, and membrane isolation studies carried out in the 1970s provided evidence that a small portion of the total phytochrome is tightly bound to various organellar membranes.

These findings led some workers to suggest that membrane-bound phytochrome represents the physiologically active fraction, and that all the effects of phytochrome on gene expression are initiated by changes in membrane permeability. On the basis of sequence analysis, however, it is now clear that phytochrome is a hydrophilic protein without membrane-spanning domains. The current view is that phytochrome in *Mougeotia* is associated with microtubules located directly beneath the plasma membrane.

If phytochrome exerts its effects on membranes from some distance, no matter how small, involvement of a "second messenger" would be implied. For example, rapid changes in cytosolic free calcium have been implicated as second messengers in several signal transduction pathways, and there is evidence that calcium plays a role in chloroplast rotation in *Mougeotia* . More recently, however, the role of calcium as a second messenger in phytochrome action has been questioned, so calcium-independent mechanisms may also be involved.

11.5 Phytochrome Regulation of Gene Expression

We have seen that phytochrome-mediated light responses at the cellular level can involve rapid effects on membrane properties. The stimulation and repression of transcription by light can also be very rapid, with lag times as short as 5 minutes. Such early-gene expression is likely to be regulated by the direct interaction of transcription factors by one or more phytochrome-initiated signal transduction pathways.

Within the nucleus, these transcription factors stimulate the transcription of specific genes. Some of the gene products of the early genes are transcription factors themselves, which activate the expression of other genes. Expression of the early genes, also called primary response genes, is independent of protein synthesis; expression of the late genes, or secondary response genes, requires the synthesis of new proteins.

As with cytokinin-regulated gene expression , the photoregulation of gene expression has focused on the nuclear genes that encode messages for chloroplast proteins: the small subunit (SSU) of ribulose-1,6-bisphosphate carboxylase/ oxygenase (rubisco) and the major light-harvesting chlorophyll *a/b*-binding proteins associated with the light-harvesting complex of photosystem II (LHCIIb proteins). These proteins play important roles in chloroplast development and greening; hence their regulation by phytochrome has been studied in detail. The genes for both of these proteins—*RBCS* and *LHCB* (also called *CAB* in some studies)—are present in multiple copies in the genome.

The abundance of these mRNAs has been shown to increase in response to light and phytochrome action. The abundance of other mRNAs, however, can be decreased by phytochrome action; this latter class includes mRNAs that encode oat phyA and the chlorophyll biosynthesis enzyme NADPH–protochlorophyllide oxidoreductase . Although the abundance of most of these mRNAs is regulated by light given in the low-fluence range, some *LHCB* mRNAs also respond to light in the very-low-fluence range.

We can demonstrate phytochrome-regulated induction of mRNA abundance (for example *RBCS* mRNAs) experimentally by giving etiolated plants a brief pulse of low-fluence red or far-red light, returning them to darkness to allow the signal transduction pathway to operate, and then measuring the abundance of specific mRNAs in total RNA prepared from each set of plants. If its abundance is regulated by phytochrome, the mRNA is absent or present at low levels in etiolated plants, but is increased by red light. The red light–induced increase in expression can be reversed by immediate treatment with far-red light, but far-red light alone has little effect on mRNA abundance.

In the case of mRNAs (e.g., *LHCB*) that respond to very-low-fluence red light, expression increases in response to both red and far-red light, and additional experiments are needed to confirm the involvement of phytochrome. In the case of mRNAs whose expression is negatively regulated by phytochrome, for example oat *PHYA* mRNA, levels are *high* in darkness, *reduced* by a pulse of red light, and *restored* to dark levels by red/far-red light treatment. However, none of these experiments prove the action of phytochrome on transcription; instead they show the effect of phytochrome only on overall mRNA abundance in total RNA, which reflects both transcription and degradation processes. As mentioned earlier, the abundance of oat *PHYA* mRNA in total RNA is the result of both transcription and degradation events.

The involvement of phytochrome in transcriptional regulation of mRNA abundance has been demonstrated by two approaches. In the first approach, etiolated plants are treated as was described for the mRNA studies, and then transcriptionally active nuclei are isolated and allowed to elongate transcripts initiated under the different in vivo light treatments in the presence of radioactive precursors. If the relative incorporation into an mRNA is similar to that described for the overall abundance in total RNA, then transcription of the mRNA is thought to be under phytochrome control.

11.6 Phytochrome and Shade Avoidance

To identify components of the shade-avoidance response pathway, microarray analysis was conducted in Arabidopsis, comparing plants grown in low and high R:FR light environments. These studies have identified both early- and late-acting genes that regulate responses to shading by other plants. Characterizations of two Arabidopsis transcription factors (*ATHB-2* and *PIL1*) also suggests that shade avoidance is initiated by rapid transcriptional changes following a decrease in the ratio of red to far-red light. These profiling studies strengthen the previous genetic studies that show that phyA acts to antagonize phyB-mediated responses under prolonged low R:FR light conditions through the transcriptional control of target genes that include components of the auxin transport and signaling pathways. Interestingly, it appears that the circadian clock is capable of regulating or "gating" physiological responses to R:FR through a modulation of gene expression. That is, changes in gene expression due to a change in the R:FR only occur at certain phases of the circadian rhythm.

11.7 Effects of Phytochromes A and B on Flowering

Phytochrome is encoded by members of a multigene family, and it is possible that the effects of red and far-red light are mediated by different types of phytochrome (A and B). In LDPs such as *Arabidopsis*, pea (*Pisum sativum*), and sorghum (*Sorghum bicolor*), phytochrome B (phyB) appears to be an *inhibitor* of flowering, since mutations in the gene *PHYB* that eliminate or reduce the amount of the phyB protein cause more rapid flowering.

The promotion of flowering by far-red light may result from a decrease in the amount of the Pfr form of phyB, which would have the same effect on the phyB Pfr levels as the mutations that reduce the total amount of phyB. As already discussed, red-light interruptions of a dark period prevent flowering in SDPs; therefore, the Pfr form of phyB may be an *inhibitor* of flowering in SDPs as well.

In the LDPs *Arabidopsis* and pea, mutations in the *PHYA* gene interfere with the promotion of flowering by long days , thus the Pfr form of phyA may promote flowering in LDPs. The effect of the *phyA* mutation is quite strong in pea (the mutant plants are essentially unable to respond to inductive photoperiods) whereas the *phyA* mutation in *Arabidopsis* has only a slight effect on the ability to respond to inductive photoperiods. The species difference in the effect of *phyA* mutations is probably due to the fact that pea relies entirely on phyA to sense inductive photoperiods, whereas *Arabidopsis* also uses a blue light photoreceptor for this purpose.

Chapter 12

Blue-Light Responses

12.1 Effect of Blue Light on Guard Cell Osmoregulation

Studies have shown that blue light influences respiratory rate, which are associated with a drastic shift in metabolism, which can also be observed in experiments using blue and red light.

Red light stimulates photosynthetic carbon fixation in photosynthetic algae, and the fixed carbon is incorporated into glucose and sucrose. The metabolic shift is observed when low photon fluxes of blue light are added to the red-light treatment; glucose and sucrose are no longer synthesized, and the fixed carbon is used in the biosynthesis of organic acids, amino acids, and proteins.

Blue light is also a powerful regulator of guard cell metabolism. Guard cells irradiated with red light accumulate sucrose ; if low photon fluxes of blue light are added to the red-light irradiation, sucrose stops accumulating, and guard cells accumulate K^+ and malate^{2-} (figure 12.1.). Thus, the blue light–stimulated shift to the synthesis of the organic acid malate in guard cells parallels the shift to the synthesis of organic acids and amino acids in photosynthetic algae.

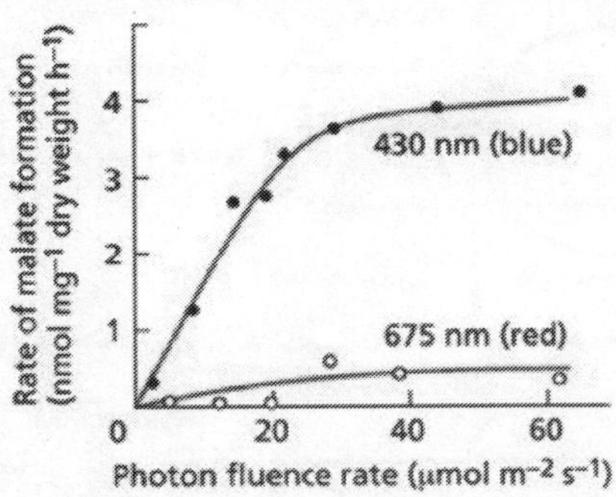

Figure 12.1. Blue light stimulates malate synthesis in guard cells. Under red light, guard cells synthesize mostly glucose and sucrose; this metabolic pathway is deactivated by blue light, and metabolism switches to starch degradation and malate biosynthesis. (Ogava *et el* ,1978, *planta* 142: 61-65)

The experimental use of blue and red light has been useful for investigations of osmoregulatory pathways associated with stomatal movements (Figure 12.2). The rapid activation of proton pumping by blue light generates a proton gradient

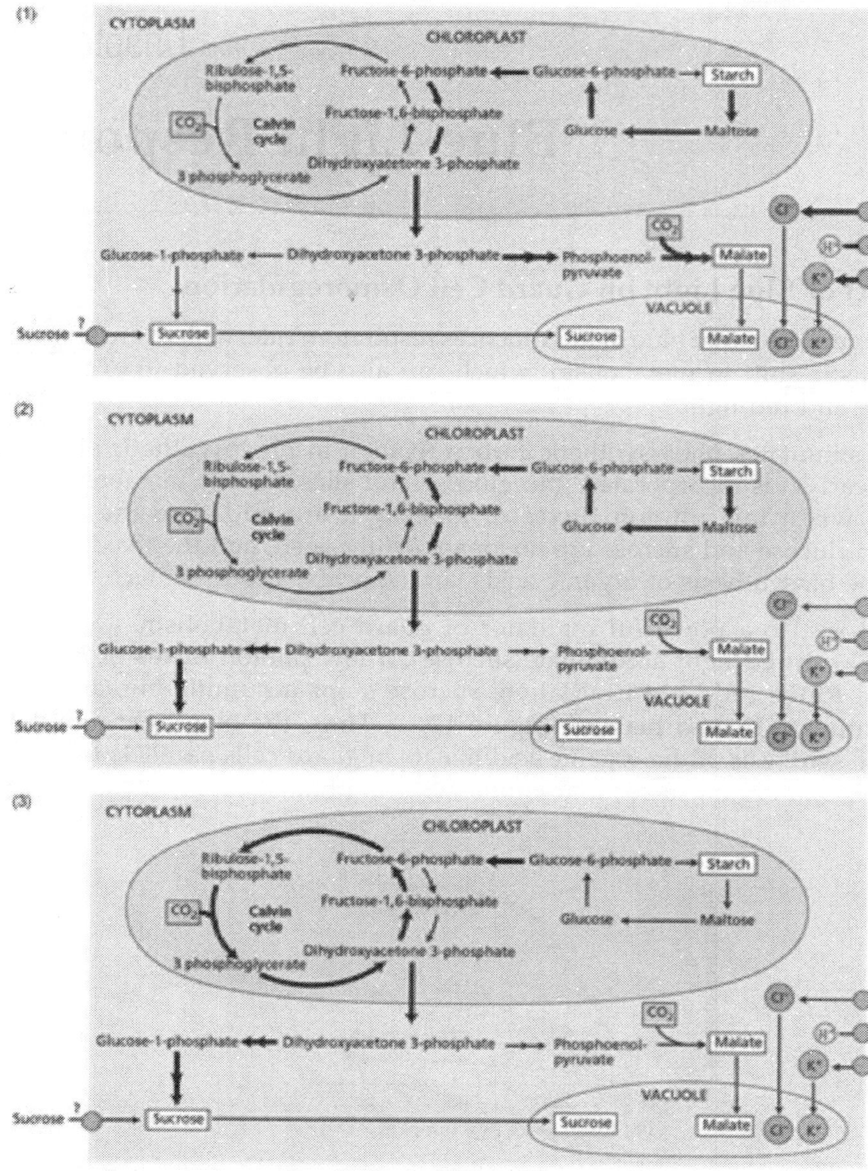

Figure 12.2 Three distinct osmoregulatory pathways in guard cells. The dark arrows identify the major metabolic steps of each pathway that lead to the accumulation of osmotically active solutes in the guard cells. (1) Potassium and its counterions. Potassium and chloride are taken up in secondary transport processes driven by a proton gradient; malate is formed from the hydrolysis of starch. (2) Accumulation of sucrose from starch hydrolysis. (3) Accumulation of sucrose from photosynthetic carbon fixation. The possible uptake of apoplastic sucrose is also indicated.

at the guard cell plasma membrane that drives potassium and chloride uptake. Blue light also stimulates starch hydrolysis and malate biosynthesis. Prolonged incubations of guard cells in detached epidermis under low fluence rates of blue light result in large stomatal apertures accompanied by extensive starch hydrolysis.

Guard cells also show a sucrose-dependent osmoregulatory phase . Sucrose can originate from starch hydrolysis (see Figure 12.2, Part 2). On the other hand, guard cells from detached epidermis irradiated with red light under ambient CO_2 concentrations operate a third metabolic pathway, showing no detectable potassium uptake or starch hydrolysis (see Figure 12.2., Part 3) . In these conditions, guard cells slowly accumulate sucrose, and this sucrose accumulation is blocked by DCMU (dichlorophenyldimethylurea), an inhibitor of photosynthetic electron transport . A recent study of guard cell osmoregulation under red light illumination and CO_2-free air showed that under these conditions, in which photosynthesis is inhibited by the absence of CO_2, red light activates the potassium-dependent pathway (Figure 12.2., Part 1) . These studies underscore the remarkable metabolic flexibility of guard cells.

A study of the effects of calcium and abscisic acid on stomatal apertures under red and blue light found that increasing calcium or ABA concentrations inhibited blue light–stimulated stomatal opening in a concentration-dependent fashion, but had no effect on red light-stimulated opening . These contrasting responses to blue and red light can be explained by the effect of the treatments on guard cell osmoregulation. ABA and high calcium concentrations have been shown to inhibit proton pumping and potassium uptake , which are central to blue light–stimulated stomatal opening. Red light, on the other hand, stimulates guard cell photosynthesis and sucrose accumulation, and this osmoregulatory pathway appears to be insensitive to ABA and calcium.

Unlike visual systems, the color information in the blue light and red light treatments does not appear to have functional significance for photosynthetic organisms. Rather, their dramatic experimental effects appear to be a result of a differential stimulation of carotenoids and chlorophylls that are selectively excited by blue light applied under a saturating background of red light, or by red light alone, respectively.

Under solar radiation, regulatory pathways activated by the excitation of these pigments are used by the chloroplast to sense key features in their environment and to control their metabolism.

12.2 The Coleoptile Chloroplast

Guard cell chloroplasts are a conspicuous structural feature of guard cells (, and their physiology and biochemistry have been studied for several decades. In contrast, the chloroplasts of phototropically sensitive coleoptiles have only recently been analyzed in detail. The coleoptile chloroplast had been suggested as a possible site for photoreception in oat coleoptiles by Kenneth V. Thimann and his coworkers in 1960, but interest in the flavin hypothesis later in that decade shifted attention away from coleoptile chloroplasts and these organelles have been virtually ignored for several decades.

Recent studies have shown that the coleoptile chloroplast evolves oxygen, fixes CO_2 photosynthetically, and operates the xanthophyll cycle . Dark-grown, etiolated coleoptile tips have violaxanthin and antheraxanthin but no zeaxanthin, and they accumulate zeaxanthin when irradiated.

When dark-grown coleoptiles are pretreated with red light for different lengths of time, their zeaxanthin content is proportional to the length of the red-light treatment (Figure 12.3). Exposure of coleoptiles with different zeaxanthin content to a brief pulse of blue light elicitis phototropic bending that is linearly related to the zeaxanthin content of the coleoptiles

(Figure 12.3). Most interestingly, the regression line through the data points extrapolates to zero (Figure 12.3, Part 1), suggesting that there is no phototropic response in the absence of zeaxanthin .

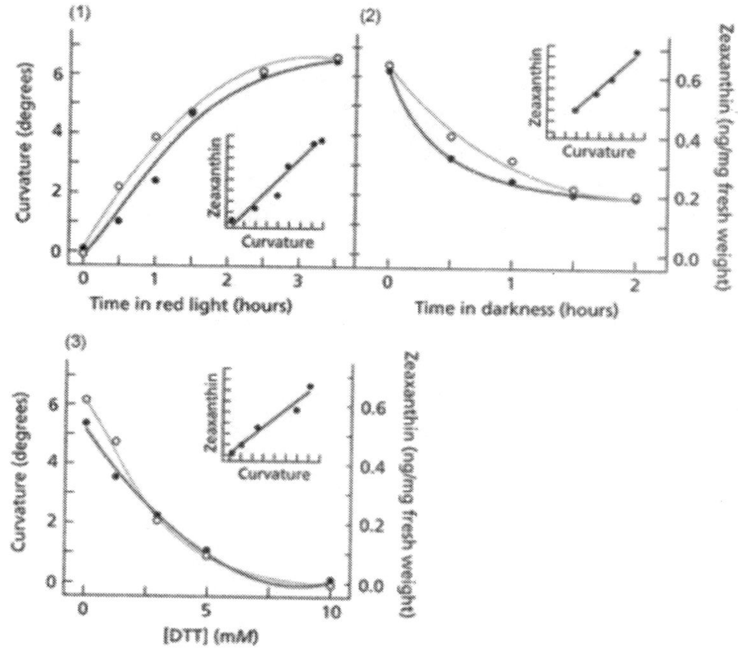

Figure 12.3 The relationship between the zeaxanthin content of corn coleoptile tips and bending in response to a pulse of blue light. Zeaxanthin content was altered by preillumination time under red light (1), transfer of zeaxanthin-containing coleoptiles to the dark (2), and a saturating red-light pretreatment in the presence of increasing concentrations of DTT, an inhibitor of zeaxanthin formation (3). (Quinones and Zeiger, 1994 *Science* 264:558-561)

Red light-treated coleoptiles with a high zeaxanthin content convert their zeaxanthin to violaxanthin when transferred to the dark. Phototropic bending in response to a pulse of blue light given at different time intervals after the transfer to the dark is linearly related to zeaxanthin content (Figure 12.3., Part 2).

When corn coleoptiles are given a saturating red-light pretreatment and a blue-light pulse in the presence of increasing concentrations of DTT (an inhibitor of zeaxanthin formation), bending and zeaxanthin content decrease linearly as a function of DTT concentration (Figure 12.3, Part 3).

Thus the relationship between the zeaxanthin content of coleoptile chloroplasts and blue light-dependent phototropic sensitivity is very similar to the relationship between the zeaxanthin content of the guard cell chloroplast and the blue-light sensitivity of stomatal opening.

Coleoptile and guard cell chloroplasts share some properties that are clearly distinct from mesophyll chloroplasts, including high rates of oxygen evolution and low rates of photosynthetic carbon fixation, a high sensitivity of their xanthophyll cycle to low levels of incident photon fluxes, and a high starch content . The coleoptile and guard cell chloroplasts also show a specific blue light response with an action spectrum that closely matches the action spectra for blue light-stimulated stomatal opening and coleoptile phototropism .

These remarkable similarities suggest that these distinct characteristics are typical of chloroplasts specialized in signal transduction, as opposed to the classic role of the mesophyll chloroplast, specialized in photosynthetic carbon fixation.

Chapter 13

Auxin: Some New Findings

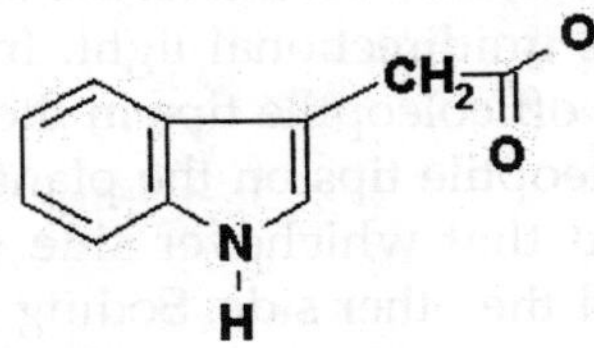

13.1 Nature of Auxins

The term auxin is derived from the Greek word auxein which means to grow. Compounds are generally considered auxins if they can be characterized by their ability to induce cell elongation in stems and otherwise resemble indoleacetic acid (the first auxin isolated) in physiological activity. Auxins usually affect other processes in addition to cell elongation of stem cells but this characteristic is considered critical of all auxins and thus "helps" define the hormone.

13.2 History of Auxins and Pioneering Experiments

Auxins were the first plant hormones discovered. Charles Darwin was among the first scientists to dabble in plant hormone research. In his book "The Power of Movement in Plants" presented in 1880, he first describes the effects of light on movement of canary grass (Phalaris canariensis) coleoptiles. The coleoptile is a specialized leaf originating from the first node which sheaths the epicotyl in the plants seedling stage protecting it until it emerges from the ground. When unidirectional light shines on the coleoptile, it bends in the direction of the light. If the tip of the coleoptile was covered with aluminum foil, no bending would occur towards the unidirectional light. However if the tip of the coleoptile was left uncovered but the portion just below the tip was covered, exposure to unidirectional light resulted in curvature toward the light. Darwin's experiment suggested that the tip of the coleoptile was the tissue responsible for perceiving the light and producing some signal which was transported to the lower part of the coleoptile where the physiological response of bending occurred. He then cut off the tip of the coleoptile and exposed the rest of the coleoptile to unidirectional light to see if curving occurred. Curvature did not occur confirming the results of his first experiment.

It was in 1885 that Salkowski discovered indole-3-acetic acid (IAA) in fermentation media. The isolation of the same product from plant tissues would not be found in plant tissues for almost 50 years. IAA is the major auxin involved in many of the physiological processes in plants. In 1907, Fitting studied the effect of making incisions on either the light or dark side of the plant. His results were aimed at understanding if translocation of the signal occurred on a particular side of the plant but his results were inconclusive because the signal was capable of crossing or going around the incision . In 1913, Boysen-Jensen modified Fritting's experiment by inserting pieces of mica to block the transport of the signal and showed that transport of auxin toward the base occurs on the dark side of the plant as opposed to the side exposed to the unidirectional light. In 1918, Paal confirmed Boysen-Jensen's results by cutting off coleoptile tips in the dark, exposing only the tips to the light, replacing the coleoptile tips on the plant but off centered to one side or the other. Results showed that whichever side was exposed to the coleoptile, curvature occurred toward the other side. Soding was the next scientist to extend auxin research by extending on Paal's idea. He showed that if tips were cut off there was a reduction in growth but if they were cut off and then replaced growth continued to occur.

In 1926, a graduate student from Holland by the name of Fritz Went published a report describing how he isolated a plant growth substance by placing agar blocks under coleoptile tips for a period of time then removing them and placing them on decapitated Avena stems. After placement of the agar, the stems resumed growth. In 1928, Went developed a method of quantifying this plant growth substance. His results suggested that the curvatures of stems were proportional to the amount of growth substance in the agar. This test was called the avena curvature test.

Much of our current knowledge of auxin was obtained from its applications. Went's work had a great influence in stimulating plant growth substance research. He is often credited with dubbing the term auxin but it was actually Kogl and Haagen-Smit who purified the compound auxentriolic acid (auxin A) from human urine in 1931. Later Kogl isolated other compounds from urine which were similar in structure and function to auxin A, one of which was indole-3 acetic acid (IAA) initially discovered by Salkowski in 1985. In 1954 a committee of plant physiologists was set up to characterize the group auxins. The term comes from the Greek auxein meaning "to grow." Compounds are generally considered auxins if they are synthesized by the plant and are substances which share similar activity to IAA (the first auxin to be isolated from plants) .

13.3 Functions of Auxin

The following are some of the responses that auxin is known to cause.

- Stimulates cell elongation
- Stimulates cell division in the cambium and, in combination with cytokinins in tissue culture

- Stimulates differentiation of phloem and xylem
- Stimulates root initiation on stem cuttings and lateral root development in tissue culture
- Mediates the tropistic response of bending in response to gravity and light
- The auxin supply from the apical bud suppresses growth of lateral buds
- Delays leaf senescence
- Can inhibit or promote (via ethylene stimulation) leaf and fruit abscission
- Can induce fruit setting and growth in some plants
- Involved in assimilate movement toward auxin possibly by an effect on phloem transport
- Delays fruit ripening
- Promotes flowering in Bromeliads
- Stimulates growth of flower parts
- Promotes (via ethylene production) femaleness in dioecious flowers
- Stimulates the production of ethylene at high concentrations

13.4 Additional Synthetic Auxins

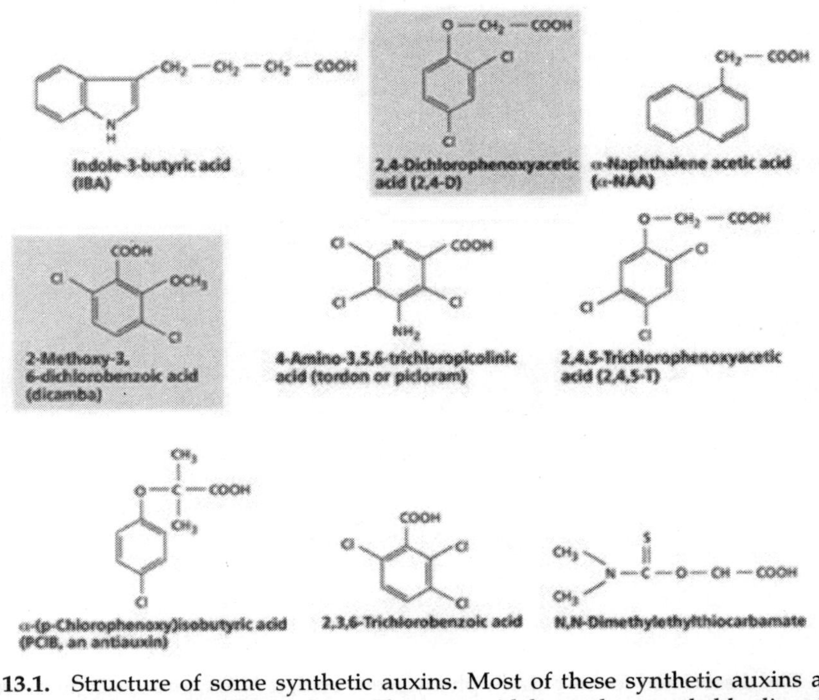

Figure 13.1. Structure of some synthetic auxins. Most of these synthetic auxins are used as herbicides in horticulture and agriculture. The most widely used are probably dicamba and 2,4-D, which are not subject to breakdown by the plant and are very stable.

13.5 The Structural Requirements for Auxin Activity

Antiauxins are another class of synthetic auxin analogs. These compounds such as x-(p-chlorophenoxy)isobutyric acid, or PCIB , have little or no auxin activity but specifically inhibit the effects of auxin. When applied to plants, antiauxins may compete with IAA for specific receptors without triggering an auxin response, thus inhibiting normal auxin action. It is possible to overcome the inhibition of an antiauxin by adding excess IAA.

How can such a diverse group of chemicals all be active auxins? No other plant hormone, with the possible exception of the cytokinins, has such a wide variety of active analogs. For example, few people would have predicted that either 2,4-dichlorophenoxyacetic acid (2,4-D) or α-naphthalene acetic acid (α-NAA), both of which bear little resemblance to IAA, would be able to stimulate coleoptile elongation or have activity in other auxin bioassays.

A comparison of the compounds that possess auxin activity reveals that at neutral pH they all have a strong negative charge on the carBoxyl group of the side chain that is separated from a weaker positive charge on the ring structure by a distance of about 0.5 nm (Figure 13.2.) . This charge separation may be an essential structural requirement for auxin activity.

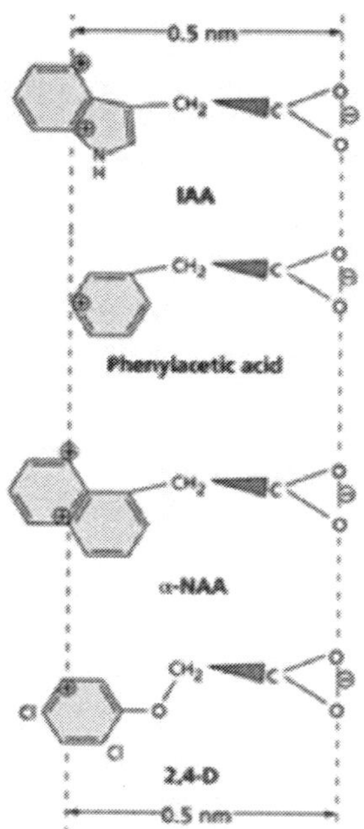

Figure 13.2. The dissociated forms of IAA, phenylacetic acid, α-NAA, and 2,4-D (in descending order), showing the negative charge on the carboxyl group and the fractional positive charge on the ring, separated by a distance of about 0.5 nm.

The main part of the molecule, the indole ring, is not essential for activity, although an aromatic or fused aromatic ring of a certain size range is required. Recently a set of molecular requirements for auxin activity is proposed based on studies of the binding of various auxin analogs to a protein, auxin-binding protein 1 (ABP1), which may be the auxin receptor. Their model defines three essential regions of the binding site: a planar aromatic ring–binding platform, a carboxylic acid–binding site, and a hydrophobic transition region that separates the two binding sites.

13.6 Auxin Measurement

A radioimmunoassay (RIA) allows the measurement of physiological levels (10^{-9} g = 1 ng) of IAA in plant tissues. An RIA (Figure 13.3.), requires two chemicals: (1) specific antibodies that recognize IAA; and (2) a radioactively labeled IAA in which several hydrogen atoms have been replaced by radioactive tritium. Because of the sensitivity and selectivity of the antibodies used in this technique, RIA is as sensitive as some of the best physicochemical methods . Radioimmunoassays also allow one to quantify a specific auxin. In a modified RIA, the radiolabeled IAA is replaced by an auxin conjugated to an enzyme, generally an alkaline phosphatase. This enzyme-linked immunosorbent assay (ELISA) affords sensitivities that are similar or better than the RIA.

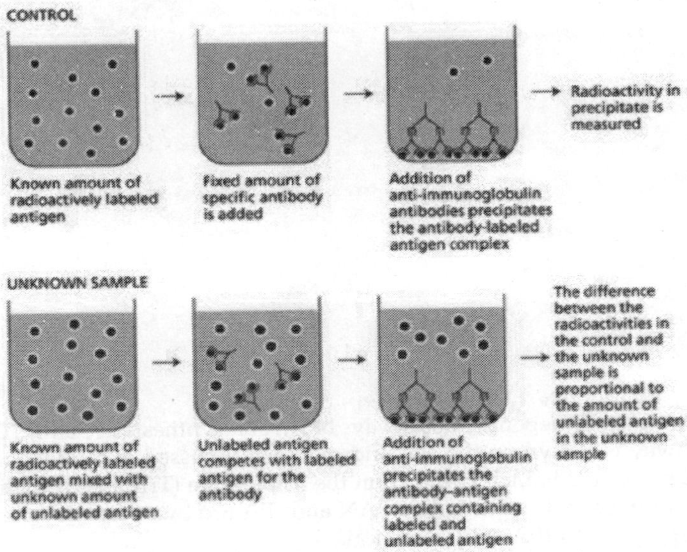

Figure 13.3. Radioimmunoassay for the amount of auxin in a plant tissue extract. In the preparation of auxin antibodies, IAA is first conjugated to a large protein and injected into a mouse or rabbit. The animal produces antibodies to all of the antigenic determinants on the protein, including the IAA. The anti-IAA antibodies can then be purified for use in the RIA. Unlabeled antigen (IAA) competes with a known amount of radioactively labeled antigen—e.g., [^{14}C]IAA—for binding to antibodies. The more unlabeled antigen present, the less radioactivity from the radioactive antigen will be found in the antibody precipitate.

13.7 Biosynthesis of IAA

Further evidence was provided by labeling studies with tryptophan auxotrophic mutants of *Arabidopsis* . Together, these studies established that the branch point for IAA biosynthesis is either indole or its precursor, indole-3-glycerol phosphate (Figure 13.4).

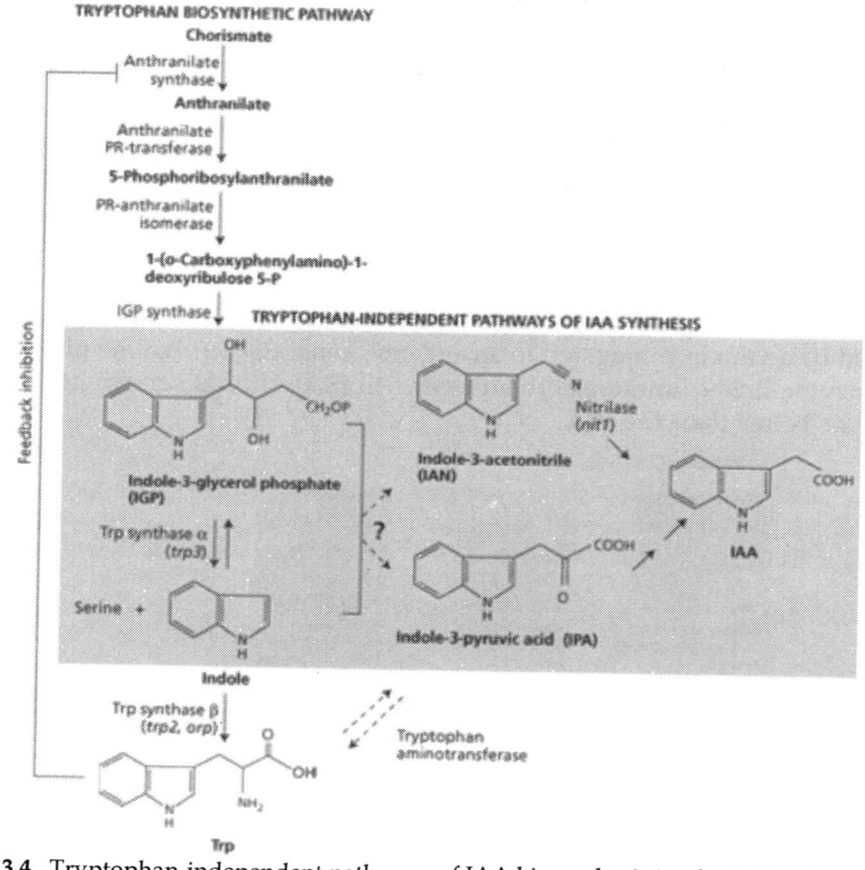

Figure 13.4 Tryptophan-independent pathways of IAA biosynthesis in plants. The tryptophan biosynthetic pathway is shown on the left, and mutants discussed in the text are indicated in parentheses. The suggested branch points from the tryptophan (Trp) biosynthetic pathway are at indole-3-glycerol phosphate or at indole. IAN and IPA are two possible intermediates. The conversion of tryptophan to IPA is hypothetical.

Studies with the *Arabidopsis* mutants *trp2* and *trp3* , which are blocked in the last two steps of tryptophan biosynthesis, have shown that IAN accumulates up to 11-fold in the mutants compared to the wild type . In tomato, IPA has been shown to be synthesized independently of tryptophan . Depending on the species, then, either IAN or IPA may serve as the intermediate between either indole-3-glycerol phosphate or indole, respectively, and IAA.

The discovery of the tryptophan-independent pathway has drastically altered our view of IAA biosynthesis, but the relative importance of the two pathways (tryptophan-dependent versus tryptophan-independent) is poorly understood. Some plants, such as bean seedlings, synthesize IAA primarily by the tryptophan-dependent pathway. The same is true of embryogenic carrot cell suspension cultures. However, upon removal from the medium containing 2,4-D, which initiates embryogenesis in these cultures, the cells switch over to the tryptophan-independent pathway of IAA biosynthesis. This result suggests that the specific pathway of IAA biosynthesis that is utilized may be under developmental control.

13.8 Regulating Steady-State IAA Levels

The steady-state level of free IAA in the cytosol is tightly regulated and is determined by several interconnected processes (Figue 13.5). Auxin metabolism includes the synthesis, degradation, and conjugation of IAA. In addition to IAA metabolism, its compartmentation in the choroplast and transport (discussed below) also play important roles in regulating the level of free IAA.

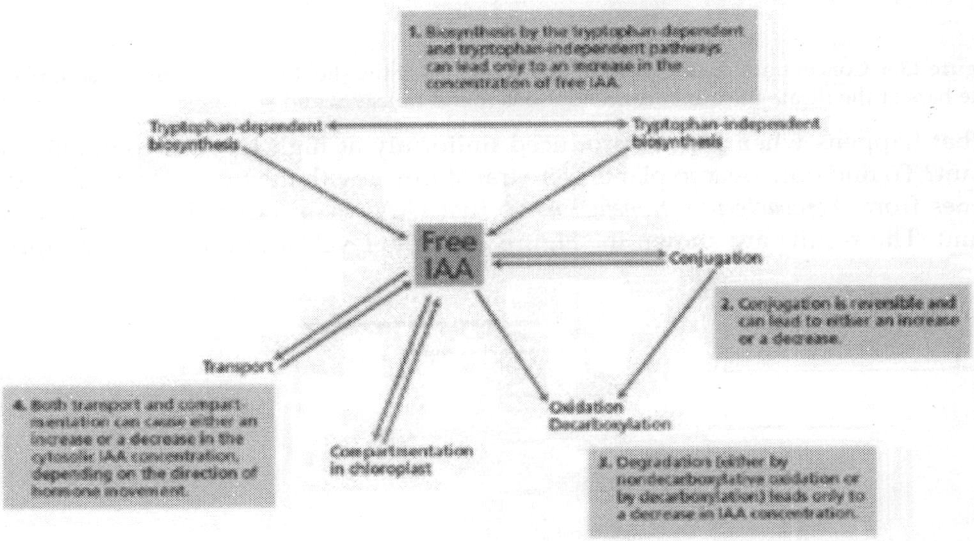

Figure 13.5 Factors that influence the steady-state levels of free IAA in plant cells. Biosynthesis by the tryptophan-dependent and tryptophan-independent pathways can lead only to an increase in the concentration of free IAA. Degradation (either by nondecarboxylative oxidation or by decarboxylation) leads only to a decrease in IAA concentration, while conjugation is reversible and can therefore lead to either an increase or a decrease. Both transport and compartmentation can cause either an increase or a decrease in the cytosolic IAA concentration, depending on the direction of hormone movement.(Normanly *et el, Proc Natl. Acad. Sci.* USA. 90:10355-10359)

The amounts of IAA in different regions of a normal tobacco plant are shown in Figure 13.6 The highest concentration of free IAA is in the apical bud. The level of free IAA in the mature regions of the stem is nearly constant. Since there is little IAA synthesis in the mature regions of the shoot, we can infer that there must be little catabolism or conjugation in these regions; otherwise the free IAA concentrations would continue to decline.

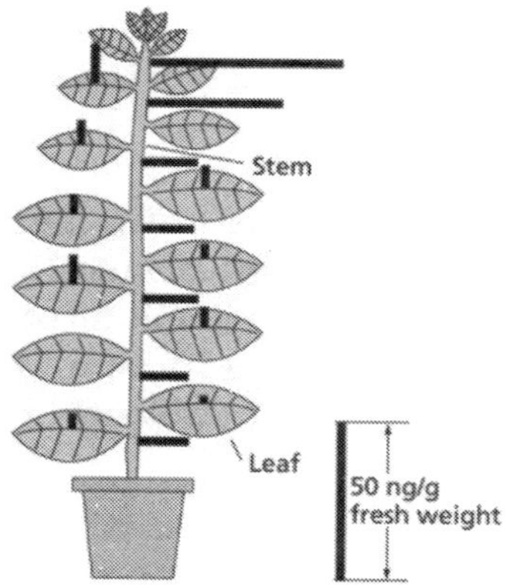

Figure 13.6 Concentrations of IAA in different regions of the shoot of a wild-type tobacco plant. The bars at the dicate the auxin concentrations found in leaves and stems.)

What happens when IAA is produced uniformly at high levels throughout the plant? To find out, tobacco plants were transformed with the two IAA biosynthesis genes from *Agrobacterium tumefaciens* so that IAA was produced throughout the plant. The results are shown in Figure 13.7 The level of free IAA in the upper

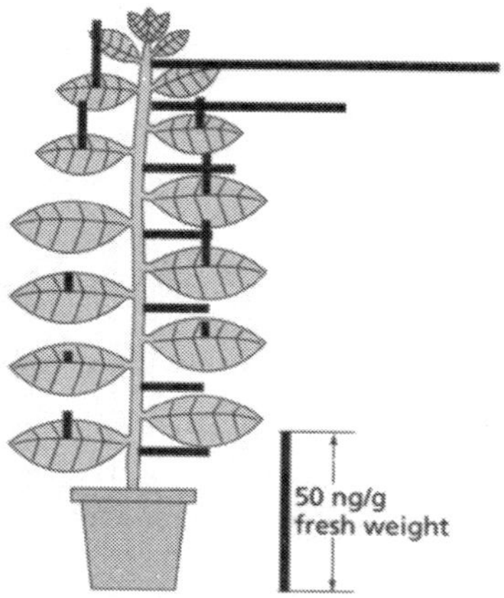

Figure 13.7 Concentrations of IAA in different regions of the shoot of a transformed tobacco plant expressing the bacterial genes that encode the two enzymes involved in the IAM pathway of IAA biosynthesis: Trp monoxygenase and IAM hydrolase. The bars at the different regions indicate the auxin concentrations found in leaves and stems.

part of the shoot increased considerably, while the free IAA concentrations in the mature parts of the shoot hardly changed. The measured levels of conjugated IAA were several times higher in the transformed tobacco plant than in the control. We can infer that processes that decrease the amount of free IAA must have been activated in the mature shoot so that the concentration of free IAA was maintained close to its normal level.

13.9 Fusicocci –A super Auxin

Our examination of the auxin response would be incomplete without a discussion of the mechanism of fusicoccin (FC) . The structure of fusicoccin A, a phytotoxin produced by the fungus *Fusicoccum amygdali*. This phytotoxin is produced by the fungus *Fusicoccum amygdali*, a parasite of peach and almond trees. Once released into the leaves, the toxin stimulates the H^+-ATPases of the guard cells. At the biochemical level it induces a rapid (10 to 30 seconds) hyperpolarization of the plasma membrane, accompanied by an acidification of the cell wall. This triggers an irreversible opening of the stomata , resulting in wilting of the leaves and, eventually, the death of the tree.

FC causes membrane hyperpolarization and proton extrusion in nearly all plant tissues. Treatment of coleoptiles and stem sections with FC leads to a transient growth response, a result that provides support for the acid growth hypothesis for auxin-induced growth. But whereas auxin-induced proton extrusion has a lag time of about 10 minutes and is inhibited by cycloheximide, FC-induced proton extrusion begins after only 1 to 2 minutes, and the response is insensitive to cycloheximide. Moreover, FC-treated cells can acidify down to a much lower extracellular pH than auxin-treated cells can (pH 3 versus pH 4).

Because of these effects, FC has sometimes been referred to as a "super auxin." However, FC acts by a mechanism entirely different from that of auxin. For example, FC does not stimulate the expression of any of the auxin-induced genes, nor can it mimic the effects of IAA on other developmental processes, such as cell division. Thus the effects of FC on plants are much more limited than those of IAA, which is not surprising since FC is a toxin, not an endogenous plant hormone. In the case of proton extrusion, FC is known to activate preexisting H^+-ATPases on the plasma membrane, probably by displacing the autoinhibitory C-terminal domain of the enzyme from the catalytic site . In contrast, auxin may promote proton extrusion in part by activation of preexisting H^+-ATPases (perhaps via acidification of the cytosol) and in part by causing the de novo synthesis of the H^+-ATPase.

Fusicoccin was first identified as a phytotoxin in the 1960s, but it was not until 1977 that a radioactively labeled version of this toxin was available that could be used to characterize its ligand. Studies with radioactively labeled FC have shown that it binds tightly to a protein on the plasma membrane. Ultimately, these studies led to the purification and identification of the fusicoccin-binding protein (FCBP).

Initial attempts to purify FCBP after photoaffinity labeling led to conflicting results. Some laboratories identified a homodimer, composed of two 30 kDa subunits. The ability of the purified 30 kDa polypeptides to stimulate the plasma membrane H^+-ATPase in the presence of FC was tested in in vitro proton-pumping experiments. The 30 kDa polypeptide was reconstituted along with purified plasma membrane ATPase into artificial membrane vesicles called proteoliposomes. The addition of FC to the reconstituted proteoliposomes caused an increase in their ATP-driven H^+-pumping activity.

Cloning and sequencing of the gene for the 30 kDa polypeptide revealed it to be a member of the so-called 14-3-3 family of regulatory proteins . Regulatory proteins of the 14-3-3 family are widespread among plants and animals. Originally identified in mammalian brain tissue (and named after the positions they occupy in chromatographs), 14-3-3 proteins make up a heterogeneous family of soluble proteins ranging from 25 to 32 kDa, which associate to form dimers . The precise function of 14-3-3 proteins has not yet been elucidated. However, the numerous functions that have been attributed to them indicate that they play multiple roles in signal transduction pathways.

Although it was initially believed that the 30 kDa 14-3-3 protein represented the complete FCBP, another laboratory reported that a 90 kDa polypeptide consistently

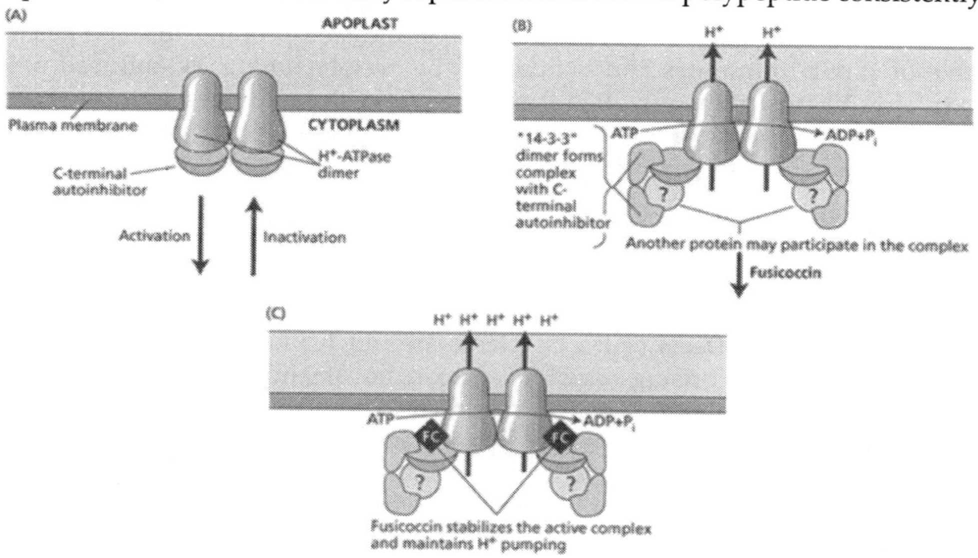

Figure 13.8. Model for the mechanism of fusicoccin activation of H^+-ATPases. (A) A pair of plasma membrane H^+-ATPases associate to form an active dimer. The activity of the pump is limited by the autoinhibitory effect of the C terminus of each monomer. (B) Dimers of 14-3-3 proteins form a transient complex with the C terminus of the H^+-ATPase and possibly another protein. When the autoinhibitory C terminus is bound to the 14-3-3 protein, the activity of the H^+-ATPase increases, but the effect is transient because of the instability of the complex. (C) The binding of fusicoccin to the complex stabilizes the complex, locking the H^+-ATPase into an active state. (Oecking *et el* , FEBS *Lett* 352:163-166)

copurified with the two 30 kDa polypeptides . More recently it has been demonstrated that the 14-3-3 protein is unable to bind FC by itself. To bind FC, the 14-3-3 protein requires the presence of H^+-ATPase . This finding confirms the earlier report that a 90 kDa polypeptide (the H^+-ATPase) copurifies with the 14-3-3 protein. According to the current model, FC activates the H^+-ATPase by stabilizing a transient complex that forms between the autoinhibitory C-terminal domain of the H^+-ATPase and the 14-3-3 dimer (Figure 13.8) locking the enzyme into its most active state.

13.10 The Fluence Response of Phototropism

Phototropism involves two fundamental processes: blue-light perception and differential cell elongation. . There is now compelling evidence that two flavoproteins associated with the plasma membrane (PHOT1 and PHOT2) are the photoreceptors for phototropism . These proteins become phosphorylated in response to blue light. , differential cell elongation in response to unilateral light is thought to be caused by the lateral redistribution of auxin to the shaded side.

If phototropism were a simple photochemical reaction, one would expect the fluence response curve to rise from a threshold level of light to a plateau at which the photoreceptor becomes saturated. However, this is not what happens. .

Starting at the light threshold, there is a bell-shaped curve that has been termed the first positive curvature. The first positive curvature is followed by a neutral zone in which little or no measurable response is observed. Under certain conditions, *Avena* coleoptiles may actually bend away from the light in this range, a phenomenon referred to as the first negative curvature. Both the first positive and the first negative curvatures (neutral zone) are restricted to the coleoptile tip. At higher light fluences the curve ascends again, and this second bending response is termed the second positive curvature. The second positive curvature differs from the first positive curvature in that the bending occurs at the base rather than at the tip of the coleoptile.

Plant physiologists have long puzzled over the meaning of this complex relationship between light fluence and phototropism. Most studies of phototropism have been carried out in the range of the first positive curvature. The first positive curvature and the neutral zone obey the Bunsen–Roscoe reciprocity law, which states that the response will be the same for a brief exposure to bright light as for a longer exposure to dim light, as long as the product of time and the photon fluence rate are the same . Thus the first positive curvature is a response to short light pulses.

In contrast to the first positive curvature, the second positive curvature is a response to irradiations of long duration. The second positive curvature does not obey the reciprocity law; instead, it is proportional to the duration and the photon fluence rate. For the second positive response to occur, both the time threshold and the fluence threshold must be exceeded . Since these are the conditions that prevail normally in nature, phototropism in nature usually represents the second positive curvature.

What causes the gradient in auxin to form at the three fluences? It has long been assumed that the light gradient that is established across the coleoptile during unilateral irradiation with blue light must generate a biochemical gradient between the cells of the shaded and the irradiated sides, but the nature of this biochemical gradient has, until recently, eluded detection. With the discovery of blue light–induced phosphorylation of the PHOT1 and PHOT2 plasma membrane proteins, much work has focused on these photoreceptors .

Recently, Michael Salomon and his colleagues at the University of Munich demonstrated that unilateral blue light causes a lateral gradient in the phosphorylation of PHOT1 from the irradiated to the shaded side . Moreover, the concentration of PHOT1 declines exponentially from the tip of the coleoptile to the base. Studies on the rates of phosphorylation of the 116 kDa protein PHOT1 at the tip versus at the base of oat coleoptiles have led to a new model for the fluence response curve of phototropism.

According to this new model proposed by Salomon and colleagues, the gradient in concentration of NPH1 from tip to base makes the tip more sensitive to light than the base is (Figure 13.9). Thus, at threshold fluences, a gradient in

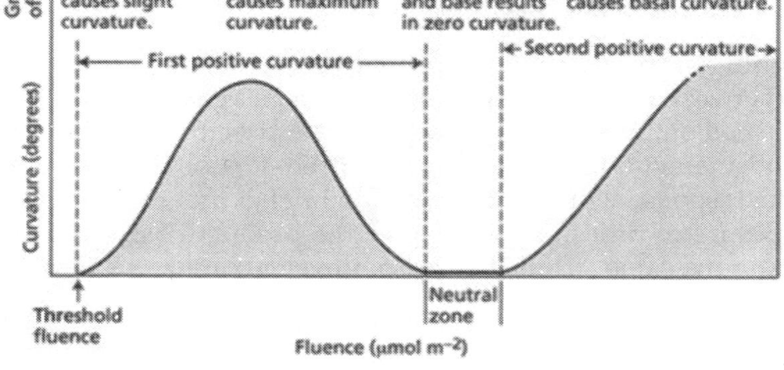

Figure 13.9 typical fluence response curve for phototropism. The phosphorylation model for the first positive curvature, the neutral zone, and the second positive curvature of coleoptiles is shown above the curve.

phosphorylation occurs only at the tip, leading to a small bending response. Increasing the fluence increases the gradient of phosphorylation at the tip, until phosphorylation on the irradiated side saturates and a maximum phosphorylation gradient is obtained. This maximum gradient represents the peak of the first positive curvature.

As the fluence is increased beyond this point, phosphorylation increases on the shaded side due to the penetration of light, reducing the gradient. When phosphorylation of NPH1 is equal on both sides of the tip, no bending occurs (the neutral zone).

However, with continued exposure to light of sufficient fluence, phosphorylation increases at the base on the irradiated side.. Phosphorylation at the base is dependent on both fluence and time, unlike phosphorylation at the tip which is strictly dependent on fluence. Thus, phosphorylation at the base appears to involve some type of time-dependent biochemical process. Since recent immunological studies have shown that the total amount of the 116 kD protein remains constant during phototropism, the time-dependent increase in phosphorylation at the base does not involve synthesis of new PHOT1 protein. Instead, a second photoreceptor (PHOT2), initially implicated in chloroplast movement in response to light appears to function as an additional photoreceptor in second positive curvature conditions.

The model proposed by Salomon and colleagues requires further testing, and other interpretations are certainly possible. Nevertheless, the new model provides the first plausible explanation for the fluence response curve of phototropism. However, even if this model is confirmed, the question remains as to how a gradient in phosphorylation can bring about a gradient in auxin.

13.11 Gravity Perception without Statoliths in *Chara*

An alternative mechanism of gravity perception that does not involve statoliths has been documented in the giant-celled fresh water alga *Chara*. Despite the absence of any discernible sedimenting statoliths, *Chara* filaments show negative gravitropism. To explain this phenomenon, it has been suggested that the entire protoplast acts like a statolith, and that gravity is perceived at the interface between plasma membrane and cell wall.

This model has also been applied to the cells of higher plants, and an alternative to the starch-statolith hypothesis has been proposed. According to the **Plasmalemma Central Control** (PCC) model, stretch-activated calcium channels are clustered around attachment centers connecting the cytoskeleton and the cell wall. These calcium channels are thought to open in response to tensions at the membrane caused by gravity-induced changes in the distribution of forces exerted by the protoplast, the cytoskeleton, or the cell wall. Something like the PCC model may explain how *Chara* and the starchless mutants of *Arabidopsis* are able to perceive and respond to gravity.

Chapter 14

Gibberellins

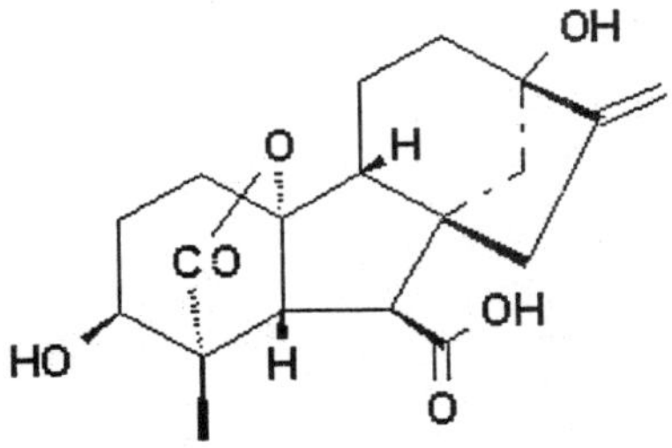

14.1 The Nature of Gibberellins

Unlike the classification of auxins which are classified on the basis of function, gibberellins are classified on the basis of structure as well as function. All gibberellins are derived from the ent-gibberellane skeleton. The structure of this skeleton derivative along with the structure of a few of the active gibberellins are shown above. The gibberellins are named GA1....GAn in order of discovery. Gibberellic acid, which was the first gibberellin to be structurally characterised , is GA3. There are currently 136 GAs identified from plants, fungi and bacteria.

GA's are widespread and so far ubiquitous in both flowering (angiosperms) and non-flowering (gymnosperms) plants as well as ferns.

Gibberellins (GAs) are a group of diterpenoid acids that function as plant growth regulators inflencing a range of developmental processes in higher plants including stem elongation, germination, dormancy, flowering, sex expression, enzyme induction and leaf and fruit senescence. The origin of research into gibberellins can be traced to Japanese plant pathologists who were investigating the causes of the "bakanae" (foolish seedling) disease which seriously lowered the yield of rice crops in Japan, Taiwan and throughout the Asian continent.

Symptoms of the disease are pale yellow, elongated seedlings with slender leaves and stunted roots. Severely diseased plants die whereas plants with slight symptoms survive but produce poorly developed grain, or none at all.

Bakanae is now easily prevented by treatment of seeds with fungicides prior to sowing. The first paper on the cause of bakanae was published in 1898 by Shotaro Hori who demonstrated that the symptoms were induced by infection with a fungus belonging to the genus Fusarium, probably Fusarium heterosporium Nees.

In 1912, Sawada published a paper in the Formosan Agricultural Review entitled "The Diseases of Crops in Taiwan" in which he suggested that the elongation in rice seedlings infected with bakanae fungus might be due to a stimulus derived from fungal hyphae.

Subsequently, Eiichi Kurosawa (1926) found that culture filtrates from dried rice seedlings caused marked elongation in rice and other sub-tropical grasses. He concluded that bakanae fungus secretes a chemical that stimulates shoot elongation, inhibits chlorophyll formation and suppresses root growth.

Although there has been controversy among plant pathologists over the nomenclature of bakanae fungus, in the 1930s, the imperfect stage of the fungus was named Fusarium moniliforme (Sheldon) and the perfect stage, Gibberella fujikuroi (Saw.) Wr. by H.W. Wollenweber. The terms "Fujikuroi" and "Saw." in Gibberella fujikuroi (Saw.) Wr. were derived from the names of two distinguished Japanese plant pathologists, Yosaburo Fujikuro and Kenkichi Sawada.

Teijiro Yabuta initiated work on the isolation of the active component using the fungal strains provided by Kurosawa. In 1934, Yabuta isolated a crystalline compound from the fungal culture filtrate that inhibited growth of rice seedlings at all concentrations tested. The structure of the inhibitor was shown to be 5-n-butylpicolinic acid or fusaric acid. The formation of fusaric acid in culture filtrates was suppressed by changing the composition of the culture medium. As a result, a non-crystalline solid was obtained from the culture filtrate that stimulated the growth of rice seedlings. This compound was named gibberellin by Yabuta in 1935; the first use of the term "gibberellin" in the scientific literature.

In 1938, Yabuta and his associate Yusuke Sumiki finally succeeded in crystallizing a pale yellow solid to yield gibberellin A and gibberellin B (The names were subsequently interchanged in 1941 and the original gibberellin A was found to be inactive.) Determination of the structure of the active gibberellin was hampered by a shortage of pure crystalline sample. By current standards the productivity of their fungal strain was extremely poor and they did not know that their sample of gibberellin A was not pure, but a mixture of structurally-related gibberellins.

In the United States, the first research on gibberellins began after the Second World War by a research unit at Camp Dietrick, Maryland. In 1950, John E. Mitchell reported optimal fermentation procedures for the fungus, as well as the effects of fungal extracts on the growth of bean (Vicia faba) seedlings. Work also began at the Northern USDA Regional Research Laboratories in Peoria, Illinois in the USA using the strain provided by Mitchell. Large scale fermentations were carried out with the purpose of producing pure gibberellin A for agricultural uses but initial fermentations were inactive.

In 1951, Sumiki visited the United States and met Frank H. Stodola. After returning to Japan he sent new cultures to the USA but these also proved inactive. The problem was traced to the lack of magnesium in the culture medium and good yields of gibberellin were obtained when the culture medium was suppiemented with magnesium sulphate. The physical properties of gibberellin isolated from these fermentations were found to be surprisingly different from those reported by the Japanese and the new compound was named gibberellin-x.

At about the same time in the UK, a team of researchers at Akers Research Laboratories (ICI) isolated a new gibberellin which was given the name "gibberellic acid".This compound had physical properties different from the Japanese gibberellin A(Curtis & Cross, 1954). Samples were exchanged between Stodola and Grove and "gibberellic acid" and gibberellin-X were found to have identical chemical and physical properties and the name gibberellic acid was accepted by both groups. A structure for gibberellic acid was proposed in 1956 but later revised see Grove 1961.

In 1955, members of Sumuki group, succeeded in separating the methyl ester of gibberellin A into three components, from which corresponding free acids were obtained and named gibberellins A1, A2, and A3. Gibberellin A3 was found to be identical to gibberellic acid. In 1957, Takahashi *et al.* isolated a new gibberellin named gibberellin A4 as a minor component from the culture filtrate.

14.2 Functions of Gibberellins

Active gibberellins show many physiological effects, each depending on the type of gibberellin present as well as the species of plant. Some of the physiological processes stimulated by gibberellins are outlined below.

- Stimulate stem elongation by stimulating cell division and elongation.
- Stimulates bolting/flowering in response to long days.
- Breaks seed dormancy in some plants which require stratification or light to induce germination.
- Stimulates enzyme production (a-amylase) in germinating cereal grains for mobilization of seed reserves.
- Induces maleness in dioecious flowers (sex expression).
- Can cause parthenocarpic (seedless) fruit development.
- Can delay senescence in leaves and citrus fruits.

14.3 Structures of Some Important Gibberellins

By definition, all GAs possess either a tetracyclic (four-ringed) *ent*-gibberellane skeleton containing 20 carbon atoms (Figure 14.1), or a 20-nor-*ent*-gibberellane skeleton, which has only 19 carbon atoms because carbon-20 (C-20) has been lost.

Gibberellins that contain all 20 carbon atoms are referred to as **C_{20}-GAs.** The first-formed GA in plants is GA_{12}, a C_{20}-GA whose structure is shown below (Figure 14.2 *top*). The other GAs have only 19 carbons because they have lost C-20 (blue, on structure of the C_{20}-GA) by metabolism. These are referred to as **C_{19}-GAs**, and the structure of one of them, GA_9, is shown below (see Figure 14.2, *bottom*). In nearly all C_{19}-GAs, the carboxyl at C-4 forms a lactone at C-10 (red, on structure of GA_9).

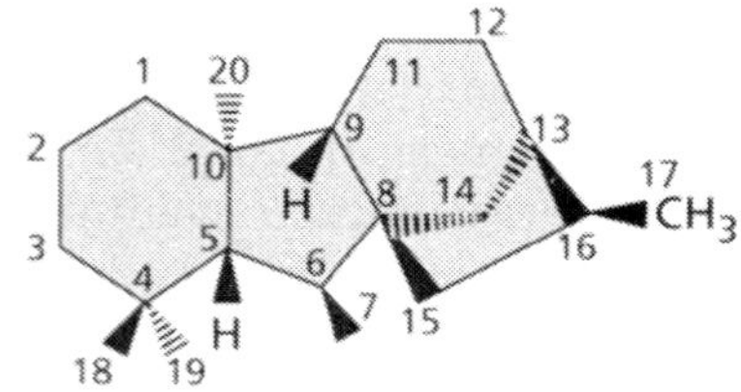

Figure 14.1 The structure of *ent*-gibberellane.

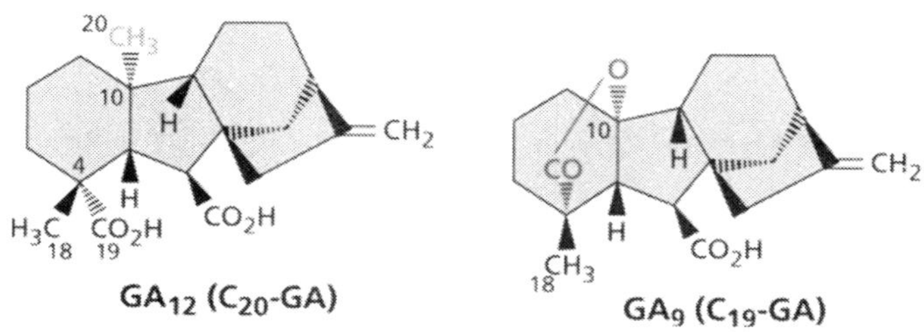

Figure14.2 The structures of GA_{12} and GA_9.

Other structural modifications include the insertion of additional functional groups, the *position* and *stereochemistry* of which can have *profound* effects on biological activity. For example, in a C_{19}-GA, the presence of a hydroxyl group at C-3 in the β-orientation (shown by a darkened wedge-shaped bond on structures) is required for high biological activity in bioassays that rely on stem growth (Figure14. 2). A C_{19}-GA possessing a hydroxyl group at C-3 in the α-orientation (α-orientation is denoted by a dotted wedge) has substantially less biological activity than its 3β-OH counterpart. Substitution at C-2 can also have very important consequences for biological activity. For instance, a biologically active C_{19}-GA that acquires a hydroxyl group at C-2 will become inactive if the hydroxyl is in β-orientation, whereas 2α-hydroxylation is less detrimental to bioactivity. **Therefore, 3β-hydroxylation activates a C_{19}-GA, whereas 2β-hydroxylation is a deactivating mechanism.**

It is assumed that the strict structural requirements for bioactivity reflect the necessary size and shape needed for binding to the receptor protein.

Interestingly—and not surprisingly—the most biologically active GAs were among the first to be discovered, namely GA_1, GA_3, GA_4, and GA_7. These GAs, all of which have intrinsic or inherent stem growth-promoting activity, are C_{19}-GAs. They all possess a 4,10-lactone, a carboxylic acid (-COOH) at C-6, a hydroxyl group at C-3 in β-orientation, and do *not* have a hydroxyl group at C-2 in β-orientation. (Figure 14.3)

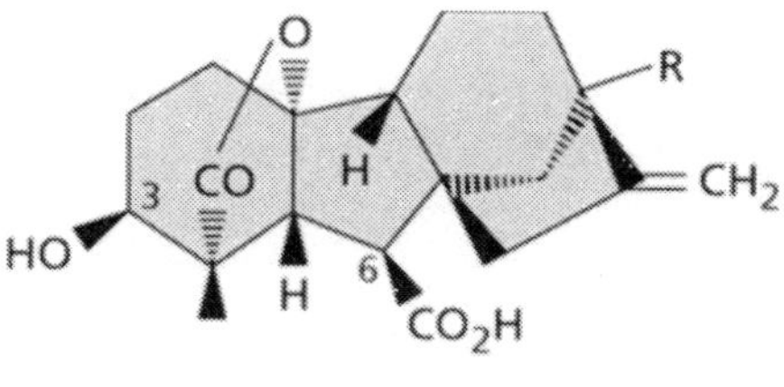

GA$_4$ R = H
GA$_1$ R = OH

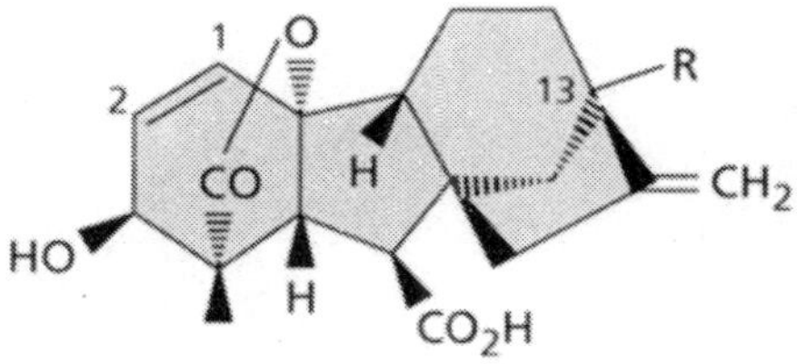

GA$_7$ R = H
GA$_3$ R = OH

Figure 14.3 The structures of GA$_4$, GA$_1$, GA$_7$, and GA$_3$.

The presence of a double bond between C-1 and C-2 (as in GA$_7$ and GA$_3$) may account for enhanced bioactivity of GA$_7$ relative to GA$_4$, and of GA$_3$ relative to GA$_1$, because GA$_7$ and GA$_3$ will not be metabolically deactivated by 2β-hydroxylation. Hence, GA$_7$ and GA$_3$ are expected to have more sustained growth-promoting activity than GA$_4$ and GA$_1$, respectively.

Gibberellins can also be converted into **conjugates**, most commonly with glucose. The linkage of the GA moiety to glucose occurs either via a hydroxyl group to produce a GA-*O*-glucosyl ether, or via the COOH at C-6 yielding a GA-glucosyl ether. When applied to bioassay plants, the GA-*O*-glucosyl ethers show little to no activity, indicating that they are deactivation products. The intact GA-glucosyl esters have no bioactivity, yet they may show biological activity if they are hydrolyzed within a bioassay plant yielding a free biologically active GA. Hence, the formation of glucosyl ester conjugates at a specific stage in the life cycle of a plant may be a mechanism for sequestration of bioactive GA that can be released later, as required.

Chemicals, either natural or synthetic, that interfere with GA biosynthesis usually retard growth. These types of chemicals, which are referred to as **growth retardants**, have been used for many years to reduce the height of plants in order to enhance yield of fruit or grain. The utility of growth retardants depends on their effectiveness to inhibit GA biosynthesis, their lack of toxicity, and on their specificity.

Varios enzymes are involved in different stages of GA biosynthesis. Growth retardants have been produced that can target most of these different classes of enzymes. For example, compounds that contain a quaternary ammonium group have been developed to target the terpene synthases in *stage 1*, of which *ent*-copalyl diphosphate synthase (CPS) is an example. **Chlormequat chloride** (also known as cycocel), **mepiquat chloride** and **AMO-1618** (Figure 14.4) are thus all growth retardants blocking the synthesis of the GA-precursor *ent*-copalyl diphosphate. Chlormequat chloride is used extensively to reduce lodging in wheat, and to reduce vegetative growth in cotton.

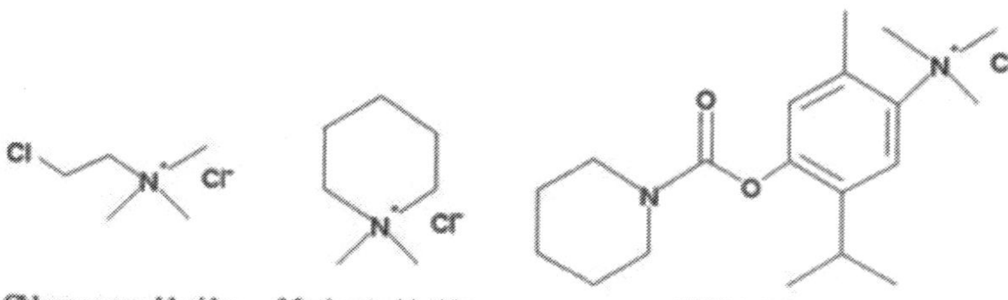

Chlormequat chloride Mepiquat chloride AMO-1618

Figure 14.4 The structures of chlormequat chloride (also known as cycocel), mepiquat chloride and AMO-1618.

Another type of GA biosynthesis inhibitor, the triazoles, of which **paclobutrazol** and **uniconazole** are examples (Figure14.5), inhibit the **monooxygenase** enzymes that catalyze ***stage 2*** in GA biosynthesis . These growth retardants target *ent*-kaurene oxidase, thereby inhibiting the formation of *ent*-kaurenoic acid. They are used commercially to restrict the stem growth of fruit trees and ornamental plants.

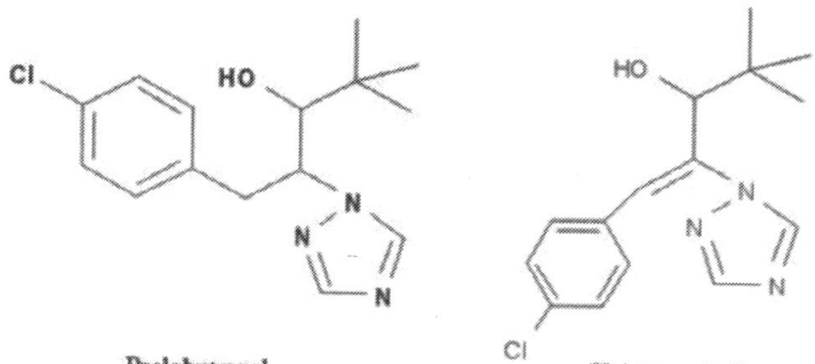

Paclobutrazol Uniconazole-P

Figure 14.5 The structures of paclobutrazol and uniconazole-P.

Dioxygenase enzymes catalyze the oxidative steps between individual C_{20}- and C_{19}-GAs in ***stage 3*** . These enzymes utilize 2-oxoglutarate as a co-substrate, and so are competitively inhibited by compounds that resemble 2-oxoglutarate. Examples of this type of inhibitor are **prohexadione calcium** (BX-112) and **daminozide** (Figure 14.6). Most commonly, these inhibitors interfere with oxidation at C-20 or at C-3, and hence reduce the formation of bioactive GA. They are therefore effective growth retardants and can be used to retard the growth of potted ornamental plants.

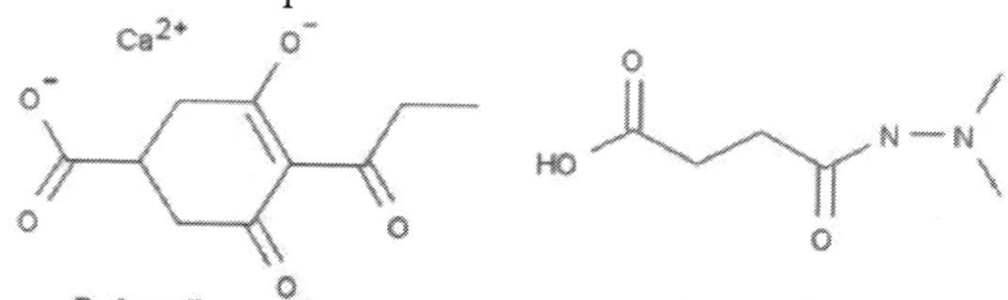

Prohexadione calcium Daminozide

Figure 14.6 The structures of prohexadione calcium (BX-112) and daminozide.

The last type of growth retardant to be considered here are analogs that interfere with the activity of dioxygenases by mimicking the GA-substrate rather than 2-oxoglutarate. One example of this type of inhibitor is a group of GAs in which the exocyclic methylene (=CH_2) at C-16 is replaced by a methyl group. This type of substitution typically reduces activity of biologically active GA, and can also make them effective inhibitors of dioxygenase enzymes. For example, *exo*-16,17-dihydro-GA_5, which is normally applied as the 13-acetate (Figure14. 7), is a very effective growth retardant for grasses but not for dicot species.

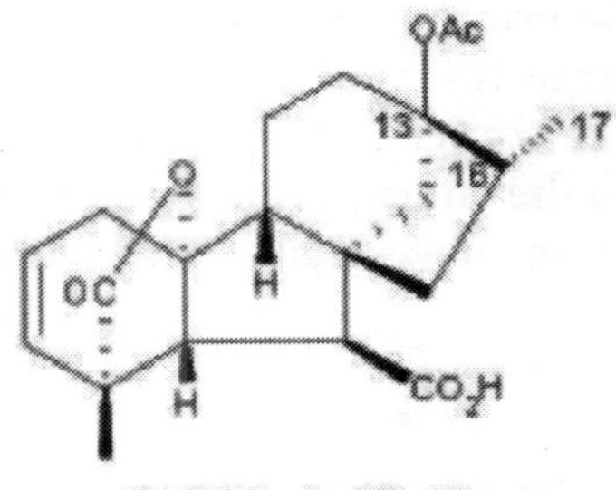

Figure 14..7. The structure of *exo*-16,17-dihydro-GA_5 13-acetate.

14.4 Gibberellin Biosynthesis

Terpenoids are compounds made up of 5-carbon **isoprenoid** building blocks, joined head to tail. The GAs are diterpenoids that are formed from *four* such isoprenoid units. They are assembled to give a 20-carbon linear molecule, geranylgeranyl diphosphate (GGPP). The stepwise conversion of GGPP to the first-formed GA, GA_{12}, is essentially similar in all plant species studied so far.

The subsequent formation of many different GAs from GA_{12} differs not only from species to species, but even in different organs within one species, or in response to different environmental conditions. The GA biosynthetic pathway can be divided into three stages, each residing in a different cellular compartment: the plastid, the endoplasmic reticulum, and the cytosol .

Stage 1: Biochemical reactions can be catagorized into three stages

The basic biological isoprenoid unit is isopentenyl diphosphate (IPP) (Figure 14.9). For many years it was assumed that the early stages of the terpenoid biosynthetic pathway were the same for all types of terpenoids. This is now known to be untrue. In plants, IPP can be synthesized in either the cytosol or in the stroma of plastids, using different precursors at each site. The cytosolic pathway utilizes acetate and is named the **mevalonate (MVA) pathway** after the important intermediate, mevalonic acid. This is the only pathway that operates for terpenoid biosynthesis in fungi and animals. The mevalonate pathway is thus the source of IPP for GA biosynthesis in the fungus *Gibberella zeae*.

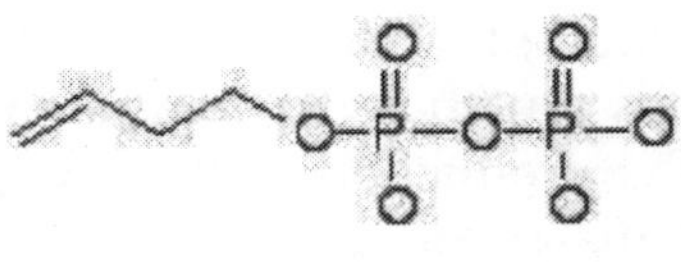

IPP

Figure 14 .8

The plastidic pathway in the green parts of plants and in algae uses glyceraldehyde 3-phosphate and pyruvate, and again the pathway is named for an important intermediate, **methyl erythritol phosphate (MEP)** . In Arabidopsis, it has been determined that IPP used in GA biosynthesis in green tissues comes predominantly from IPP synthesized in plastids by the MEP pathway , and this is thought to be the source of IPP for GA biosynthesis in most plants. (One exception is known— in the endosperm of pumpkin seeds, IPP that is used for GA biosynthesis is formed from mevalonate.) An overview of the two pathways is shown in Figure 14.9.

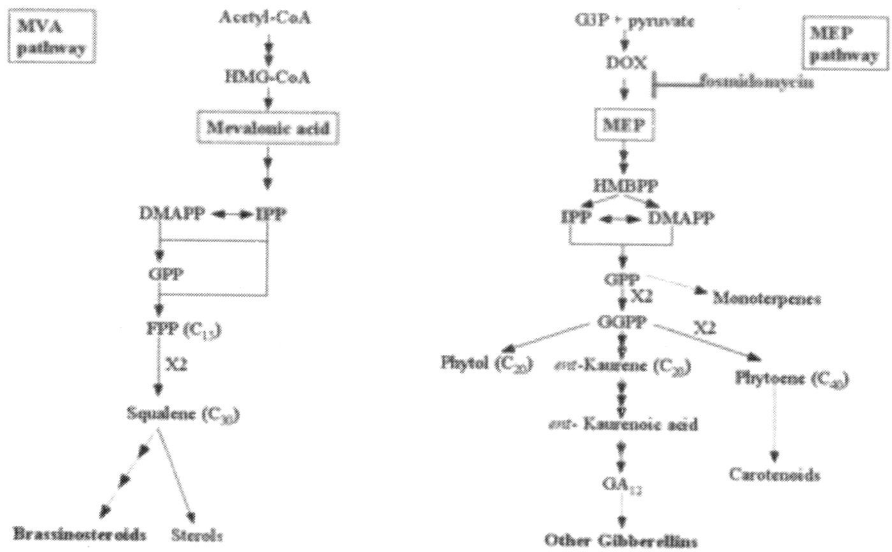

Figure 14.9 Two biosynthetic pathways to terpenoids in plants. The MVA pathway is in the cytosol, the MEP pathway in the stroma of plastids.

Once synthesized, the 5-carbon units condense to produce intermediates of 10 carbons (geranyl diphosphate, GPP), 15 carbons (farnesyl diphosphate, FPP), and 20 carbons (geranylgeranyl diphosphate, GGPP) as shown in Figure 2. GGPP is a precursor of many diterpenoid compounds, including the phytol side chain of chlorophyll, and tetraterpenoids (40 carbons), including carotenoids. Therefore, any mutation or chemical that blocks the MEP pathway prior to GGPP will prevent the production of these pigments giving albino plants, and is most likely to be lethal/toxic. In fact, fosmidomycin, which blocks early in the MEP pathway, can be used as a herbicide. It is only after GGPP that, in most plants, the pathway is dedicated for GA biosynthesis.

There are two cyclization reactions that convert linear GGPP to *ent*-kaurene (Figure 14.10). The formation of a bicyclic intermediate *ent*-copalyl-diphosphate (CPP) is catalyzed by *ent*-copalyl diphosphate synthase (CPS), and the conversion of CPP to the tetracyclic *ent*-kaurene is catalysed by *ent*-kaurene synthase (KS). (The precursor of GAs is the enantiomeric form of kaurene, hence the prefix "*ent*." However, in this Topic, the prefix is dropped after its first use with each compound.)

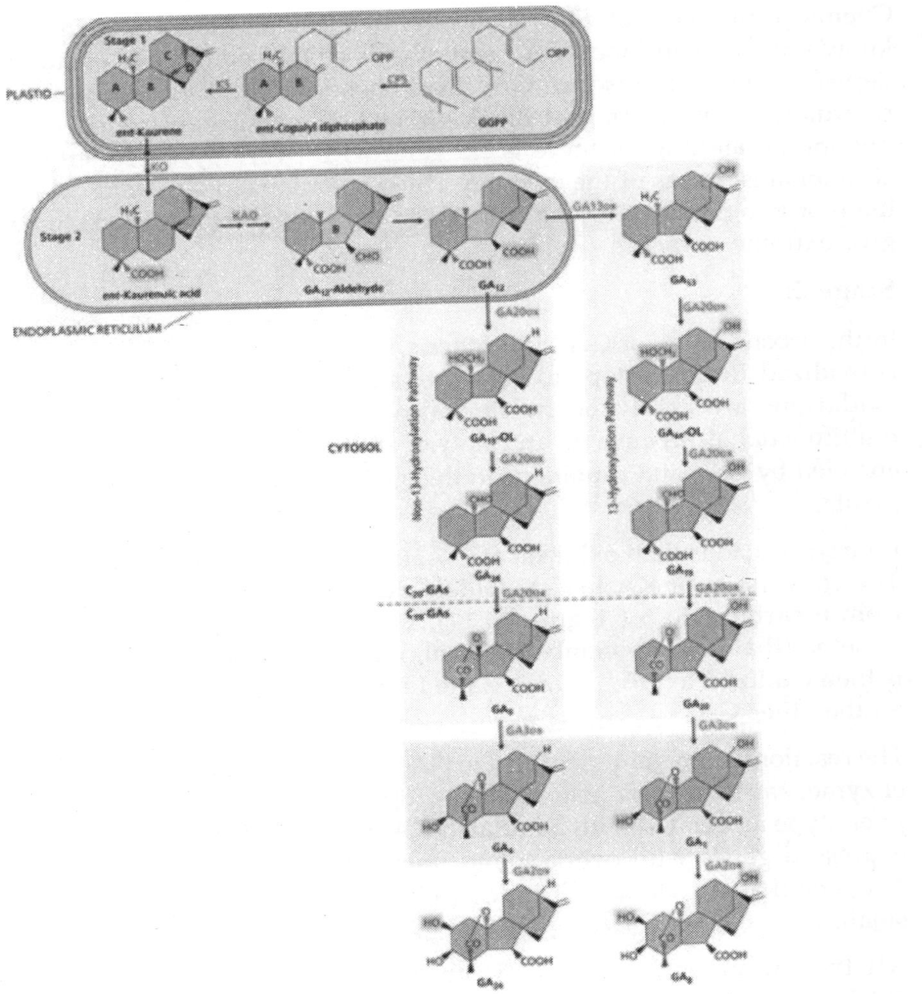

Figure 14.10 The three stages of GA biosynthesis. In stage 1, isopentenyl diphosphate is converted (not shown) to geranylgeranyl diphosphate (GGPP), which is then converted to *ent*-kaurene via *ent*-copalyl diphosphate in plastids. In stage 2, which takes place on the plastid envelope and endoplasmic reticulum, *ent*-kaurene is converted to GA_{12}. In many plants, GA_{12} is converted to GA_{53} by hydroxylation at C-13. In most plants, the 13-hydroxylation pathway predominates, although in Arabidopsis and some others, the non-13-OH-pathway is the main pathway. In stage 3, in the cytosol GA_{12} or GA_{53} is converted, via parallel pathways, to other GAs. This conversion proceeds with a series of oxidations at C-20, resulting in the eventual loss of C-20 and the formation of C_{19}-GAs. In the non-13-hydroxylation pathway, this leads to the production of GA_9. GA_9 is then oxidized to the bioactive GA_4 by a 3β-hydroxylation reaction. In the 13-hydroxylation pathway, GA_{53} is sequentially oxidized at C-20, leading to GA_{20}, which is then 3β-hydroxylated to yield bioactive GA_1. Finally, hydroxylation at C-2 converts GA_4 and GA_1 to the inactive forms, GA_{34} and GA_8, respectively. CPS = *ent*-copalyl diphosphate synthase; KS = *ent*-kaurene synthase; KO = *ent*-kaurene oxidase; KAO = *ent*-kaurenoic acid oxidase; GA13ox = GA 13-oxidase; GA20ox = GA 20-oxidase; GA3ox = GA 3-oxidase; GA2ox = GA 2-oxidase; OL = open lactone.

Chemical inhibitors of CPS, such as AMO-1618 and chlormequat chloride (also known as cycocel) block GA biosynthesis at this point . This leads to severely depleted levels of bioactive GA; these chemicals are therefore powerful growth retardants Mutations that block either CPS or KS cause a severely dwarfed phenotype as long as there is no gene with overlapping function to provide additional isoforms of the enzyme. Thus, mutations in *GA1* and *GA2* that encode the first two enzymes in the committed part of GA biosynthesis in Arabidopsis, give extreme dwarfs.

Stage 2:

In the second stage of GA biosynthesis, the C-19 methyl (CH_3) group of kaurene is oxidized in three steps to give *ent*-kaurenoic acid (KA) (see Figure 3). These oxidations are catalyzed by *ent*-kaurene oxidase (KO), which is said to be multifunctional because it can catalyze all three reactions. In Arabidopsis, KO is encoded by GA_3, and mutations in this gene will again produce severely dwarfed plants.

Kaurenoic acid is then oxidized at C-7 and C-6 in two steps to give GA_{12}-aldehyde. The conversion of KA to GA_{12}-aldehyde involves the contraction of the B-ring from 6 carbons to 5 (Figure 14.11), which distinguishes the *ent*-gibberellane skeleton (that has a 6-membered B-ring) from that of *ent*-kaurene. GA_{12}-aldehyde is then oxidized to GA_{12}, which is the first-formed GA, and thus the precursor, of all the other GAs.

The reactions from kaurenoic acid to GA_{12} are catalyzed by another multifunctional enzyme, *ent*-kaurenoic acid oxidase (KAO). In Arabidopsis, there is no known phenotype associated with a mutated KAO, probably because there are two stem-expressed genes with overlapping functions. , mutations in the stem-expressed *NA* gene that encodes KAO in pea yields a so-called *nana* phenotype that is very small.

All the enzymes involved in the second phase of GA biosynthesis are **monooxygenases** that utilize cytochrome P450 in their reactions, and are typically associated with membranes. As mentioned earlier, enzymes that catalyze the first part of the pathway are in the chloroplast stroma. Kaurene must therefore exit the chloroplast, because in Arabidopsis, KO is located on the plastid envelope while KAO is located in the endoplasmic reticulum . Actually, kaurene can be volatile, and not only does it exit the chloroplast, it can also diffuse from the plant and affect the growth of neighboring plants .

Paclobutrazol and other inhibitors of P450 monooxygenases specifically inhibit this second stage of GA biosynthesis . These compounds are growth retardants and are used for ornamental plants—especially potted ones—to produce shorter plants that are easier to manage and more attractive.

Stage 3:

All subsequent steps from GA_{12} are carried out by a group of soluble **dioxygenases** in the cytosol. These enzymes require 2-oxoglutarate and molecular oxygen as co-substrates, and they use Fe^{2+} and ascorbate as cofactors.

The specific steps in the modification of GA_{12} vary from species to species, and between organs of the same species. Typically, the reactions in vegetative tissue (young seedlings and stems) are more conserved between species and follow a similar basic plan of C-20 oxidation along with hydroxylation at C-3, and eventually, at C-2. Hydroxylation at C-13 may or may not occur.

In contrast, in maturing seeds there are many additional modifications of GAs—for example, hydroxylations at carbons 1, 11, 12, or 15 that consequently lead to the production of the large number of polyhydroxylated GAs. Many of the 136 known GAs have been identified only in seeds, and closely related species often have seed GAs with similar patterns of hydroxylation. In vegetative tissues, we will discuss only the *major* GA pathway in each of three examples: pea, maize and Arabidopsis. Often minor pathways are also present, giving a metabolic network rather than a single pathway, but only the *major* pathway in specific plants is described here. The basic pathway from GA_{12} in vegetative tissues is as follows:

1. **Hydroxylation at C-13.** This is a major metabolic event in pea and maize pathways (but *not* in the main pathway in Arabidopsis). 13-Hydroxylation converts GA_{12} to GA_{53}.
2. **Successive oxidation at C-20.** This occurs in four steps, and in the final step C-20 is lost, resulting in the first-formed, C_{19}-GA. , this loss of C-20 results in the formation of a lactone between C-4 and C-10. In Arabidopsis, this sequence is GA_{12} (CH_3 at C-20) to GA_{15} (CH_2OH at C-20 in the open-lactone form) to GA_{24} (CHO at C-20) to GA_9 (C-20 has been lost) (see Figure 3). In pea and maize, the sequence is identical, giving the 13-hydroxylated counterparts of the Arabidopsis GAs, namely GA_{53} (CH_3 at C-20) to GA_{44} (CH_2OH at C-20 in the open-lactone form) to GA_{19} (CHO at C-20) to GA_{20} (C-20 has been lost). The four sequential oxidations are all catalyzed by the same multifunctional GA 20-oxidase (GA20ox).
3. **3β-Hydroxylation.** In Arabidopsis, this converts GA_9 to the bioactive GA, GA_4. In pea and maize, the 13-OH counterpart of GA_9, namely GA_{20}, is converted to bioactive GA_1. Both GA_4 and GA_1 are C_{19}-GAs with a 4,10-lactone, a carboxylic acid group at C-6, and a 3β-OH group.
4. **2β-Hydroxylation.** The bioactive GAs can be deactivated by hydroxylation at C-2, converting GA_4 to GA_{34} in Arabidopsis, and GA_1 to GA_8 in pea/maize.

As we can see from Figure 14.11, in Arabidopsis there are 12 reactions in the linear pathway from the earliest committed precursor (GGPP) to the bioactive GA, GA_4. In pea and maize, there are 13 steps to the bioactive GA, GA_1, because of the additional hydroxylation step at C-13.

Inhibitors of the third stage of the GA biosynthetic pathway interfere with enzymes that utilize 2-oxoglutarate as a co-substrate. Among these, **prohexadione calcium** (**BX-112**) is especially useful because it specifically inhibits GA 3-oxidase, and it is a growth retardant. It is used to control the canopy size of apple and orange trees and reduce the amount of pruning required.

Gibberellins can be converted into **conjugates**, most commonly with glucose. The linkage of the GA moiety to glucose occurs either via a hydroxyl group to produce a GA-*O*-glucosyl ether, or via the COOH at C-6, yielding a GA-glucosyl ether. Gibberellin conjugates are frequently found in *Phaseolus vulgaris* (bean) seeds during the later stages of seed maturation.

When applied to bioassay plants, the GA-*O*-glucosyl ethers show little-to-no activity, indicating that they are deactivation products. In contrast, the GA-glucosyl esters may be hydrolyzed within a bioassay plant, yielding the free GA. Hence, the formation of glucosyl ester conjugates may be a mechanism for sequestration of bioactive GA that can be released later, as required by the plant.

14.5 Commercial Uses of Gibberellins

Gibberellic acid (GA_3) is the GA most often used commercially, since it can be readily obtained in large quantities from fermentations of the fungus ***Gibberella fujikuroi***. The global (excluding China) use of GA_3 per annum is approximately 50 tons. Other GAs, for example GA_4 and/or GA_7, are used for specific crops or specific purposes for which they are more effective than GA_3, though $GA_{4/7}$ are produced in lower yields by commercial fermentations and are therefore more expensive than GA_3.

The major commercial uses of GAs are to promote the growth of a variety of fruit

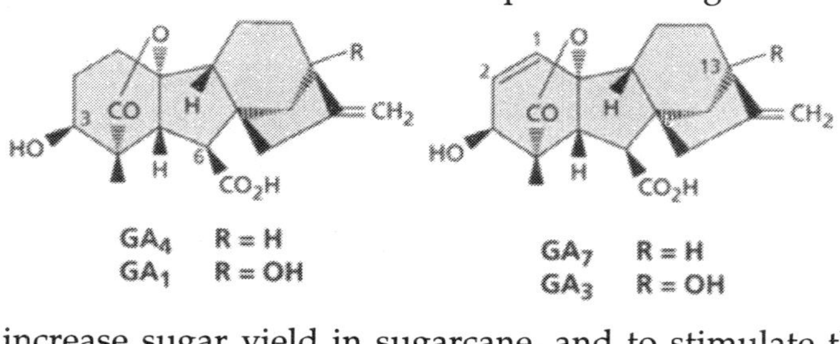

crops, to increase sugar yield in sugarcane, and to stimulate the barley-malting process in the beer-brewing industry. Details of additional uses of GAs in horticulture, albeit on a smaller scale than those mentioned below, can be found in Gianfagna .

Many of the table grapes grown in the United States are a genetically seedless variety that would naturally produce small fruit on very compact clusters. Almost all seedless grapes on the market are treated with GA_3. It substitutes for the presence of seeds, which would normally be the source of native GAs for fruit growth. Repeated spraying with GA_3 increases both rachis length (producing looser clusters) and fruit size . The increased rachis length prevents the cluster from being too compact, and this reduces the chance of fungal growth inside the cluster. Two to three additional applications of GA_3 during fruit development are thought to increase berry size by enhancing the import of carbohydrates into the developing fruit. In excess of 8 tons of GA_3 are used in the California grape industry annually.

Gibberellic acid is also used to boost cherry production. Sweet, bing cherries are sprayed 4 to 6 weeks before harvest to increase fruit size. Application of GA_3 to tart cherries increases yield through enhanced bearing.

Gibberellin A_4 (GA_4) is used to promote the fruit set of apple and pear trees. For example, in some apple cultivars the amount of fruit produced is often limited by biennial bearing, a phenomenon whereby the production of a heavy crop of fruit one year inhibits the subsequent production of flower buds, and hence, the yield of fruit the following year. The alternate bearing of some cultivars can be overcome by applying GA_4 in the "off" year to promote the formation of flower buds, and subsequent fruit set. In regions of Europe where fruit set of apple and pear trees is often reduced by inclement weather at the time of pollination, the application of a hormone mixture can promote the production and subsequent growth of parthenocarpic (seedless) fruit. $GA_{4/7}$ is also used on Golden Delicious apples to prevent abnormal cell divisions in the epidermal layer that produce "russetting." The russetted appearance of the apple fruit is considered by many consumers to be undesirable.

Gibberellic acid is also applied to citrus crops, though the actual use depends on the particular crop. For example GA_3 is sprayed onto oranges and tangerines to delay or prevent rind-aging, so that fruit can be harvested later without adverse effects on rind quality and appearance. For lemons and limes, GA_3 synchronizes ripening and enhances fruit size.

Gibberellic acid is used extensively to increase the sucrose yield of sugarcane. Sugarcane, a normally fast-growing C_4 member of the Poaceae (grass) family, is sensitive to cooler winter temperatures, which reduce internode elongation and subsequent sucrose yield. The adverse effects of cooler temperatures can be counteracted by the application of GA_3.

The biochemistry of GA-induced α-amylase synthesis in the aleurone of cereal grains has been studied extensively and is one of the best-understood GA response pathways. Gibberellins from the embryo of germinating grains are necessary for the synthesis of α-amylase by the cells of the aleurone layer, which, in turn is necessary for the hydrolysis of starch within the endosperm (Figure 14.11). In the brewing industry, the production of beer relies on this hydrolytic breakdown of starch in barley grains to yield fermentable sugars, principally maltose, which are then subjected to fermentation by yeast. During fermentation, glycolytic enzymes from yeast break down the sugars, resulting in ethanol.

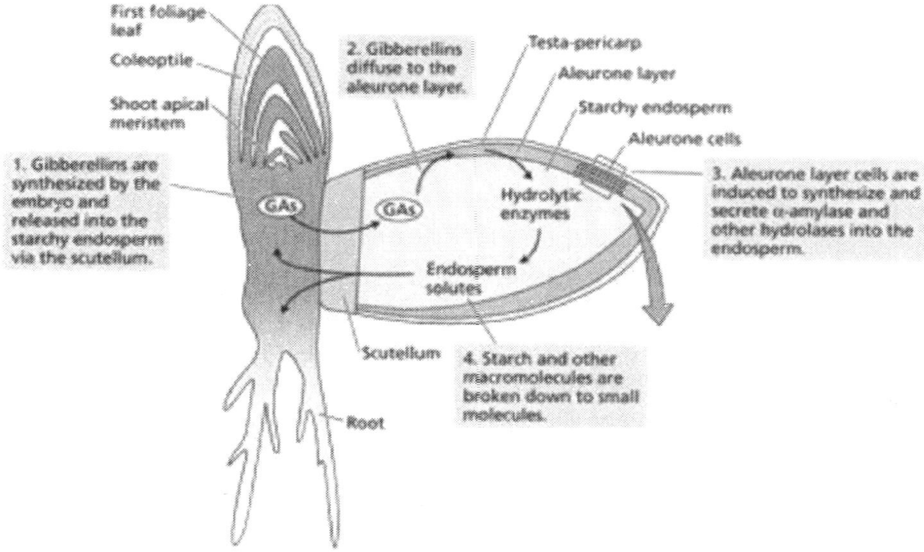

Figure 14.11 Structure of a germinating barley grain and the biochemical processes that occur during the "modification" part of the malting process.

In the multistep malting process, mature barley grains are steeped or soaked to allow them to imbibe water. Next, the grains are spread out to germinate, during which time the starch within the endosperm will be hydrolyzed by α-amylase allowing the embryo to begin to grow. This process of starch breakdown is referred to as "modification." Gibberellic acid may be applied during this time and will supplement the native GAs in the grain, enhance the production of α-amylase, and consequently, speed up the hydrolysis of starch. . The germinated grains, which show a well-developed root and have a shoot (termed an "acrospire"), which is approximately 75% of the length of the grain, are then kilned. This is a two-step process, first to dry the grain, and second, to cure or heat it. The temperature to which the modified malt is cured will determine whether light-colored (low-temperature curing) or dark beer (high-temperature curing) is to be produced. The modified and kilned malt provides the raw material for the fermentation.

14.6 Gas-Chromatography Mass-Spectrometry of Gibberellins

Fundamental to our understanding of how GAs control growth and development is a need to be able to identify the GAs present in our experimental plants. Systems of measurement that use a biological response, called bioassays, were originally important for detecting GA-like activity in partly purified extracts. Fractions of extracts were applied to specific plants—for example, dwarf rice or maize seedlings—and the biological response, such as leaf sheath elongation, was measured after several days. The amount of bioactivity in extracts was estimated by comparing samples with a standard curve generated using GA_1 or GA_3 (Figure 14.12).

Figure 14 .12 Gibberellin causes elongation of the leaf sheath of rice seedlings, and this response is used in the dwarf rice leaf sheath bioassay. Here, 4-day-old seedlings were treated with increasing amounts of GA_3 and allowed to grow for another 5 days.

Bioassays are sensitive and specific for a particular type of hormone, but they cannot provide conclusive identification of the GAs within the extract, just an indication that the extract contains GA-like substances. Moreover, extracts frequently contain many GAs, and these may not necessarily have been separated one from another during the fractionation process. Also, extracts almost always contain additional substances that could interfere with the bioassay. For these reasons, use of bioassays for impure plant extracts has declined with the development of highly sensitive physical techniques, such as mass spectrometry, that allow precise identification of specific GAs from small amounts of tissue.

To identify GAs, plants or plant parts are normally homogenized in aqueous methanol to yield an extract. Extracts need to be subjected to several different purification procedures (e.g., anion exchange chromatography) to separate some of the contaminating materials, such as pigments, from the GAs. Extracts may then be fractionated again, often using high-pressure liquid chromatography. Individual fractions are normally treated with reagents to convert the GAs to derivatives that will be appropriate for gas chromatographic separation. This usually entails methylating carboxylic acid functions and converting the hydroxyl groups to trimethylsilyl (TMS) derivatives. These procedures not only facilitate gas chromatography, they also impart structural features to the GAs that can be useful in analyzing their mass spectra.

Mass spectra can readily be obtained from pure compounds, but this option is not available unless enough plant material has been extracted to isolate individual GAs. Instead, a purified extract (that will contain a mixture of compounds) is fractionated into constituent components by gas chromatography (GC) before transferring those components directly into the source of a mass spectrometer (MS). Fragmentation of each GA will produce a mass spectrum that records the abundance of ions of a particular mass to charge (m/e) ratio. Spectra of all characterized GAs, as well as their Me and/or MeTMS derivatives, can be obtained at the Plant Hormones website *www.plant-hormones.info/ga1info.htm*. GC retention time, relative to alkane standards, gives a retention index (RI) which is also important for confirming the identity of GA isomers that may have similar mass spectra.

Methods for analyzing extracts by liquid chromatography coupled to MS (LC-MS) have also been developed. This technique has the advantage that GAs do not need to be derivatized; but it has the disadvantage that a spectral library for nonderivatized GAs is not yet publicly available.

There are 136 known GAs, numbered GA_1 to GA_{136} in chronological order of their discovery. The website *www.plant-hormones.info/occurrence_of_gas_in_plants.htm* contains very comprehensive information on the plant species in which each one occurs, together with other useful information.

14.7 Effect of Environment on Gibberellin Biosynthesis

Gibberellins play an important role in mediating the effects of environmental stimuli on plant development. Environmental factors such as light and temperature can alter the levels of active GAs by affecting gene transcription for specific steps in the biosynthetic pathway . This topic focuses on the effects of light on GA biosynthesis and metabolism.

Light quality, quantity and duration (day length) can all influence GA biosynthesis and catabolism. Effects can vary depending on the species and on the developmental event being examined. Generally, enhanced or decreased production of bioactive GA is correlated with a specific growth response and/or a switch in development that is/are triggered by changes in the plant's light environment. We focus here on the effect of light on seed germination, the greening of dark-grown seedlings (a process known as de-etiolation), the regulation of stem length, and tuberization.

In dicots such as lettuce and Arabidopsis, **seed germination** is stimulated by exposure to light, which acts to promote production of bioactive GAs. Germination in these species has an absolute requirement for GAs, which have to be newly synthesized in the germinating seed. Gibberellins induce the production of enzymes that are needed to digest the endosperm, which would otherwise form a mechanical barrier to root emergence. The wavelength of light is critically important, implying that specific photoreceptor(s) are involved. **Red light** acts to promote the 3□-hydroxylation step in the GA pathway, whereby inactive C_{19}-GAs are converted to GAs that have intrinsic biological activity. **Phytochrome B** is one of the photoreceptors involved in this response. Enhanced transcription of two genes encoding 3□-hydroxylase enzymes is observed within 4 hours of red light treatment to imbibing seeds.

Plants growing in very-low-light conditions or in complete darkness are characterized by an **etiolated** growth habit with long, thin stems and small, yellow cotyledons and leaves . On exposure to normal light conditions, the growth rate of the stem is reduced, the stems thicken, and leaves expand and green. This process is known as **de-etiolation** After years of controversy as to whether or not de-etiolation involves changes in GA content, it has now been clearly established that exposure of etiolated pea seedlings to light results in a rapid, but temporary, reduction in the content of GA_1, which is the major bioactive GA in pea stems. The light, which is detected by **phytochrome A** and possibly a blue light receptor, causes a down-regulation of 3□-hydroxylation and an up-regulation of 2□-hydroxylation within 30 minutes of exposure. The consequence of this altered metabolism—a transient reduction in the steady-state level of GA_1—is thought to be important in establishing the types of growth changes that accompany **photomorphogenesis** . More prolonged changes in GA status affect the *sensitivity* of light-grown tissues to GA_1, rather than the actual level of GA_1.

Plants use changes in day-length **(photoperiod)** as cues for the initiation of numerous other development changes. Gibberellins have been shown to act as secondary messengers in processes induced by **long photoperiods,** such as flowering in long-day rosette plants, the breaking of bud dormancy in woody plants, and the induction of stolons, rather than tubers, in potato.

Long-day rosette plants grow vegetatively in short photoperiods and are induced to flower when the photoperiod extends beyond a critical value. Flowering in such species is accompanied by rapid stem extension, a process known as **bolting.** In spinach (*Spinacia oleracea*), which has an absolute requirement for long days, petiole extension and bolting requires the presence of bioactive GA(s), synthesis of which is induced when plants are transferred from short days to long days. Enhanced GA biosynthesis in long days is associated with higher levels of *ent*-kaurene production and of GA 20-oxidase activity leading to a fivefold increase in GA_1 that induces stem growth.

In Arabidopsis, a facultative long-day plant, GA 20-oxidase gene expression is also enhanced in long days, but flower induction is not absolutely dependent on GAs under these conditions, although it is delayed when GAs are absent. However, GAs are necessary for stem elongation and normal development of flowers, and for flower induction in short days. In this species, GAs participate in one of several pathways to flower induction and appear to operate independently of the main photoperiod-dependent pathway.

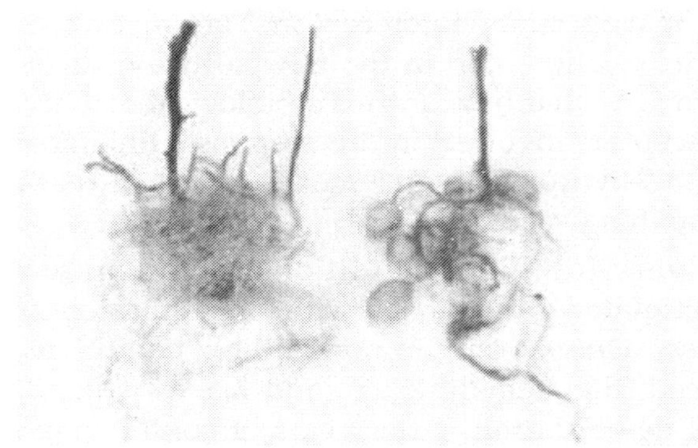

Figure 14.13 Tuberization of potatoes is promoted by short days. Potato (*Solanum tuberosum* spp. *Andigena*) plants were grown under either long days or short days. The formation of tubers in short days is associated with a decline in GA_1 levels.

Tuberization is another process regulated by photoperiod (Figure 14.13). Tubers form on wild potatoes only in **short days** (although the requirement for short days has been bred-out of many cultivated varieties), and this tuberization can be blocked by applications of GA. The transcription of GA 20-oxidase was found to fluctuate during the light–dark cycle, leading to lower levels of GA_1 in short days. Potato plants overexpressing the *GA20ox* gene showed delayed tuberization, whereas transformation with the antisense gene for *GA20ox* promoted tuberization. This result demonstrates the importance of the transcription of this gene in regulating the level of GA_1, which must be *below* a particular threshold level for tuberization to occur

14.8 Second Messengers in GA Signaling

Components of the GA signal transduction chain in aleurone cells have been identified in the plasma membrane, cytosol and nucleus (Figure 14.14). Several lines of evidence indicate that bioactive GA, perceived at the plasma membrane and bound to a hypothetical receptor protein, interacts with a **membrane-localized heterotrimeric G protein** . Treatment of oat aleurone protoplasts with a peptide called Mas7, which stimulates GTP/GDP exchange by G-proteins, was found to induce *α-amylase* gene expression and to stimulate α-amylase secretion, suggesting that such a GTP/GDP exchange on the cell membrane is necessary for the induction of α-amylase biosynthesis by GA. In addition, GA-induced *α-amylase* gene expression and enzyme secretion were inhibited by a guanine nucleotide analog that binds to the α subunit of heterotrimeric G-proteins and inhibits GTP/GDP exchange, further supporting the preceding conclusion.

Genetic studies have provided further support for the role of G-proteins as intermediates in the GA signal transduction pathway. The rice dwarf mutant

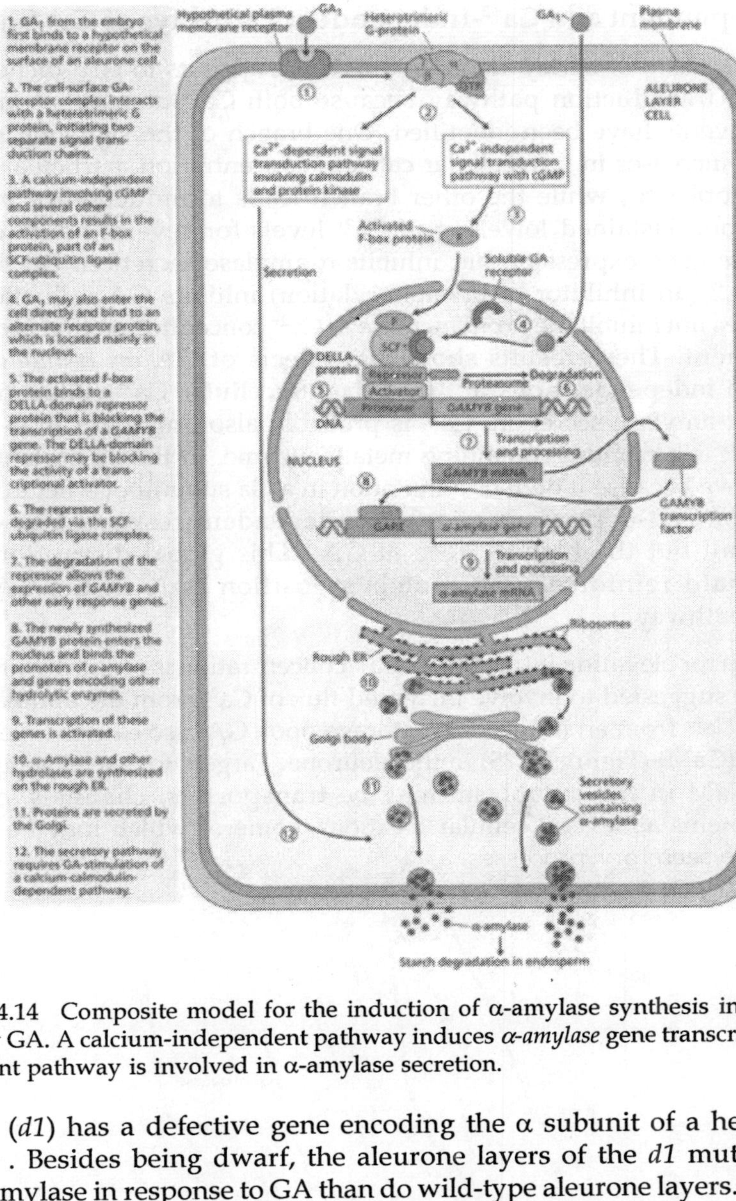

Figure 14.14 Composite model for the induction of α-amylase synthesis in barley aleurone layers by GA. A calcium-independent pathway induces *α-amylase* gene transcription; a calcium-dependent pathway is involved in α-amylase secretion.

dwarf 1 (*d1*) has a defective gene encoding the α subunit of a heterotrimeric G protein . Besides being dwarf, the aleurone layers of the *d1* mutant synthesize less α-amylase in response to GA than do wild-type aleurone layers. This reduction in α-amylase production by the *d1* mutant demonstrates that G-proteins are one of the components of the GA signal transduction pathway involved in both the growth response and the production of α-amylase. However, the difference in α-amylase production between the mutant and the wild type goes away with increasing GA concentration, suggesting that GA can also stimulate α-amylase production by a G-protein-independent pathway .

14.9 Ca^{+2}-Dependent and Ca^{+2}-Independent Pathways of GA Signaling

Within the cytoplasm of barley aleurone cells there appears to be a dichotomy in the GA signal transduction pathway because both Ca^{2+}-dependent and Ca^{2+}-independent events have been identified. One branch of the signaling pathway leads to rapid increases in intracellular calcium concentration, particularly in the peripheral cytoplasm , while the other branch leads to induction of *α-amylase* gene expression. Sustained lowering of Ca^{2+} levels for several hours does *not* affect *a-amylase* gene expression but inhibits α-amylase secretion. On the other hand, syntide 2 (an inhibitor of phosphorylation) inhibits GA induction of *α-amylase* but does not inhibit the rapid increase in Ca^{2+} concentration that is observed on GA treatment. These results show that effects of GA on *α-amylase* gene expression are independent of Ca^{2+}, though intracellular Ca^{2+} concentration is important in α-amylase secretion. Ca^{2+} is probably also important in α-amylase formation, as it is a calcium containing metalloenzyme. In the *d1* mutant of rice, mentioned above because it defines a mutation in a Ga subunit of a heterotrimeric G protein, *both* Ca^{2+}-dependent and Ca^{2+}-independent responses to GA_3 are perturbed at all but the highest dose of GA_3. This global disruption of GA responses would reinforce a Ga protein's position *early* in the GA signal transduction pathway.

The mechanism for elevating intracellular Ca^{2+} concentration is presently unknown, but it has been suggested to invòlve increased flux of Ca^{2+} from the apoplast and/or a release of Ca^{2+} from an intracellular storage pool. GA also causes an elevation of calmodulin (CaM) (Figure 14.15) in the aleurone. Targets for the elevated levels of Ca^{2+} and CaM in the cytoplasm may be transporters, channels, or other membrane proteins at several cellular locations, some of which may be directly involved in the secretory processes.

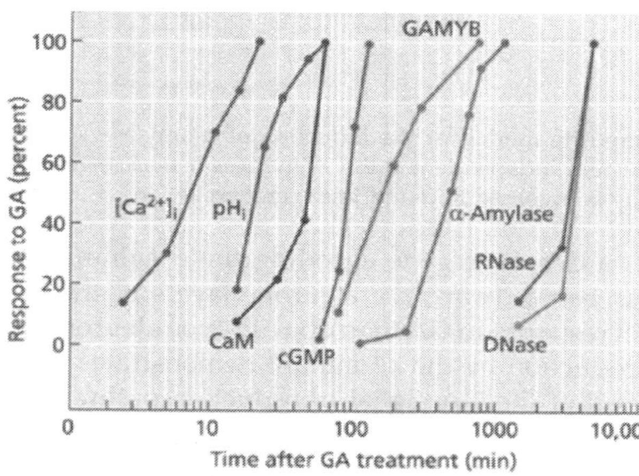

Figure 14.15 Following the addition of GA to barley aleurone protoplasts, a multiple-component signaling pathway is initiated. The timing of some of these events is shown. CaM = calmodulin.

Use of guanylyl cyclase inhibitors prevents a transient increase in **cyclic GMP (cGMP)** normally seen in barley aleurone layers two hours after GA treatment (see Figure 14.15). These inhibitors also reduce the accumulation of *α-amylase* mRNA, and that of the transcription factor, *GAMYB*. For this and other reasons, it has been suggested that cGMP may be a component of the transduction chain, provisionally placed downstream of the G proteins and upstream of MYB transcription factors.

Reversible **phosphorylation** is also involved in the GA signal transduction chain. The aforementioned syntide 2, which essentially blocks phosphorylation by competing with natural substrates, inhibits GA induction of gene expression. In contrast, okadaic acid, which is a phosphatase inhibitor, negates the effects of GA on Ca^{2+} accumulation, on gene expression, and on α-amylase secretion. This inhibitor therefore must act early (i.e., before the transduction chain branches) or, if it acts later, it must act at two sites to inhibit both branches of the signal transduction chain.

14.10 Effects of GAs on Flowering Genes

Flowering in Arabidopsis, a plant for which more tools exist to study the molecular genetics of the process than in any other species, is regulated by four separate pathways: (a) the **photoperiodic** (long day) pathway, which operates in the leaves; (b) the convergent **autonomous** (leaf number)/**vernalization** (low temperature) pathway; (c) the **carbohydrate** (sucrose) pathway; and (d) the **gibberellin** pathway. The latter three pathways all operate in the shoot apical meristem. The four pathways converge on a number of floral pathway integrators that together regulate floral initiation.

The recent identification of ***FLOWERING LOCUS T*** (*FT*) as the gene expressed in leaves that encodes a phloem-mobile mRNA or protein that fits the description of a universal flowering stimulus, **florigen**, has been an exciting development after the decades-long search for this elusive flowering signal . Additional work is seeking to further define the integrative networks that exist between the different pathways. Figure 14.16 , provides a complete consideration of this subject. This topic briefly addresses the GA pathway, when it operates, and what is known about how it is integrated with the other pathways.

In Arabidopsis, flowering occurs in long days (LD) even in the GA-deficient *ga1-3* mutant, although it is somewhat delayed compared to wild-type plants. However, this mutant will *not* flower in short days (SD) unless treated with GA. Double mutants of *ga1-3* and *co*, (the wild-type *CONSTANS* is normally expressed in leaves and is upstream of FT in the LD pathway) will not normally flower, either in SD or LD. From this and other information it is proposed that the LD and GA pathways are independent, with the GA pathway being essential for flowering in noninductive conditions. Input from the LD pathway (transduced

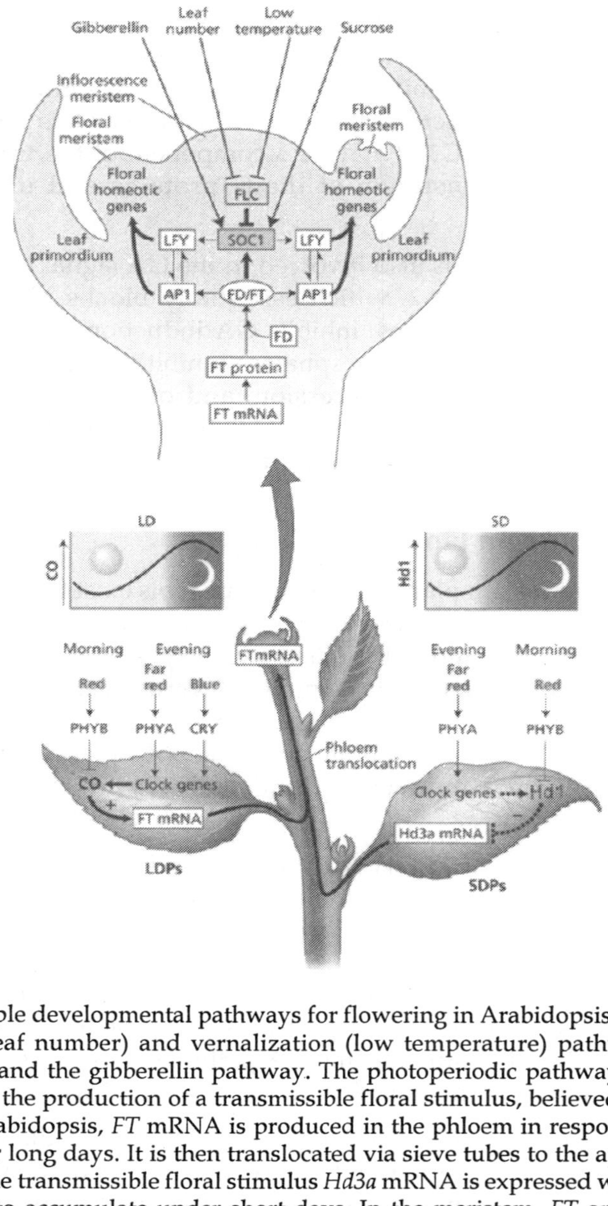

Figure 14.16 Multiple developmental pathways for flowering in Arabidopsis: photoperiodism, the autonomous (leaf number) and vernalization (low temperature) pathways, the energy (sucrose) pathway, and the gibberellin pathway. The photoperiodic pathway is located in the leaves and involves the production of a transmissible floral stimulus, believed to be *FT* mRNA. In LDPs such as Arabidopsis, *FT* mRNA is produced in the phloem in response to CO protein accumulation under long days. It is then translocated via sieve tubes to the apical meristem. In SDPs such as rice, the transmissible floral stimulus *Hd3a* mRNA is expressed when the repressor protein, Hd1, fails to accumulate under short days. In the meristem, *FT* or *Hd3a* mRNA are translated to proteins. In Arabidopsis, FT interacts with another protein, FD. The FT/FD complex then activates the *AP1* and *SOC1* genes, which trigger *LFY* gene expression. *LFY* and *AP1* then trigger the expression of the floral homeotic genes. The autonomous (leaf number) and vernalization (low temperature) pathways act in the apical meristem to negatively regulate *FLC*, a negative regulator of *SOC1*. The sucrose and gibberellin pathways, also localized to the meristem, promote *SOC1* expression.

through FT) and from the GA pathway converge on SOC 1 (SUPPRESSOR OF OVEREXPRESSION OF CO 1), that is expressed in the apical meristem. Thus, GA treatment to *ga1-3* mutants in SD causes both enhanced expression of SOC in the apical meristem, and flowering. Overexpression of *SOC1* in *ga1-3* mutants allows them to flower in SD without GA treatment.

An additional floral pathway integrator is LFY, the product of a floral meristem identity gene that is expressed in apical meristems. *LFY* expression, which is high in LD-grown plants, is only observed in SD-grown *ga1-3* mutants that have been treated with GA. Constitutive *LFY* expression will also allow SD-grown *ga1-3* to flower without GA treatment . Using deletion analysis, it was identified an 8 bp motif in the *LFY* promoter that is necessary for GA responsiveness, but not for response to daylength. This sequence is a potential target for MYB transcription factors, and candidate MYBs whose expression increases in the shoot apex in response to exogenous GA_4, or to elevated levels of native GA_4, have been identified in both *Lolium temulentum* and Arabidopsis . Interestingly, LDs—but not GA treatment—cause a rapid increase in expression of the *L. temulentum* ortholog of *FT* (*LtFT*), indicating in this plant too that the LD and GA pathways converge downstream of FT .

14.11 Auxin-GA Interaction

The first indications that auxin may be required for GA signaling in rootswas obsearved with the studies of *ga1-3* (GA-deficient) Arabidopsis seedlings that had either been decapitated or treated with an auxin transport inhibitor are described . Both treatments reduced the response of roots to applied GA_3, and delayed or prevented the degradation of the RGA-repressor, as measured by RGA-GFP abundance. Additionally, silencing the expression of *PIN1*, which is required to ensure polar transport of endogenous IAA from the stem apex to the roots, or the presence of an *axr1-12* mutation, which confers resistance to IAA, also prevented GA-induced RGA degradation . Thus auxin appears to be necessary for the GA-induced degradation of a DELLA repressor, and hence, auxin can be regarded as a positive regulator of GA signaling in this system.

Chapter 15

Cytokinins

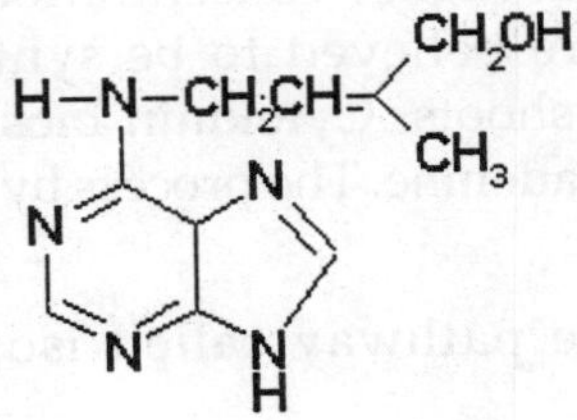

15.1 Nature of Cytokinins

Cytokinins are compounds with a structure resembling adenine which promote cell division and have other similar functions to kinetin. Kinetin was the first cytokinin discovered and so named because of the compound's ability to promote cytokinesis (cell division). Though it is a natural compound, It is not made in plants, and is therefore usually considered a "synthetic" cytokinin (meaning that the hormone is synthesized somewhere other than in a plant). The most common form of naturally occurring cytokinin in plants today is called zeatin which was isolated from corn (Zea mays).

Cytokinins have been found in almost all higher plants as well as mosses, fungi, bacteria, and also in tRNA of many prokaryotes and eukaryotes. Today there are more than 200 natural and synthetic cytokinins combined. Cytokinin concentrations are highest in meristematic regions and areas of continuous growth potential such as roots, young leaves, developing fruits, and seeds.

15.2 History of Cytokinins

In 1913, Gottlieb Haberlandt discovered that a compound found in phloem had the ability to stimulate cell division. In 1941, Johannes van Overbeek discovered that the milky endosperm from coconut also had this ability. He also showed that various other plant species had compounds which stimulated cell division. In 1954, Jablonski and Skoog extended the work of Haberlandt showing that vascular tissues contained compounds which promote cell division. The first cytokinin was isolated from herring sperm in 1955 by Miller and his associates. This compound was named kinetin because of its ability to promote cytokinesis. Hall and deRopp reported that kinetin could be formed from DNA degradation products in 1955. The first naturally occurring cytokinin was isolated from corn in 1961 by Miller. It was later called zeatin. Almost simultaneous with Miller

Letham published a report on zeatin as a factor inducing cell division and later described its chemical properties. It is Miller and Letham that are credited with the simultaneous discovery of zeatin. Since that time, many more naturally occurring cytokinins have been isolated and the compound is ubiquitous to all plant species in one form or another.

15.3 Biosynthesis and Metabolism of Cytokinins

Cytokinin is generally found in higher concentrations in meristematic regions and growing tissues. They are believed to be synthesized in the roots and translocated via the xylem to shoots. Cytokinin biosynthesis happens through the biochemical modification of adenine. The process by which they are synthesized is as follows:

A product of the mevalonate pathway called isopentyl pyrophosphate is isomerized.

This isomer can then react with adenosine monophosphate with the aid of an enzyme called isopentenyl AMP synthase.

The result is isopentenyl adenosine-5'-phosphate (isopentenyl AMP). This product can then be converted to isopentenyl adenosine by removal of the phosphate by a phosphatase and further converted to isopentenyl adenine by removal of the ribose group.

Isopentenyl adenine can be converted to the three major forms of naturally occurring cytokinins.

Other pathways or slight alterations of this one probably lead to the other forms. Degradation of cytokinins occurs largely due to the enzyme cytokinin oxidase. This enzyme removes the side chain and releases adenine. Derivitives can also be made but the pathways are more complex and poorly understood.

15.4 Cytokinin Functions

A list of some of the known physiological effects caused by cytokinins are listed below. The response will vary depending on the type of cytokinin and plant species.

- Stimulates cell division.
- Stimulates morphogenesis (shoot initiation/bud formation) in tissue culture.
- Stimulates the growth of lateral buds-release of apical dominance.
- Stimulates leaf expansion resulting from cell enlargement.
- May enhance stomatal opening in some species.
- Promotes the conversion of etioplasts into chloroplasts via stimulation of chlorophyll synthesis.

15.5 Idientification of Cytokinins

Naturally occurring molecules with cytokinin activity may be detected and identified by a combination of physical methods and bioassays. The introduction of immunoaffinity purification methods combined with liquid chromatography

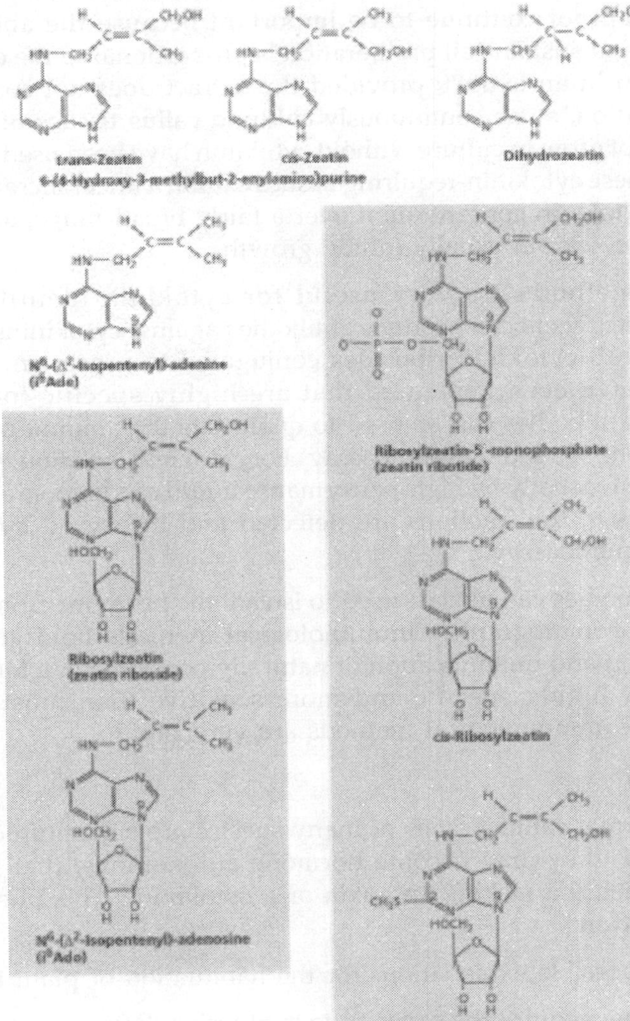

Figure 15.1 Structures of some naturally occurring cytokinins. Most plants have trans-zeatin as the principal free cytokinin, but dihydrozeatin and isopentenyl adenine (i6Ade) are also native plant cytokinins. Free cytokinins also include the ribosides and ribotides of zeatin, dihydrozeatin, and isopentenyladenosine, although these may be active as cytokinins by conversion to the respective bases. Free cytokinins from bacteria include 2-methylthio-cis-ribosylzeatin, as well as cis- or trans-zeatin, and their ribosides and ribotides.

and mass spectrometry, along with the use of isotopically labeled cytokinins as internal standards, has resulted in major advances in the measurement of endogenous cytokinins .

Historically, bioassay was the only method available for the identification of molecules that act as cytokinins. Cell proliferation bioassays using tobacco pith tissue or carrot taproot continue to be important because the ability of these tissues to initiate and sustain cell proliferation is proportional to the concentration of active cytokinin in an extract, provided the extract does not contain growth inhibitors. In addition, some continuously cultured callus tissues of tobacco and soybean that cannot grow in culture without cytokinin have been used for cytokinin bioassays. All of these cytokinin-requiring tissues exhibit a linear increase in growth with increasing cytokinin concentration over a fairly broad range, although high concentrations of cytokinin usually inhibit growth .

Immunological methods are very useful for cytokinin identification and quantification. Researchers can produce antibodies against cytokinins by injecting rabbits or mice with cytokinin ribosides conjugated to a protein. Monoclonal antibodies also have been generated that are highly specific for individual cytokinins. These antibodies can be used to quantitate the amount of a cytokinin in a sample by means of a radioimmunoassay . For cytokinin isolation, plant extracts are first fractionated, usually by high-performance liquid chromatography (HPLC), and the cytokinins in the fractions are detected and measured by means of a cytokinin radioimmunoassay.

The cytokinin antibodies can also be used to isolate the hormone from extracts by immunoaffinity chromatography Immunological methods hold great promise for the identification and quantification of naturally occurring cytokinins because the antibodies are highly specific and more sensitive than most bioassays Furthermore, these immunological methods are very rapid.

15.6 Habituation

When cultured normal callus tissues of many species are subcultured repeatedly over a long period, they may become hormone autonomous; that is, they will grow on culture medium lacking an auxin or a cytokinin . This phenomenon is known as **habituation**.

There are several possible explanations for the habituation of plant tissues.

First, the tissues may acquire the capacity to synthesize their own cytokinin, and in some cases habituated callus tissues have been shown to contain cytokinins. Alternative explanations are that the rate of cytokinin destruction is reduced, or that the tissues become more sensitive to cytokinins, such that lower levels of the hormone are effective.

Frederick Meins and his colleagues in Basel, Switzerland, have shown that different tissues of tobacco plants exhibit different capacities to synthesize cytokinins in tissue culture. Tobacco pith tissue is cytokinin dependent but capable of being habituated. In contrast, tissues cultured from the cortex of the stem are cytokinin autonomous; they can be cultured indefinitely on medium containing auxin as the only hormone supplement. Finally, tobacco leaf tissue has an absolute requirement for cytokinins and cannot be habituated .

These observations indicate that the ability to synthesize cytokinins can be switched on or off during development, and that the on and off states can be stably maintained during cell divisions. Such developmental control of cytokinin synthesis may be important in morphogenesis.

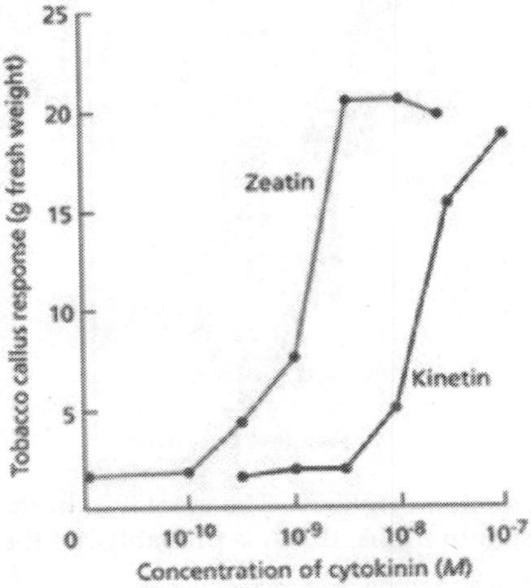

Figure 15.2 Tobacco tissues depend on cytokinin for their growth in culture. Cultured tobacco callus was transferred to fresh medium that contained an auxin and either zeatin or kinetin at the indicated concentrations. The tissues were weighed after growing for 1 month on the various media. The results show that zeatin is more effective than kinetin in supporting tobacco tissue growth. The maximum response was obtained with 5×10^{-8} *M* zeatin, and higher zeatin concentrations either did not stimulate additional growth or were slightly inhibitory. The kinetin concentration had to be at least tenfold higher to achieve the same growth stimulation. (Leorard *et el* ,1968. *Proc. Natl. Acad. Sci.* USA 581:15-21)

15.7 Presence of Cytokinins in tRNAs

Most organisms contain a variety of modified bases in their tRNA. The normal RNA bases are incorporated into the newly synthesized tRNA chain, and then specific enzyme modify a small subset of the bases. As with the free cytokinins, Ä²-isopentenyl groups are transferred to the adenine molecules. The enzyme catalyzing this reaction is called tRNA-IPT, which is weakly similar in its sequence to the plant cytokinin biosynthetic enzyme. The similarity was used to help identify the cytokinin biosynthetic enzyme from plants. The tRNA-IPT must be able to recognize a specific base sequence in the tRNA and transfer the isopentenyl group to the adenosine nearest the 3' end of the anticodon (Figure 15.3.). It does not utilize free AMP as a substrate.

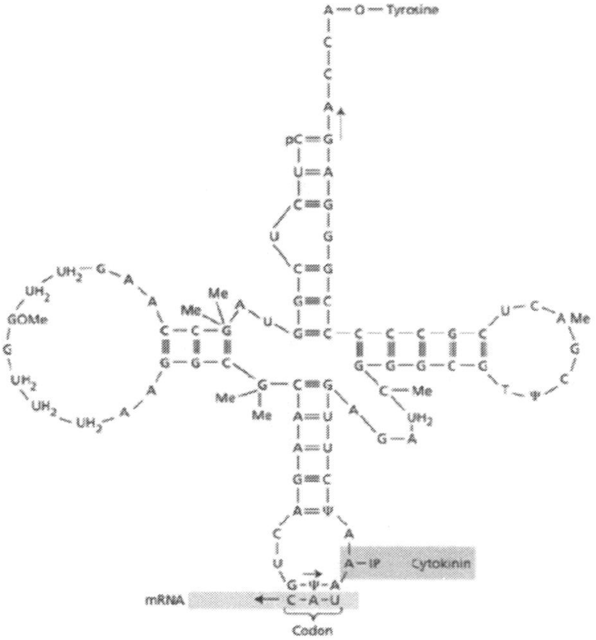

Yeast tRNA for tyrosine

Figure 15.3 Some tRNAs contain a cytokinin. The cytokinin in the examples shown is [^{9}R] iP. Although cytokinins are present in tRNA, tRNA is probably not the source of free cytokinins in plants.

The possibility that the free cytokinins are derived from tRNA has been explored extensively. Although the tRNA-bound cytokinins can act as hormonal signals for plant cells if the tRNA is degraded and fed back to the cells, it is unlikely that any significant amount of the free hormonal cytokinin in plants is derived from the turnover of tRNA.

15.8 Role of Cytokinin in Development

The *det* mutants of *Arabidopsis* develop many of the characteristics of light-grown seedlings when they are germinated in the dark. The *det* mutants have short, unhooked hypocotyls and expanded cotyledons, and the apical meristem initiates leaves that subsequently expand without light. Although the *det* mutants do not develop full photosynthetic competence in the dark , there is partial development of the chloroplasts, including the expression of genes encoding many photosynthetic enzymes that are not present in the chloroplasts of dark-grown wild-type seedlings.

When wild-type *Arabidopsis* seedlings are germinated in darkness in the presence of cytokinin, they develop many of the characteristics of the *det* mutants: Hypocotyls shorten, cotyledons expand, the apical meristem initiates leaves, and there is partial development of the chloroplasts, including the synthesis of some

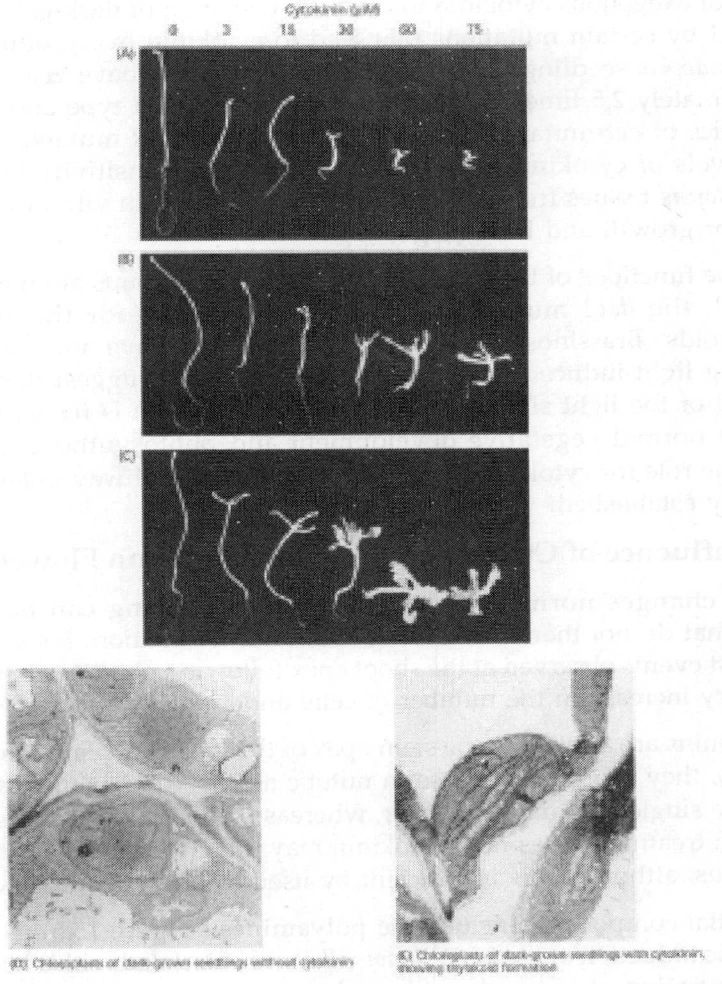

Figure 15.4. The effects of cytokinin on the development of wild-type *Arabidopsis* seedlings grown in darkness. (A–C) The appearance of the seedlings after 1, 2, and 3 weeks, respectively, of growth in the dark with increasing concentrations of cytokinin. The control (no cytokinin) is on the left in each case. The next five seedlings were treated with 3, 15, 30, 60, and 75 μ*M* of cytokinin, respectively. As the cytokinin concentration was increased, the inhibition of hypocotyl elongation became more pronounced, while the cotyledons expanded somewhat and leaves were initiated from the shoot apical meristem. At the higher cytokinin concentrations the seedlings were phenocopies of *det* mutants. Cytokinin treatment also resulted in thylakoid formation in the plastids of dark-grown seedlings (E) as compared to the development of plastids as etioplasts in the untreated, dark-grown wild-type control (D). (Chory *et el*, 1994, *Plant Physiol* 104:339-347)

photosynthetic enzymes. As Figure 15.4.A shows, this effect is proportional to the cytokinin dosage given to the seedlings . Cytokinin is said to produce a *phenocopy* of the *det* mutations.

The ability of exogenous cytokinin to cause de-etiolation of dark-grown seedlings is mimicked by certain mutations that lead to cytokinin overproduction. Dark-grown *Arabidopsis* seedlings carrying the *amp1* mutation have zeatin levels that are approximately 2.5 times higher than those of the wild type and develop the characteristics of *det* mutant seedlings . Although the *det* mutants do not have elevated levels of cytokinins, they may have altered sensitivity to cytokinins, since *Arabidopsis* tissues from *det1* plants, when cultured in vitro, do not require cytokinin for growth and become green.

Although the functions of the genes altered by the *det* mutants are not completely understood, the *det2* mutation is in a gene required for the synthesis of brassinosteroids Brassinosteroids and cytokinins may have antagonistic effects in regulating light-induced development. These results suggest that cytokinins may be part of the light signal transduction pathway that is responsible for the initiation of normal vegetative development and photosynthetic competence. However, the role for cytokinins in the developmental pathway initiated by light is not clearly established.

15.9 The Influence of Cytokinins and Polyamines on Flowering

Sometimes changes normally associated with flowering can be caused by treatments that do not themselves bring about floral initiation. For example, one of the earliest events observed at the shoot apex following photoperiodic induction is a transitory increase in the number of cells undergoing mitosis.

When cytokinins are applied to the stem apex of the long-day plant white mustard (*Sinapis alba*), they cause an increase in mitotic activity similar to that caused by exposure to a single long day. However, whereas the long day induces flowering, the cytokinin treatment does not. Cytokinin may therefore be a component of the floral stimulus, although it is insufficient by itself to bring about flowering.

Other potential components include the **polyamines** (Figure 15.5). For example, photoperiodic induction of the LDP *Sinapis alba* was correlated with a large increase in the concentration of putrescine, the major polyamine, in the phloem exudate from the leaves. Spraying the leaves with an inhibitor of putrescine biosynthesis lowered the level of putrescine in the phloem exudate and also inhibited flowering. This result suggests that putrescine may be a component of the floral stimulus in *S. alba*.

In addition to receiving signals from the leaves, the shoot apex may accumulate flowering signals from the roots. Classic experiments with cuttings have shown that the initiation and growth of adventitious roots inhibit flowering in the shoot. In many species, root removal promotes flowering. Thus, at least in certain species, the root appears to produce one or more flowering inhibitors. In other cases, the roots may contribute floral activators.

(A) $^{+}H_3N—(CH_2)_4—NH_3^{+}$

Putrescine

(B) $^{+}H_3N—(CH_2)_3—\overset{+}{N}H_2—(CH_2)_4—NH_3^{+}$

Spermidine

(C) $^{+}H_3N—(CH_2)_3—\overset{+}{N}H_2—(CH_2)_4—\overset{+}{N}H_2—(CH_2)_3—NH_3^{+}$

Spermine

Figure 15.5 Polyamine structures. (A) Putrescine. (B) Spermidine. (C) Spermine. At neutral pH, polyamines are polycations and can bind to polyanions in the cell, such as DNA, RNA, and phospholipids. The binding of polyamines to nucleic acids and membranes may stabilize these macromolecules. In addition to playing a role in the control of flowering, polyamines have been implicated in the regulation of embryogenesis and senescence.

For example, both GA and cytokinin are synthesized in the root and transported to the shoot via the xylem. During photoperiodic induction in *Sinapis alba* , the amount of cytokinin present in the xylem exudate from the root increases considerably . This finding indicates that signals emanating from the leaf are transported both to the shoot apex and to the root during photoperiodic induction. In response to the signal from the leaves, the roots produce and export additional chemical signals that modulate the flowering response.

Chapter **16**

Ethylene

$$H_2C = CH_2$$

16.1 Nature of Ethylene

Ethylene, unlike the rest of the plant hormone compounds is a gaseous hormone. Like abscisic acid, it is the only member of its class. Of all the known plant growth substance, ethylene has the simplest structure. It is produced in all higher plants and is usually associated with fruit ripening and the tripple response.

16.2 History of Discovery in Plants

Ethylene has been used in practice since the ancient Egyptians, who would gas figs in order to stimulate ripening. The ancient Chinese would burn incense in closed rooms to enhance the ripening of pears. It was in 1864, that leaks of gas from street lights showed stunting of growth, twisting of plants, and abnormal thickening of stems (the triple response). In 1901, a russian scientist named Dimitry Neljubow showed that the active component was ethylene. Doubt discovered that ethylene stimulated abscission in 1917. It wasn't until 1934 that Gane reported that plants synthesize ethylene. In 1935, Crocker proposed that ethylene was the plant hormone responsible for fruit ripening as well as inhibition of vegetative tissues. Ethylene is now known to have many other functions as well.

16.3 Biosynthesis and Metabolism

Ethylene is produced in all higher plants and is produced from methionine in essentially all tissues. Production of ethylene varies with the type of tissue, the plant species, and also the stage of development. The mechanism by which ethylene is produced from methionine is a 3 step process.

ATP is an essential component in the synthesis of ethylene from methionine. ATP and water are added to methionine resulting in loss of the three phosphates and S-adenosyl methionine.

1-amino-cyclopropane-1-carboxylic acid synthase (ACC-synthase) facilitates the production of ACC from SAM.

Oxygen is then needed in order ro oxidize ACC and produce ethylene. This reaction is catalyzed by an oxidative enzyme called ethylene forming enzyme. The control of ethylene production has received considerable study. Study of ethylene has focused around the synthesis promoting effects of auxin, wounding, and drought as well as aspects of fruit-ripening. ACC synthase is the rate limiting step for ethylene production and it is this enzyme that is manipulated in biotechnology to delay fruit ripening in the "flavor saver" tomatoes.

16.4 Functions of Ethylene

Ethylene is known to affect the following plant processes:

- Stimulates the release of dormancy.
- Stimulates shoot and root growth and differentiation (triple response)
- May have a role in adventitious root formation.
- Stimulates leaf and fruit abscission.
- Stimulates Bromiliad flower induction.
- Induction of femaleness in dioecious flowers.
- Stimulates flower opening.
- Stimulates flower and leaf senescence.
- Stimulates fruit ripening.

16.5 Cloning of the Gene That Encodes ACC Synthase

A major hurdle in the understanding of ethylene biosynthesis was the difficulty in purifying the low abundance, unstable protein and the cloning of the corresponding gene. Researchers overcame this obstacle by raising antisera against partly purified ACC synthase from zucchini fruit slices that were induced to produce large amounts of ethylene. The resulting antisera contained a mixture of antibodies, including antibodies to ACC. The investigators then removed the contaminating antibodies by allowing them to bind to proteins extracted from zucchini tissue that lacked the enzyme. The key to the success of this procedure was the availability of tissue largely lacking the enzyme with which to subtract the non-ACS antibodies from the polyclonal antibody mixture. This circumvented the need for highly purified ACC synthase protein. The purified ACC synthase antibodies were then used to isolate the gene encoding the enzyme .

Once the gene encoding ACS was isolated, it could be used to express the ACC synthase protein in the bacterium *E. coli* at high levels and isolate purified ACC synthase in large amounts . The cloned zucchini gene also allowed genes encoding ACC synthase to be isolated from numerous other plant species, using low stringency hybridization and degenerate PCR. The deduced amino acid sequences of ACC synthase genes were found to show a high degree of similarity to the *aminotransferase* superfamily of enzymes. These enzymes catalyze the transfer of amino groups among amino acids and require pyridoxal phosphate as a cofactor.

16.6 Demonstration of the Dimeric Nature of ACC Synthase

The deduced crystal structure of both apple and tomato ACS reveals a homodimeric protein with shared active sites, which is similar to aminotransferases . The dimeric nature of ACC synthase has been confirmed by an elegant study involving functional complementation. In this study, researchers examined potential dimerization of ACS enzymes by expressing in *E. coli* two ACS proteins that harbor distinct active-site mutations, neither of which has activity on its own . If ACS acts as a monomer, then expressing the two different mutant versions should not result in an active enzyme, because each is inactive. Conversely, if the two inactive ACS subunits come together as a dimer, the different amino acid changes can complement each other, forming a functional enzyme. The researchers indeed found that functional ACS enzyme was formed when various pairs of mutants were expressed, which provides compelling evidence that ACS enzymes are able to act as both homo- and heterodimers with shared active sites.

16.7 ACC Synthase Gene Expression and Biotechnology

As we have seen, ACC synthase is a cytosolic enzyme that catalyzes the first committed step in ethylene biosynthesis in higher plants. It is a key regulatory enzyme that catalyzes the production of the ethylene precursor ACC from AdoMet (*S*-adenosylmethionine). However, this enzyme is difficult to purify because it is labile and present in low abundance in plant tissues. The molecular cloning and functional expression of ACC synthase in *E. coli* and yeasts have facilitated biochemical and structural studies of this enzyme.

ACC synthase cDNAs and genomic sequences have been cloned from numerous plant species. The emerging picture from the study of these genes is that ACC synthase is encoded by members of a divergent multigene family in which each gene is differentially regulated by various environmental and developmental factors during the growth of a plant. *Arabidopsis*, for example, has twelve and tomato has at least nine gene members, which are differentially regulated by inducers such as auxin, fruit ripening, and wounding .

The protein synthesis inhibitor cycloheximide is a potent inducer of many ACC synthase genes. Cycloheximide inducibility is a characteristic of *primary response genes* such as the early auxin-inducible genes . Two mechanisms have been proposed to explain this inducibility.

- First, the transcription of these genes may be under the control of a short-lived repressor protein whose synthesis is prevented by the addition of cycloheximide;
- Second, the mRNA transcripts of these genes may be short-lived and are stabilized when cycloheximide prevents the synthesis of a nuclease.

The induction of at least one of the genes for ACC synthase (*ACS2*) in *Arabidopsis* by cycloheximide appears to be due to transcriptional activation , and thus is likely regulated by a short-lived transcriptional repressor.

A clue to the signal transduction mechanism involved in the regulation of ACC synthase gene expression has come from experiments with lithium (Li^+). Li^+ is known to interfere with the phosphoinositide signaling pathway by inhibiting the activity of inositolphosphate phosphatase. Li^+ is also one of the strongest inducers of ACC synthase activity in plants . This finding suggests that the regulation of at least some ACC synthase genes may involve Ca^{2+} mobilization mediated by inositol trisphosphate (IP_3) .

Because ethylene plays such an important role in fruit ripening, the cloning of the genes for ACC synthase and ACC oxidase has enabled scientists to use biotechnology to reduce ethylene production during fruit ripening, and thus to prolong the storage life of fruit.

In the case of tomato, two different strategies have been employed to control ethylene production.

- The first strategy was to use the Ti plasmid of Agrobacterium tumefaciens to transform tomato tissue with antisense DNA for ACC synthase or ACC oxidase. When the genes for the two enzymes are expressed in the reverse direction in the cell, the resulting "antisense transcript" binds to and inactivates the "sense transcript" produced by the wild-type gene, effectively shutting down expression of the wild-type gene . After transformation plants can be regenerated in tissue culture.
- In the second strategy for reducing ethylene production, tomato tissue was transformed (using A. tumefaciens) with genes encoding enzymes that metabolize ACC or AdoMet . In this case, the ACC that is produced is rapidly degraded, and no ethylene can be synthesized.

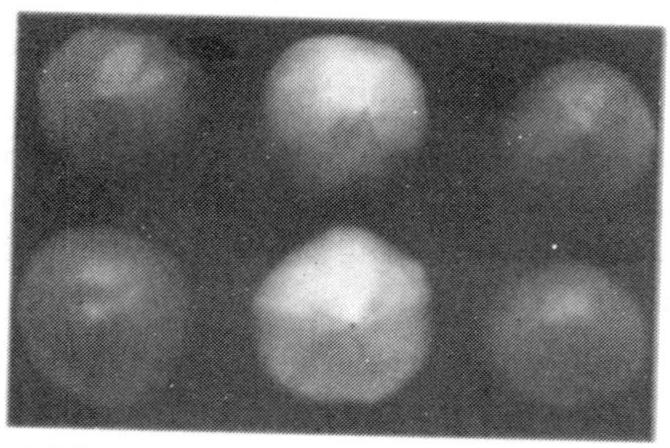

Figure 16.1 Inhibition of tomato fruit ripening in detached tomato fruits by antisense ACC synthase RNA. Antisense tomato fruits (center) remain firm and green unless treated by ethylene. Ethylene-treated antisense fruits (left) are indistinguishable from naturally ripened fruits (right) in terms of color, flavor, aroma, and texture. (Oellar *et el*, 1991; *Science* 24: 437-439)

Regardless of the strategy used, all fruits from these genetically engineered plants show significant delays in ripening. Tomato fruits from plants transformed with the ACC synthase antisense DNA fail to ripen unless treated with ethylene (Figure 16.1).

Similar approaches are now being carried out in other agronomically important crops, such as melon and banana. In addition, these approaches are being used to retard floral senescence in carnations, thereby reducing postharvest losses. The research on ACC synthase and other ethylene biosynthesis genes thus provides an excellent example of how basic research can lead to improvements in agricultural and horticultural crops, the full benefits of which we are only beginning to realize.

16.8 Conservation of Ethylene Signaling Components in Other Plant Species

Genes that are similar to the *Arabidopsis* ethylene signaling components have been found in other plant species. For example, the tomato genome, like that of *Arabidopsis*, encodes a family of ethylene receptor genes, including *Never-ripe*, which are differentially expressed throughout development. The tomato ethylene receptor genes are homologous to ETR1. A tobacco homolog of EIN3 has been found, and genes similar to all the *Arabidopsis* signaling components can be found in the rice genomic sequence. These and other studies suggest that the components of these proteins probably represent ubiquitous mediators of ethylene signaling in plants.

Chapter **17**

Abscisic Acid

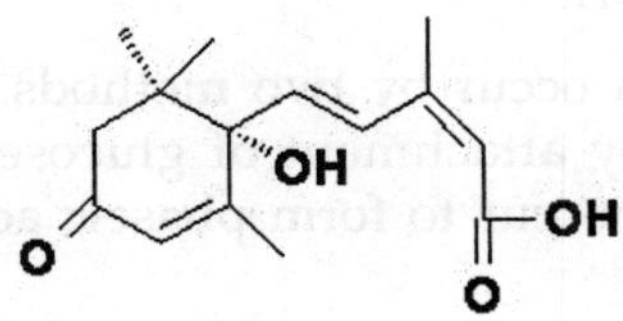

17.1 Nature of Abscisic Acid

Abscisic acid is a single compound unlike the auxins, gibberellins, and cytokinins. It was called "abscisin II" originally because it was thought to play a major role in abscission of fruits. At about the same time another group was calling it "dormin" because they thought it had a major role in bud dormancy. The name abscisic acid (ABA) was coined by a compromise between the two groups. Though ABA generally is thought to play mostly inhibitory roles, it has many promoting functions as well.

17.2 History of Abscisic Acid

In 1963, Abscisic acid was first identified and characterized by Frederick Addicott and his associates. They were studying compounds responsible for the abscission of fruits (Cotton). Two compounds were isolated and called Abscisin I and Abscisin II. Abscisin II is presently called Abscisic acid (ABA). Two others groups at about the same time discovered the same compound.One group headed by Philip Wareing was studding bud dormancy in woody plants. The other group led by Van Steveninck was studying abscission of flowers and fruits from lupine. Plant Physiologists agreed to call the compound Abscisic Acid.

17.3 Biosynthesis and Metabolism

ABA is a naturally occurring compound in plants. It is a sesquiterpenoid (15-carbon) which is partially produced via the mevalonic pathway in chloroplasts and other plastids. Because it is sythesized partially in the chloroplasts, it makes sense that biosynthesis primarily occurs in the leaves. The production of ABA is accentuated by stresses such as water loss and freezing temperatures. It is believed that biosynthesis occurs indirectly through the production of carotenoids. Carotenoids are pigments produced by the chloroplast which have 40 carbons. Breakdown of these carotenoids occurs by the following mechanism: Violaxanthin is a carotenoid which has forty carbons.

It is isomerized and then split via an isomerase reaction followed by an oxidation reaction.

One molecule of xanthonin is produced from one molecule of violaxanthonin and it is uncertain what happens to the remaining biproduct.

The one molecule of xanthonin produced is unstable and spontaneously changed to ABA aldehyde.

Further oxidation results in ABA.

Activation of the molecule can occur by two methods. In the first method, an ABA-glucose ester can form by attachment of glucose to ABA. In the second method, oxidation of ABA can occur to form phaseic acid and dihyhdrophaseic acid.

The transport of ABA can occur in both xylem and phloem tissues. It can also be translocated through paranchyma cells. The movement of abscisic acid in plants does not exhibit polarity like auxins. ABA is capable of moving both up and down the stem.

17.4 Functions of Abscisic Acid

The following are some of the physiological responses known to be associated with abscisic acid.

- Stimulates the closure of stomata (water stress brings about an increase in ABA synthesis).
- Inhibits shoot growth but will not have as much affect on roots or may even promote growth of roots.
- Induces seeds to synthesize storage proteins.
- Inhibits the effect of gibberellins on stimulating *de novo* synthesis of α-amylase.
- Has some effect on induction and maintanance of dormancy.
- Induces gene transcription especially for proteinase inhibitors in response to wounding which may explain an apparent role in pathogen defense.

17.5 The Structure Of Lunularic Acid from Liverworts

In liverworts, a compound similar to ABA, named **lunularic acid** appears to play a physiological role similar to that of ABA in higher plants (Figure 17..1.).

17.6 Structural Requirements for Biological Activity of Abscisic Acid

The features shown to be essential for biological activity include the carboxyl group, the tertiary hydroxyl group, the 2-*cis*, 4-*trans*-pentadienoic side chain, a 4'-ketone, and a double bond in the cyclohexane ring. Catabolic products of ABA that are present in the tissue and are missing any of these groups are biologically inactive.

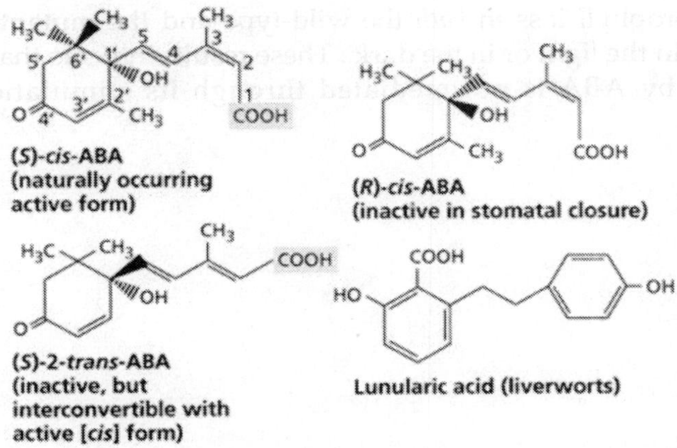

Figure 17.1. The chemical structures of the *S* (counterclockwise array) and *R* (clockwise array) forms of *cis*-ABA, the (*S*)-2-*trans* form of ABA, and lunularic acid. The numbers in the diagram of (*S*)-*cis*-ABA identify the carbon atoms.)

17.7 The Bioassay of ABA

Coleoptile growth, the classic bioassay devised for auxins , is also used for ABA detection in plant extracts by measurement of coleoptile growth inhibition. This bioassay has adequate sensitivity (minimum detectable level is 10^{-7} *M*) and shows a linear response in the range of 10^{-7} to 10^{-5} *M*, but it has some disadvantages. Plant extracts have other promoters and inhibitors, which decrease the specificity of the bioassay and make preliminary purification steps necessary.

Other available bioassays for ABA include inhibition of germination, inhibition of gibberellic acid- induced α-amylase synthesis in cereal aleurone layers, induction of leaf abscission, and stomatal closure. Stomatal closure is highly specific for ABA because it is affected little by other plant growth regulators. Additional advantages of this bioassay include a fast response of guard cells to ABA, high sensitivity (minimum detectable level 10^{-9} *M*), and a linear response over a wide range of concentrations.

17.8 ABA-Induced Senescence and Ethylene

Ethylene has also been found to play a role in the senescence of oat leaf segments, but ABA appears to be the initiating agent, whereas ethylene appears to exert its effects at a later stage. To determine whether the ABA effects are mediated by ethylene, the effects of ABA and ethylene on senescence of leaf discs from wild-type and ethylene-insensitive mutants were compared. Chlorophyll loss from untreated leaf discs of the ethylene-insensitive mutant was much slower than that from the wild type, in the light and in the dark. Ethylene accelerated the chlorophyll loss from the wild-type leaf discs, but it had no effect on the yellowing of the ethylene-insensitive mutant discs. In contrast, treatment with ABA stimulated chlorophyll loss in *both* the wild-type and the mutant discs, whether they were held in the light or in the dark . These results indicate that the promotion of senescence by ABA is not mediated through its stimulation of ethylene production.

Chapter **18**

Seed Dormancy

18.1 Types of Seed Dormancy

During seed maturation, the embryo enters a quiescent phase in response to desiccation. Seed germination can be defined as the resumption of growth of the embryo of the mature seed; it depends on the same environmental conditions as vegetative growth does. Water and oxygen must be available, the temperature must be suitable, and there must be no inhibitory substances present.

However, in many cases a viable (living) seed will not germinate even though all the necessary environmental conditions for growth are satisfied. This phenomenon is termed **seed dormancy.** Seed dormancy introduces a temporal delay in the germination process that provides additional time for seed dispersal over greater geographical distances. It also maximizes seedling survival by preventing germination under unfavorable conditions.

Two types of seed dormancy have been recognized, coat-imposed dormancy and embryo dormancy. **Coat-imposed dormancy** is dormancy imposed on the embryo by the seed coat and other enclosing tissues, such as endosperm, pericarp, or extrafloral organs. The embryos of such seeds will germinate readily in the presence of water and oxygen once the seed coat and other surrounding tissues are either removed or damaged. There are five basic mechanisms of coat-imposed dormancy:

1. *Prevention of water uptake.* Prevention of water uptake by the seed coat is a common cause of seed dormancy in families found in arid and semiarid regions, especially among legumes, such as clover (*Trifolium* spp.) and alfalfa (*Medicago* spp.). Waxy cuticles, suberized layers, and lignified sclereids all combine to restrict the penetration of water into the seed.
2. *Mechanical constraint.* The first visible sign of germination is typically the radicle breaking through the seed coat. In some cases, however, the seed coat may be too rigid for the radicle to penetrate. Nuts with hard, lignified shells are examples of dormancy caused by mechanical constraint. Such shells must be broken by biotic or environmental forces for the seed to germinate. Even nonlignified tissues, such as the endosperm of lettuce seeds, can suppress expansion of the embryo. For the seeds to germinate, the endosperm cell walls must be weakened by the production of cell wall–degrading enzymes.

3. *Interference with gas exchange.* Some seed coats are considerably less permeable to oxygen than an equivalent thickness of water is— e.g., less by a factor of about 10^4 in seeds of cocklebur (*Xanthium pennsylvanicum*). This lowered permeability to oxygen suggests that the seed coat inhibits germination by limiting the oxygen supply to the embryo. In support of this idea, investigators can break the dormancy of such seeds either by making a small hole in the coat with a pin (without weakening the coat mechanically), or by treating the coat with concentrated oxygen. However, other studies suggest that the oxygen consumption of the embryos from such seeds is considerably less than the amount of oxygen able to penetrate the seed coats under normal aerobic conditions. Thus the role of oxygen impermeability in seed coat dormancy remains unresolved.
4. *Retention of inhibitors.* The seed coat may prevent the escape of inhibitors from the seed. For example, growth inhibitors readily diffuse out of isolated *Xanthium* embryos, but not from intact seeds.
5. *Inhibitor production.* Seed coats and pericarps may contain relatively high concentrations of growth inhibitors that can suppress germination of the embryo. ABA is a common germination inhibitor present in these maternal tissues. In certain cases where repeated washing (leaching) removes dormancy, the effect is thought to be due to the loss of such inhibitory compounds.

The second type of seed dormancy is **embryo dormancy**, which refers to a dormancy that is inherent in the embryo and is not due to any influence of the seed coat or other surrounding tissues. In some cases, embryo dormancy can be relieved by amputation of the cotyledons. Species in which the cotyledons exert an inhibitory effect include European hazel (*Corylus avellana*) and European ash (*Fraxinus excelsior*). A fascinating demonstration of the cotyledon's ability to inhibit growth is found in species (e.g., peach) in which the isolated dormant embryos germinate but grow extremely slowly to form a dwarf plant. If the cotyledons are removed at an early stage of development, however, the plant abruptly shifts to normal growth.

Embryo dormancy is thought to be due to the presence of inhibitors, especially ABA, as well as the absence of growth promoters, such as GA (gibberellic acid). The loss of embryo dormancy is often associated with a sharp drop in the ratio of ABA to GA.

18.2 Secondary Dormancy

Different types of seed dormancy also can be distinguished on the basis of the timing of dormancy onset rather than the cause of the dormancy.

- Seeds that are released from the plant in a dormant state are said to exhibit **primary dormancy**.
- Seeds that are released from the plant in a nondormant state but which be-

come dormant if the conditions for germination are unfavorable exhibit **secondary dormancy**. For example, seeds of *Avena sativa* (oat) can become dormant in the presence of temperatures higher than the maximum for germination, whereas seeds of *Phacelia dubia* (small-flower scorpionweed) become dormant at temperatures below the minimum for germination. The mechanisms of secondary dormancy are poorly understood.

18.3 Release from Seed Dormancy

Various external factors release the seed from dormancy, and dormant seeds typically respond to more than one condition. Many seeds lose their dormancy when their moisture content is reduced to a certain level by drying. This method of breaking seed dormancy is called **after ripening,** and is usually performed in a special drying oven. On the other hand, if the seed becomes too dry (5% water content or less), the effectiveness of afterripening is diminished.

Another factor that can release seeds from dormancy is low temperature, or **chilling**. Many seeds require a period of cold (0 to 10°C) while in a fully hydrated (imbibed) state in order to germinate. In temperate-zone species, this requirement is of obvious survival value, since such seeds will germinate not in the fall, but only in the following spring. Chilling seeds to break their dormancy is a time-honored practice in horticulture and forestry and traditionally has been referred to as *stratification*. This term is derived from the old agricultural practice of allowing seeds with a chilling requirement to overwinter outdoors in layered mounds of moist soil. Today the seeds are simply stored in a refrigerator.

The third external factor that plays an important role in breaking seed dormancy is **light**. Many seeds have a light requirement for germination, which may involve only a brief exposure, as in the case of lettuce, an intermittent treatment (e.g., succulents of the genus *Kalanchoe*), or even a specific photoperiod involving short or long days. Phytochrome is the main sensor for light-regulated seed germination. Interestingly, all light-requiring seeds exhibit seed coat dormancy, and removal of the outer tissues of the seed allows the embryo to germinate in the absence of light. The effect that light has on the embryo is thus to enable the radicle to penetrate the seed coat. This penetration often involves some enzymatic weakening of the enclosing tissues.

18.4 The Longevity of Seeds

How long can a seed remain dormant and still remain viable? Seed longevity is of practical importance because of ongoing efforts to preserve plant genetic resources for future agricultural crops by setting up seed gene banks. Spectacular claims of dormant seeds that remained viable for thousands of years have been made, but are considered highly controversial. Extreme examples of ancient seeds that have been reported to retain their viability include submerged lotus seeds found in a 3,000-year-old boat near Tokyo, barley seeds from the 3,000-year-old tomb of

King Tutankhamen, and arctic lupine seeds associated with rodent burrows determined to be 14,000 years old. In most cases, however, a scientifically rigorous examination of the data has either disproved or raised serious doubts about the claimed antiquity of the seeds .

The most serious criticism is that the dates of the seeds have been inferred from the age of their immediate surroundings or associated artifacts rather than from direct measurements of the seeds themselves. This question raises the possibility that the seeds are much younger than the site at which they were found. In fact, microscopic examination of authenticated ancient cereal grains from various Middle Eastern sites has shown that cells and tissues of these seeds are no longer intact. Such seeds are clearly inviable.

Although the most sensational claims for seed longevity are almost certainly bogus, seeds of *Canna compacta* apparently can live for at least 600 years. Viable *Canna* seeds were obtained from inside a walnut in a tomb in Argentina. The *Canna* seeds had apparently been inserted into the immature seeds of a growing walnut fruit before the hard outer shell formed. Once the shell hardened and the nut dried out, the result was a rattle. Native people strung the rattles together to form a necklace. In this case, the seeds had to be at least as old as the walnut shell, and carbon dating of the shell indicated that it was about 600 years old .

Plant collections begun during the late eighteenth century in Europe—for example, at the British Museum in London and at the Museum of Natural History in Paris—have been reliable sources of seeds for viability determinations. *Cassia multijuga* (false sicklepod) seeds, collected in 1776, were still viable after being tested in 1934. In 1879, W. J. Beal initiated the longest-running experiment on seed longevity by burying the seeds of 21 different species in unstoppered bottles in a sandy hilltop near the Michigan Agricultural College in East Lansing. After 100 years, only one species, *Verbascum blattaria* (moth mullein), remained viable.

Why are some seeds able to survive in a dormant state for hundreds of years, whereas others lose viability in less than 5 years? Almost nothing is known about the mechanisms that deter-mine the longevity of seeds. If these mechanisms could be understood, we might some day be able to greatly increase the seed longevity of agriculturally important species and varieties, thereby enhancing our ability to preserve plant genetic resources for generations to come.

18.5 Genetic Mapping Of Dormancy:

Complex traits, e.g. dormancy, are affected by many different, often unlinked genes. Consequently, attempts to map them result in continuous, or quantitative, variation within a mapping population. Quantitative trait locus (QTL) mapping is used in these instances to map individual genes affecting the trait. In QTL mapping, inbred lines that differ in the trait of interest are crossed and an F_2 population is produced. The F_2 population has continuous variation for the trait of interest, and the goal of the analysis is to determine which chromosomal segments contain genes affecting the trait. Using marker loci along each chromosome (loci that have different alleles between the parental inbred lines), the genotypes of entire chromosomal segments in each F_2 individual can be inferred as being homozygous for one or the other parental lines or heterozygous (Figure 18.1.). If the inheritance of a chromosomal segment is correlated with the trait of interest, then a gene affecting the trait is likely to be between the two marker loci. By analyzing chromosomal segments along each of the chromosomes, the number and locations of genes affecting the quantitative trait can be determined.

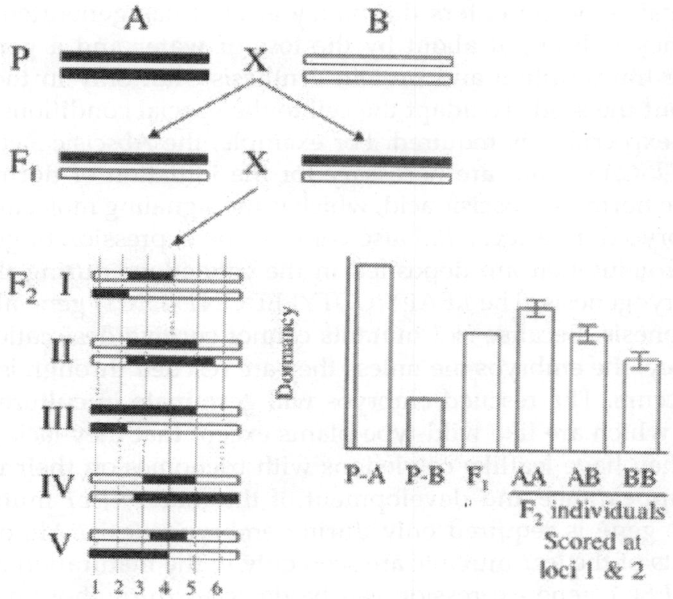

Figure 18.1 Two parental (P) inbred lines, A & B, which differ for a trait of interest (e.g. dormancy) are crossed to produce heterozygote F_1 progeny which have an intermediate degree of dormancy. Only one chromosome pair is depicted for simplicity. The F_1 progeny are crossed to produce a large segregating F_2 population (usually 100-500 individuals are analyzed, here only five individuals, I-V, are depicted). Because of recombination during F_1 meiosis, the F_2 individuals have varying genetic combinations of the original inbred lines (depicted by shading of chromosomes) and have continuous variation, from high to low, in degree of dormancy. In order to estimate whether segments of the chromosome are homozygous for one of the parental-lines or heterozygote, the F_2 individuals are scored for marker loci (1-6) along the chromosome. For example, loci 1 & 2 are identified as homozygous for A-line alleles in individual III, as heterozygous in individuals I & V and as homozygous for B-line alleles in individuals II & IV. The F_2 individuals are also scored for their dormancy. After scoring many F_2 individuals a correlation between the genotypes at locus 1 & 2 and phenotype is determined and therefore this segment likely contains at least one gene affecting dormancy (bar graph). The analysis is repeated for each chromosomal segment to identify which ones are likely to contain genes affecting dormancy.

Candidate genes are specific loci, identified by predicted biochemical function or homology to regulatory genes from other species, that are suspected of affecting the trait of interest. Molecular differences between the parental alleles at these loci can be used to determine whether they map to the same region as any of the QTLs, thereby supporting the hypothesis that a given gene contributes to regulation of the trait.

18.6 Embryonic Dormancy

Embryo Maturation Requires Specific Gene Expression

The Arabidopsis embryo enters dormancy after it has generated about 20,000 cells. Dormancy is brought about by the loss of water and a general shutting down of gene transcription and protein synthesis—not only in the embryo, but also throughout the seed. To adapt the cell to the special conditions of dormancy, specific gene expression is required. For example, the *A*bscisic *A*cid *I*nsensitive*3* (*ABI3*) and *FUSCA3* genes are necessary for the initiation of dormancy and are sensitive to the hormone abscisic acid, which is the signaling molecule that initiates seed and embryo dormancy. *ABI3* also controls the expression of genes encoding the storage proteins that are deposited in the cotyledons during the maturation phase of embryogenesis. The *LE*AFY *C*OTYLEDON*1* (*LEC1*) gene also is active in late embryogenesis. Because *lec1* mutants cannot survive desiccation and do not enter dormancy, the embryos die unless they are rescued through isolation before desiccation occurs. The rescued embryos will germinate in culture and produce fertile plants, which are like wild-type plants except that they lack the 7S storage protein and they have leaflike cotyledons with trichomes on their upper surface. The normal appearance and development of the mature *lec1* mutants indicates that the *LEC1* gene is required only during embryogenesis. Although the most obvious defects of the *lec1* mutants are seen only in the maturation phase embryo, mRNA from *LEC1* gene expression can be detected throughout embryogenesis. It has been proposed that *LEC1* is a general repressor of vegetative development and its expression is necessary throughout embryogenesis .

Chapter **19**

Cell Signaling during Cold, Drought and Salt Stress

19.1 Introduction

Low temperature, drought and high salinity are common stress conditions that adversely affect plant growth and crop production. The cellular and molecular responses of plants to environmental stress have been studied intensively. Understanding the mechanism by which plants perceive environmental signals and transmit the signals to cellular machinery to activate adaptive responses is of fundamental importance to biology. Knowledge about stress signal transduction is also vital for continued development of rational breeding and transgenic strategies to improve stress tolerance in crops. In this review, we first consider common characteristics of stress signal transduction in plants, and then examine some recently studies on the functional analysis of signaling components. Finally, we attempt to put these component and pathways into signal transduction networks that are grouped into three generalized signaling type.

19.2 General Stress Signal Transduction Pathways

A generic signal transduction pathway starts with signal perception, followed by the generation of second messengers (e.g., inositol phosphate and reactive oxygen species [ROS]). Second messengers can modulate intracellular Ca^{2+} levels, often initiating a protein phosphorylation cascade that finally targets proteins directly involved in cellular protection or transcription factors controlling specific sets of stress-regulated genes. The products of these genes may participate in the generation of regulatory molecules like the plant hormones abscisic acid (ABA), ethylene, and salicylic acid (SA). These regulatory molecules can, in turn, initiate a second round of signaling that may follow the above involved.

Signal transduction requires the proper spatial and temporal coordination of all signaling molecules. Thus, there are certain molecules involved in modification, delivery, or assemblies of signaling components, but do not directly relay the signal. They too are critical for the accurate transmission of stress signals. These proteins include protein modifier (e.g., enzymes for protein lipidation, methylation, glycosylation, and ubiquitination), scaffolds, and adaptors .

19.3 Multiplicity of Abiotic Stresses as Signals for Plants and the need for Multiple Sensors.

Low temperature, drought, and high salinity are very complex stimuli that possess many different yet related attributes, each of which may provide the plant cell with quite different information. For example, low temperature may immediately result in mechanical constrains, changes in activities of macromolecules, and reduced osmotic potential in the cellular milleu. High salinity includes both an ionic (chemical) and an osmotic (physical) component. The multiplicity of information embedded in abiotic stress signals underlies one aspects of the complexity of stress signaling.

On the basis of this multiplicity, it is unlikely that there is only one sensor that perceives the stress condition and controls all subsequent signaling. Rather, a signal sensor might only regulate branches of the signaling cascade that are initiated by one aspect of the stress condition. For example, low temperature is known to change membrane fluidity . A sensor detecting this change would initiate a signaling cascade responsive to membrane fluidity but would not necessarily control signaling initiated by an intracellular protein whose conformation/activity is directly altered by low temperature. Thus, there may be multiple primary sensors that perceive the initial stress signal.

Secondary signals (i.e. hormones and second messengers) can initiate another cascade of signaling events, which can differ from the primary signaling in time (i.e., lag behind) and in space (e.g., the signals may diffuse within or among cells, and their receptors may be in different subcellular locations from the primary sensors). These secondary signals may also differ in specificity from primary stimuli, may be shared by different stress pathways, and may underlie the interaction among signaling pathways for different stresses and stress protection. Therefore, one primary stress condition may activate multiple signaling pathways differing in time, space, and outputs. These pathways may connect or interact with one another using shared components generating intertwined networks.

19.4 Potential Sensors for Abiotic Stress Signals

Given the multiplicity of stress signals, many different sensors are expected, although none have been confirmed for cold, drought, or salinity. All three stresses have been shown to induce transient Ca^{+2} influx may represent one type of sensor for these stress signals. The activation of certain Ca^{+2} channels by cold may result from physical alteration in cellular structures. This phenomenon was demonstrated in studies showing that cold induced Ca^{+2} influx in plants occurs only following a rapid temperature drop , and that membrane fluidity and cytoskeletal reorganization are involved in early cold signaling .

Another type of membrane protein sensor for low temperature perception could be a two component histidine kinase. Evidence suggests that the *cynobacterium* histidine kinase Hik33 and the *Bacillus subtilis* histidine kinase Desk are thermosensors that regulate desaturase gene expression in response to temperature downshifts. In the genome of *Arabidopsis thaliana,* several putative two component histidine kinases have been identified , although no evidence has been reported for any of these histidine kinases as thermosensors.

In plants, cold, drought, and salt stresses all stimulate the accumulation of compatible osmolytes and antioxidants . In yeast and in animals, mitogenactivated protein kinase (MAPK) pathways are responsible for the production of compatible osmolytes and antioxidants. These MAPK pathways are activated by receptor/ sensors such as protein tyrosine kinases, G-protein-coupled receptors, and two component histidine kinases. Among these receptor type proteins, histidine kinase has been unambiguously identified in plants. An Arabidopsis histidine kinase, AtHK1, can complement mutation in the yeast two component histidine kinase sensors may be involved in osmotic stress signal transduction in plants Understanding the in vivo function of AtHK1 and other putative histidine kinases and their relationship to osmotic stress activated MAPK pathways will certainly shed light on osmotic stress signal transduction.

Pathways leading to the activation of the embryogenesis abundant (LEA)-type genes including the dehydration-responsive element (DRE)/C-repeat (CRT) class of stress responsive genes may be different from the pathways regulating osmolytes production. The activation of LEA-type genes may actually represent damage repair pathways . Because the activity of phospholipids C in plants might be regulated by G proteins, and phosphoinositols modulate the expression of these LEA like genes under cold, drought and salt stress, the protein associated receptors may exist and function in the perception of a secondary signal derived from these stresses. . G-protein associated receptors might also serve as one kind of membrane bound receptors for ABA.

19.5 Intracellular Secondary Signal Molecules

One early response to low temperature, drought and salinity stress in plant cells is a transient increase in cytosolic Ca^{+2}, derived from either influx from the apoplastic space or release from internal ca^{+2} release which is controlled by ligand sensitive ca2- channels. These ligands are second messengers that have been described in animal cells including, for example, inositol polyphosphates, cyclic ADP ribose, and nicotinic acid adenine dinucleotide phosphate. These molecules have all been found to be able to induce Ca^{+2} release in plant cells and in particular, guard cells (reviewed by Schroeder *et al.* 2001). An important feature of the role of Ca^{+2} as a signal is the presence of repetitive Ca^{+2} transients. These transients may be generated both by first round second messengers and by signaling molecules such as ABA that may themselves be produced as a result of cascade of early Ca^{+2} signals. These rounds of Ca^{+2} signals may have quite different signaling consequences and therefore physiological meaning.

1. Phospholipids

As the selective barrier between living cells and their environments, the plasma membrane plays a key role in the perception and transmission of external information. Upon osmotic stress, changes in phospholipids composition are detected in plants as well as in other organisms . However, during exposure to stress, the major role of phospholipids, the backbone of cellular membranes, may be to serve as precursors for the generation of second messenger molecules. Where as the relative cleaving enzymes are the phospholipase A2, C and D, the most studied is the phosphoinositide specific phospholipase C (PI-PLC). PI-PLC hydrolyzes phosphotidyinositol 4, 5-bisphosphate (PIP2) upon activation. PIP2 itself is a signal and may be involved in several processes, such as the recruitment of signaling complexes to specific membrane locations and their assembly hydrolysis of PIP2 I in animal cells has been shown to desensitize a G protein stimulate K+ current . Thus PIP2 could directly affect cellular ion homeostasis. During osmotic stress, plant cells may increase the production of PIP2 by upregulating the expression of PI5K , a gene that encode phosphotidylinositol 4-phosphate 5-kinase functioning in the production of PIP2. Consistent with this observation, osmotic stress was found to rapidly increase PIP2 levels in cultured Arabidopsis cells . Drought or salt stress also upregulates the mRNA levels of certain PI-PLC iso forms . This increase in PI-PLC expression could contribute to increased cleavage of PIP2 to produce two important molecules, diacylglycerol and inositol 1, 4, 5-trisphosphate (IP3). Diacylglycerol and IP3 are second messengers that can activate protein kinase C and trigger Ca2+ release, respectively.

In plants, the role of exogenous IP3 in releasing Ca^{+2} from cellular stores has been widely reported . Transient increases in IP3 were found in plants upon exposure to light, pathogen, gravity, anoxia, or several plant hormones . IP3 levels increase in *Arabidopsis* plants under salt stress and the time frame for the increase correlates with changes in cytosolic Ca^{+2} levels . Transient increase in IP3 levels were also observed in plant tissues or cultured cells during salt stress . Inhibition of PI-PLC activity eliminated transient IP3 increases and inhibited the osmotic stress induction of the stress responsive genes RD29A and COR47 . The stress hormone ABA also elicits transient increase in IP3 levels in *Vicia faba* guard cell protoplasts and in Arabidopsis seedlings.

Given the critical role of IP3 in signaling, cellular IP3 levels must be tightly regulated through both controlled production and degradation. Biochemical studies suggest that in animal cells, IP3 is degraded through either an inositol polyphosphate 3-kinase pathway or an inositol polyphosphate 5-phosphatase (ins5pase) pathway, resulting in the generation of inositol 1,3,4,5-tetraphosphate and inositol 1,4-bisphophate [ins(1,4) P2], respectively. However, information regarding the turn over of IP3 in plants is limited.

Accumulating evidence suggests that phospholipase D (PLD) is also involved in the transduction of stress signals. PLD hydrolyzes phospholipids to generate phosphatidic acid (PA), another second messenger in animal cells that can activate PI-PLC and protein kinase. PA may also serve as a messenger in plants . In guar cell protoplasts, PLD activity mediates ABA induced stomatal closure . Drought and hyperosmolarity activate PLD and lead to transient increase in PA levels in plants . PLD appears to be activated by osmotic stress through a G protein independently of ABA. However, excess PLD activity may have a negative impact on plant stress tolerance. PA is a nonbilayer lipid favoring hexagonal phase formation and many destabilize membrane at high concentrations . Drought stress induced PLD activities were found to be higher in drought sensitive than in drought tolerant cultivars of cowpea , suggesting that a high PLD activity may jeopardize membrane integrity. Consistent with this notion, Arabidopsis deficient in PLDα was found to be more tolerant to freezing stress.

2. ROS

Drought, salt and cold stress all induce the accumulation of ROS such as superoxide, hydrogen peroxide, and hydroxyl radicals . Thee ROS may be signals inducing ROS scavengers and other protective mechanisms, as well as damaging agents contributing to stress injury in plants . Because ABA was shown to induce H_2O_2 production , ROS may be immediate signal for ABA in mediating Catalase 1 gene (CAT1) expression, thermotolerance, activation of Ca2 channels in guard cell, stomatal closure and even ABA biosynthesis While it is possible that ROS may activate downstream signal cascades via Ca^{+2}, it is also possible that they can be sensed directly by key signaling proteins such as a tyrosine phosphatase through oxidation of conserved cysteine residues . In animal cells, reduced tyrosine phosphatase activity causes an increase in the output of MAPK pathways because tyrosine phosphatase inhibits MAPKs through dephosphorylation .

It is clear that ROS contribute to stress damage, as evidenced by observation that transgenic plants over expressing ROS scavengers or mutants with higher ROS scavenging ability show increased tolerance to environmental stresses . Whereas the connections between ROS signal transduction and osmotic stress signal transduction are just beginning to emerge the involvement of ROS in pathogenesis signal transduction is well documented . In hypersensitive responses, SA is thought to potentiate ROS signaling . Although it is unclear whether osmotic stress leads to an increased SA level in plants, the observation is that osmotic stress and SA activate the same MAPK.

3. Ca^{+2} Coupled Phosphoprotein Cascade

Transient increase in cytosolic Ca2- is perceived by various Ca^{+2} binding proteins. In the case of abiotic stress signaling, evidence suggests that Ca^{+2} dependent protein kinase (CDPKs) and the SOS3 family of Ca^{+2} sensors are major players in coupling this universal inorganic signal to specific protein phosphorylation cascades. CDPKs are serine/threonine protein kinases with a C-terminal calmodulin like domain that can directly bind Ca^{+2}. Indeed, CDPKs from rice (OsCPK2) and zucchini (CpCPK1) were shown to be membrane fractions . The Arabidopsis genome encodes at least 34 putative CDPKs. A number of studies have shown that CDPKs are induced or activated by abiotic stresses, suggesting that they may be involved in abiotic stress signaling . In rice plants, a membrane associated CDPK was activated by cold treatment . In addition, over expression of OsCDPK7 resulted in increased cold and osmotic stress tolerance in rice . Thus, CDPKs somehow play roles in the development of stress tolerance. A clear demonstration of the involvement of CDPK in stress signal transduction has come from experiments in which an active A1CDPK1 induced the expression of the stress responsive HVA1 promoter driven reporter gene in maize leaf protoplasts . Interestingly, a protein phosphatase type 2C (AtPP2CA) can block AtCDPK1 activation of the HVA driven reporter gene expression . It is unclear whether AtPP2CA acts directly on AtCDPK1 or modulates a downstream phosphorylation cascade.

19.6. Dependence of Stress Signaling on ABA

Salt, drought and some extent, cold stress cause an increased biosynthesis and accumulation of ABA, which can be rapidly catabolized following the relief of stress . Many stress responsive genes are upregulated by ABA . The role of ABA in Osmotic stress signal transduction was previously addressed by studying the stress induction of several of these genes in the Arabidopsis ABA deficient mutant aba1-1 and dominant ABA insensitive mutants abl1-1 and all2-1. A general conclusion from these studies was that whereas low temperature regulated gene expression is relatively independent of ABA, osmotic stress regulated genes activated through both ABA dependent and ABA independent pathways . However re cent genetic evidence suggests that stress signaling pathways for the activation of LEA like genes that are completely independent of ABA may not exist.

19.7 Regulation of ABA Biosynthesis Genes

Increased ABA levels under drought and salt stress are mainly achieved by the induction of genes coding for enzymes that catalyze ABA biosynthesis reactions. The ABA biosynthesis pathway in higher plants is understood to a great extent

ZEP (encode by ABA1 in Arabidopsis and ABA2 in tobacco; catalyzes the epoxidation of zeaxanthin and antheraxan to violaxanthin . The 9-*cis*-epoxycarotenoid dioxygenase (NCED), which catalyzes the oxidative cleavage of AAO3, and MCSU) is likely to be regulated through a common cascade that is Ca2+ dependent .

19.8 Categorized Stress Signaling Pathways: Outputs, Specificity and Interactions

Many signal transduction process occur when plants are challenged with environmental stresses. However, there has been no consensus for how to categorize these many signaling events. On the basis of the above discussion on the major signaling processes, we think that the signal transduction networks for cold, drought and salt stress can be divided into three major signaling type: (I) osmotic/oxidative stress signaling that makes use of MAPK modules, (II) Ca2+ dependent signaling that lead to the activation of LEA type gene (such as the DRE/CRT class of genes), and (III) Ca2+ dependent SOS signaling that regulates ion homeostatis. Type I signaling may contribute to the production of compatible osmolytes and antioxidans, and may also relate to cell cycle regulation under osmotic stress. Type II signaling leads to the activation of the DRE/CRT class and other types of LEA like genes, and is most extensively studied. Type III signaling appears to be relatively specific for the ionic aspect of salt stress. Target of this type of signaling are ion transporters that control ion homeostasis under salt stress. The *SOS* mutant (*sos*3, *sos*2 and *sos*1) fall into this category. These mutants are hypersensitive to salt stress, but activation of the DRE/CRT class of genes is unchanged in them . In addition, salt induced accumulation of the compatible osmolytes praline was not reduced but rather was enhanced in the *SOS* mutant . The enhanced praline production represents a compensatory response likely triggered by reduced salt tolerance in the mutants. Besides these major signaling routes, some additional pathways also exist, as discussed earlier.

One important issue regarding various stress signal transduction pathways is their specificities with respect to the input stimuli. The specificity and interaction between pathways have been addressed explicitly As discussed before, each of the stress conditions (i.e. cold, drought and high salinity) has more than one attribute. If two stress conditions have a common attribute (for example, hyperosmotic stress for drought and salinity), then the signaling arising from this common attribute might no be specific for either of the stress conditions. Additionally, it is important to distinguish the particular pathways when signaling specificity is considered. Interaction among these three signaling types is no extensive, as evidenced by the signaling types and by the results from additional transgenic studies discussed above . Likewise, although both drought and salt stress result in a transient increase in cytosolic Ca2+, Drought stress does not appear to activate the SOS pathway. It is possible that these different stresses have different Ca2+ signatures that could be decoded by their respective Ca2+ sensors. Specific Ca2+ oscillations in guard cells in the regulation of stomatal movement have been reported when a signaling component is overexpressed of ecotopically expressed, unnatural interactions among the different signaling pathways may occur.

General Glossary for Plant Sciences

A horizon : Mineral horizons which formed at the surface or below an o horizon and which exhibit obliteration of all, or much, of the original rock structure and are: (1) characterized by an accumulation of humified organic matter intimately mixed with the mineral fraction and not dominated by properties characteristic of e or b horizons; or (2) have properties resulting from cultivation, pasturing, or similar kinds of disturbance. (sssa).

Abaxial : Directed away from the stem of a plant; pertaining to the lower surface of a leaf (see adaxial).

Abiotic : Pertaining to the absence of life, as diseases not caused by living organisms.

Abscise : To separate from a plant, as leaves, flowers, and fruits do when they fall.

Abscission : The shedding of leaves or other plant parts as the result of physical weakness in a specialized layer of cells (the abscission layer) that develops at the base of the structure.

Absorption : The process of moving a substance from the exterior of a plant to the internal portion.

Acc oxidase : The enzyme in the ethylene biosynthesis pathway converting acc (1aminocyclopropane1carboxylic acid) to ethylene.

Acc synthase : The enzyme in the ethylene biosynthesis pathway converting sam (sadenosylmethionine) to acc (1aminocyclopropane1carboxylic acid).

Accase : Acetylcoenzyme A carboxylase; enzyme responsible for production of fatty acids and is the target of many herbicides that have the lipid synthesis inhibitor mode of action.

Acceptable daily intake (adi) : A measure of the level of a chemical, i.e. A pesticide residue, that is believed to be able to be consumed on a daily basis over a lifetime without harm; used by the u.s. Government to establish safe levels of potentially hazardous substances in food.

Acervulus (pl. Acervuli) : Erumpent, cushionlike fruiting body bearing conidiophores, conidia, and sometimes setae.

Acid : Having a ph of less than. 7 OR A chemical compound that can donate a proton.

Acid precipitation : Precipitation (fog, rain, snow) with a low ph, due to the presence of nitric and sulfuric acid formed by the reaction of air pollutants (n0x and s02) with water.

Acid rain : Precipitation with a low ph, due to the presence of nitric and sulfuric acid formed by the combination of air pollutants (n0x and s02) with water.

Acidic : Having a ph lower than 7, having a higher concentration of H+ ions.

Acropetal : Upward from the base to the apex of a shoot of a plant; in fungi, the production of spores in succession in the direction of the apex so that the apical spore is the youngest (see basipetal).

Acrylamide : Chemical with the basic formula C3H5NO that can be chemically induced to form a gel matrix. This matix can be used for electrophoretic separation of proteins and nucleic acids.

Actinomycete : A group of microorganisms similar to bacteria that produce long filaments.

Action site : A location in dense vegetation, that will collect water droplets and ensure a high degree of moth survival. The area where mating between male and female moths occurs. OR The location on an enzyme (protein) to which substrate molecules bind and are converted to their products. The shape of this site must be maintained for the enzyme to remain functional.

Acute: Pertaining to symptoms that develop suddenly(see chronic).

Acute toxicity: Ability of a single dose of a compound to poison (see chronic toxicity).

Adaxial: Directed toward the stem of a plant; pertaining to the upper surface of a leaf (see abaxial).

Addition : The movement into or deposition of a quantity of material.

Additive : Adjuvant; any of several categories of compounds that can be added to a herbicide solution to enhance its activity. Some examples are oils, surfactants and fertilizers.

Adhesion : A molecular attraction between a surface and solvents and solutes in contact with it.

Adjuvant : An additive. Any of several categories of compounds that can be added to a herbicide solution to enhance its activity; some examples are oils, surfactants and fertilizers.

ADPG pyrophosphorylase : Catalyzes ATP + α-D-glucose 1-phosphate $\rightarrow$ pyrophosphate + ADP-glucose.

Adpressed (syn. Appressed) : Closely flattened down or pressed against a surface.

Adsorption : The process of a substance binding to a surface.

Adult : A fully grown, mature organism.

Adventitious : Arising from other than the usual place, as roots from a stem rather than as branches of a root.

Aeciospore : Dikaryotic spore of a rust fungus produced in an aecium; in heteroecious rusts, a spore stage that infects the alternate host.

Aecium (pl. Aecia; adj. Aecial) : The fruiting body of a rust fungus in which the first dikaryotic spores (aeciospores) are produced.

Aeration : The condition and sum of all processes affecting soil porespace gaseous composition, particularly with respect to the amount and availability of oxygen for use by soil biota and/or soil chemical oxidation reactions.

Aerial : Occurring in the air.

Aerification: The act of infusing or forcing air into, for example, soil.

Aerobic: Living only in the presence of oxygen. OR Condition that has ample amount of oxygen. This condition will allow organisms that require molecular oxygen to growth and function well.

Aflatoxin : Chemical byproduct from aspergillus flavus and a. Parasiticus harmful to humans and other animals.

AFLP : Amplified fragment length polymorphism. A molecular marker method consisting of (1) DNA cleavage with two restriction enzymes, (2) ligation to synthetic linkers, and (3) two pcr amplification steps using primers 16 to 17 bases in length. The technique results in a large number of DNA bands per analysis (50 to 100 are commonly observed), and thus it can be an efficient way of detecting polymorphic loci. However, the large number of bands can be difficult to score. Aflps are usually dominant and require no prior genome information. Because aflp bands are "anonymous" (i.e., their chromosome positions are not known in advance), rflp or ssr markers are sometimes evaluated in the same population in order to "anchor" the aflps to specific chromosome arms. Aflps can be detected either with radioactive label or silver staining.

Agar: Jellylike material derived from algae and used to solidify liquid culture media; term also applied to the medium itself.

Agarose : A carbohydrate powder purified from sea weed that can be boiled and cooled into a semisolid matrix.

Aggregates : Soil particles glued or cemented together. The agents responsible for cementing particles include organic matter and calcium carbonates.

Aggressiveness: Relative ability of a plant pathogen to colonize and cause damage to plants (see virulence).

Agrobacteria : A soil bacteria that can be a plant pathogen. It works as a natural genetic engineer by invading a plant through wounds and inserting a piece of DNA into the plant's chromosomes. The plant will then begin producing the protein encoded by the inserted gene. Bacillus thuringiensis (Bt) is a strain of agrobacteria.

Agrobacterium : A soil bacteria that can be a plant pathogen. It works as a natural genetic engineer by invading a plant through wounds and inserting a piece of DNA into the plant's chromosomes. The plant will then begin producing the protein encoded by the inserted gene. Bacillus thuringiensis (Bt) is a strain of agrobacteria.

Air pollution : Any air contaminant that causes undesirable effects on living organisms or materials.

Airfilled porosity : The percent of air occupying the empty spaces of soil and is inversely related to the percent of water occupying the empty spaces of soil. Total porosity= air filled porosity plus water filled porosity.

Alate : Winged (see apterous).

Albino (n. Albinism) White or light-colored; having a marked deficiency in pigmentation.

Aldose : A sugar with a terminal aldehyde.

Algorithm : A mathematical equation used to make a prediction, for example, predicting fertilizer requirement from soil analysis and anticipated crop yield.

Alkaline : Having basic (nonacidic) properties; ph greater than 7.

Alkaloids Nitrogen-containing ring compounds produced by plants that cause physiological effects in animals.

Allele Any of one or more alternative forms of a gene. OR One of the different forms of a gene (or marker) that can exist at a single locus. A single allele for each locus is inherited separately from each parent.

Allelic : An allele is one of a pair or series of forms of a gene possible. An example of allelic variation would be a dominant vs recessive allele for one gene.

Allelopathy (adj. Allelopathic) Ability of one species to inhibit or prevent the growth of another species through the production of toxic substance(s).

Allergen : A biological or chemical substance that elicits a reaction involving the body's immune system.

Alluvium : Sediments deposited by the running water of streams or rivers. Alluvium may occur on terraces well above present streams, on the present flood plains or deltas, or as a fan at the base of a slope. (SSSA).

Alsahas : Acetolactate synthase or acetohydroxy acid synthase; an enzyme essential for amino acid production and is the target of many herbicides including the imidazolinones and sulfonylureas; this enzyme has mutated in some species to become herbicide resistant.

Alternate host : One of two kinds of plant on which a parasitic fungus (e.g. Rust) must develop to complete its life cycle.

Alternation of generations: A reproductive cycle in which a haploid phase alternates with a diploid phase.

Alternative host: A plant other than the main host that a parasite can colonize; alternative hosts are not required for completion of the developmental cycle of the parasite.

Aluminum oxides : Group name for the oxides and hydroxides of aluminum. Includes the minerals gibbsite, bauxite.

Aminization : The process by which microorganisms break down complex proteins to simpler amino acids, amides, and amines. Aminization is the first step of mineralization as part of the nitrogen cycle.

Amino acid: Organic nitrogenous acids from which protein molecules are constructed. OR The basic building blocks of proteins. The sequence of amino acids is a protein and protein function are determined by the genetic code. It is an organic molecule containing a carboxyl group and an amino group.

Amino acid Synthesis Inhibitors : A mode of action of herbicides that inhibit the production of essential amino acids.

Ammonia : One of the mineral forms of nitrogen in the gas form NH_3.

Ammonia volatilization : The process by which ammonia is lost to the atmosphere. Volatilization occurs when ammonium in the soil, because of its ph, is converted to ammonia, which dissipates as a gas.

Ammonification : The release of ammonium (NH4+)ions from organic matter via microbial action (more broadly termed mineralization).

Ammonium : One of the mineral forms of nitrogen. NH_4^+

Ammonium nitrate sulfate : A dry nitrogen fertilizer (3000 + 6.5s) manufactured by reacting anhydrous ammonia with a mixture of nitric acid and sulfuric acid; good storage and handling characteristics.

Ammonium phosphate : A material produced by reacting ammonia with phosphoric acid or a mixture of phosphoric and either sulfuric or nitric acids.

Ammonium phosphate is the leading nitrogenphosphorus product in the fertilizer industry; high analysis, high water solubility, good physical characteristics and low production costs.

Ammonium sulfate : A dry crystalline material (2100 +24s) produced by reacting anhydrous ammonia with sulfuric acid; stores well. An excellent choice when both nitrogen and sulfur are needed.

Ammonium thiosulfate solution (ats)(1200+26s) : The most commonly used source of sulfur in fluid fertilizer; weighs 11.5 lbs./gal. After application to the soil, thiosulfate is decomposed to form approximately equal amounts of sulfate sulfur and elemental sulfur.

Ammoniumnitrate : A dry granular material manufactured by reacting nitric acid with anhydrous ammonia. One half of the nitrogen in ammonium nitrate is in the nitrate form, and one half is in the ammonium form.

Amorphous wax : Smooth or glasslike wax occurring on leaf surfaces and is generally easy to wet.

Amphid : A chemosensory organ located in the anterior region of a nematode.

Amphigynous : Having an antheridium through which the oogonium grows, as in many phytophthora species (see paragynous).

Amphimixis Copulation of two unrelated cells and nuclei, e.g., egg and sperm; reproduction by a sexual process.

Amphimobile: Able to move both upward and downward in phloem, as do some pesticides.

Amplification : An increase in the number of copies of a specific DNA fragment.

Amylopectin : The constituent of starch having a polymeric, branched structure in which an α-d-1,6 bond occurs every 20–30 glucose units linked via α-d-1,4 bonds.

Amylose : The constituent of starch in which α-d-1,4 glucosidic bonds form linear chains of 200–2000 glucose units.

Anabolic : Synthetic processes resulting in the production of complex products.

Anaerobic Anamorph (adj. Anamorphic; syn. Imperfect state): Living in the absence of oxygen. The asexual form in the life cycle of a fungus, when asexual spores (such as conidia) or no spores are produced (see holomorph and teleomorph).

Anaerobic : Condition that is lacking molecular oxygen. Under this condition, organisms that do not require molecular oxygen will survive and funtion well.

Anastomosis (pl. Anastomoses) : Fusion between branches of the same or different structures (e.g. Hyphae) to make a network.

Angiosperm : A plant which flowers and forms fruits with seeds; seeds are enclosed in a pericarp.

Anhydrous ammonia : A gaseous material that is compressed and stored as a liquid. At 60of, a gallon of anhydrous ammonia weighs 5.15 pounds. Due to the fact anhydrous ammonia needs to go through the nitrification process, it is more resistant to losses from the soil by leaching or denitrification because it is converted by bacterial action to the nitrate form more slowly than are other nitrogen sources.

Anion : A negatively charged ion. OR Having a negative charge.

Annual A plant that completes its life cycle and dies within one year (see biennial, perennial).

Annual ring : Single-year growth of xylem in a woody stem.

Annulations : Grooved bands in the cuticle of some nematodes.

Anomer : Epimers that differ only in the configuration at the carbonyl carbon -(c=o)- of aldoses or ketoses.

Antagonism : A general term for interference between organisms that may include antibiosis or competition for nutrients or space; action of two or more pesticides that reduces the effectiveness of one or all (see synergism). OR Reduced activity of one or more herbicides when applied together compared to applying them individually.

Antagonist: An organism or substance that limits or counteracts the action of another. OR A nonphytotoxic chemical that can protect the crop from herbicide injury without reducing the damage imposed on the weed; the mechanism of safening can involve induction of enzymes which metabolize the herbicide in the protected crop; also known as safeners, antidotes or crop protectants.

Antagonize : To bother.

Antena : A pigment with the primary function of capturing the energy from photons and transferring that energy to other pigments within the photosystem. Most chlorophyll molecules function as antennae, with relatively few of the hundreds of chlorophyll molecules carrying out photochemistry in the reaction center. Carotenoids also function as antennae, but additionally play an important role in quenching triplet chlorophyll and singlet oxygen molecules that would cause damage to the plant cell.

Anterior: Toward the front or head (see posterior).

Anther : Pollen-bearing portion of a flower.

Antheridium (pl. Antheridia) : Male sexual organ (male gametangium) found in some fungi.

Anthers : The part of the stamen (part of the male reproductive structure) that develops and holds pollen; supported by the filament.

Anthesis : The period of the opening of a flower during which pollination can occur.

Anthracnose : Disease caused by acervuli-forming fungi (order melanconiales) and characterized by sunken lesions and necrosis.

Anthropogenic : The effects of human influence/activities on nature.

Antibiosis : An association between organisms, or between an organism and a metabolic product of another organism, that is harmful to one of them.

Antibiotic : A chemical compound produced by one microorganism that inhibits growth or kills other living organisms.

Antibody : A specific protein formed in the blood of warm-blooded animals in response to the presence of an antigen. OR Immune system protein made by mammals to specifically bind to foreign molecules allowing the mammal to fight off disease.

Antigen : Any foreign chemical (normally a protein) that induces antibody formation in warm-blooded animals. OR A molecule which trigger's an immune system to create an antibody to bind to it.

Antiserum (pl. Antisera) : Blood serum containing antibodies.

Aonocots : A subclass of Angiosperm plants based on anatomical characteristics. They tend to have:

Narrow leaves
One cotyledon
Parallel veins in the leaves
Flower parts are usually in multiples of threes
A scattered arrangement of primary vascular bundles in the stem
Fibrous root system
Ex. Grass, daylilies, corn

Apex (pl. Apices; adj. Apical) : Tip of a root or shoot, containing the apical meristem.

Aphid : Small, sucking insect of the family aphididae (order homoptera) that produces honeydew and injures plants when in large populations.

Apoplastic path : Water moves in the cell wall or nonliving region between cells without crossing any membranes. Apoplastic means 'nonliving' tissue.

Apothecium (pl. Apothecia) : Open, cuplike or saucerlike, ascus-bearing fungal fruiting body (ascocarp), often supported on a stalk.

Appressed (syn. Adpressed) : Closely flattened down or pressed against a surface.

Appressorium (pl. Appressoria) : Swollen, flattened portion of a fungal filament that adheres to the surface of a higher plant, providing anchorage for invasion by a fungus.

Apterous : Wingless (see alate).

Aqua ammonia : Anyhdrous ammonia dissolved in water. It is a low pressure solution and contains free ammonia, the amount of which increases as air temperature increases.

Arable : Able to be cultivated for agriculture.

Arbuscular mycorrhiza (abbr. Am; syn. Endomycorrhiza) : Symbiotic association between a nonpathogenic or weakly pathogenic fungus and the roots of plants in which fungal hyphae invade cortical cells of the root and produce vesicles and arbuscles (see ectomycorrhiza).

Arbuscule (adj. Arbuscular) : Branched haustorial structure of certain endomycorrhizal fungi that forms within living cells of the root

Aromatic amino acids : Amino acids whose side chains chemical structures contain an aromatic ring. Examples include phenylalanine, tyrosine and tryptophan.

Arthropod : Member of the phylum arthropoda, which consists of animals with articulated bodies and limbs and which includes insects, arachnids, and crustaceans.

Ascocarp (syn. Ascoma) : Sexual fruiting body of an ascomycetous fungus that produces asci and ascospores; e.g. Apothecium, ascostroma, cleistothecium, perithecium, pseudothecium.

Ascogenous Pertaining to ascus-producing hyphae

Ascogonium (pl. Ascogonia) A specialized cell that gives rise to the hyphae that produce asci.

Ascoma (pl. Ascomata; syn. Ascocarp) Sexual fruiting body of an ascomycetous fungus that produces asci and ascospores; e.g. Apothecium, ascostroma, cleistothecium, perithecium, pseudothecium.

Ascomycetes (adj. Ascomycetous) A group of fungi, also called the ascomycota, that produces sexual spores (ascospores) within a saclike structure called an ascus.

Ascospore Sexual spore borne in an ascus.

Ascostroma (pl. Ascostromata) A fruiting body containing bitunicate (double-walled) asci in locules (cavities); usually dark with multiple locules, but sometimes single (see pseudothecium)

Ascus (pl. Asci) Saclike structure containing ascospores (typically eight) and usually borne in a fungal fruiting body.

Aseptate Having no cross walls; lacking septa; nonseptate; coenocytic.

Asexual : Vegetative; without sex organs, gametes, or sexual spores; the imperfect or anamorphic stage of a fungus.

Asexual reproduction : Any type of reproduction not involving the union of gametes and meiosis

Aspect : The predominant direction of slope of the land.

Atp : Adenosine 5'triphosphate .

Audpc (abbr. For area under disease progress curve) : A measure of the total amount of disease over a period of time, determined from graphs of disease vs. Time, which can be used to compare epidemics quantitatively.

Autoclaving : A treatment of solutions, tools and containers with high temperature (usually at 120oc) under high pressure (usually at 1.05 kg/cm2) to eliminate organic contaminations.

Autoecious : In reference to rust fungi, producing all spore forms on one species of host plant (see heteroecious).

Autotroph : An organism that synthesizes its nutritive substances from inorganic molecules; e.g. Plants capable of photosynthesis (see heterotroph).

Autotroph : An organism that can use CO2 as its principle carbon source.

Auxin : Plant hormone (growth regulator) influencing growth through cell elongation. OR Indole3acetic acid (IAA); the first plant hormone discovered; a plant hormone involved in many plant growth processes including embryo development, root and flower development, vascular differentiation, stem elongation, apical dominance and tropic response.

Auxin Herbicde : The first selective organic herbicides developed. This family of herbicides is described as synthetic auxins, growth regulators with herbicidal action, or herbicides with growth regulatoryactivity.

Available water : (Capacity) The amount of water released between in situ field capacity and the permanent wilting point (usually estimated by water content at soil matric potential of –1.5 Mpa). Available water is not the portion of water that can be absorbed by plant roots, which is plant specific. (Soil Science Society of America, 1997, p. 9).

Avirulence (avr) gene : Gene in a pathogen that usually causes a hypersensitive reaction, is associated with active plant defense reactions in a resistant plant, and causes disease in a susceptible plant.

Avirulent (syn. Nonpathogenic) : Unable to cause disease (see virulent).

Avoidance : Principle of plant disease control in which plants are grown at times or locations where the pathogen is inactive or not present.

Awn : Bristlelike structure at the apex of the outer bract of some cereal and grass flowers.

Axenic : Culture in the absence of living bacteria or other organisms; pure culture.

Axil : The angle formed by the leaf petiole and the stem.

Axillary : Pertaining to or placed within an axil.

Axillary bud (syn. Lateral bud) : Bud that develops in the axil of a leaf.

B horizon : Horizons that formed below an a, e, or o horizon and are dominated by obliteration of all or much of the original rock structure and show one or more of the following:

1. Illuvial concentration of silicate clay, iron, aluminum, humus, carbonates, gypsum, or silica, alone or in combination;
2. Evidence of removal of carbonates;
3. Residual concentration of sesquioxides;
4. Coatings of sesquioxides that make the horizon conspicuously lower in value, higher in chroma, or redder in hue than overlying and underlying horizons without apparent illuviation of iron;
5. Alteration that forms silicate clay or liberates oxides or both and that forms granular, blocky, or prismatic structure if volume changes accompany changes in moisture content; or
6. Brittleness. (sssa)

Bacilliform : Shaped like short rods with rounded ends.

Bacillus thuringiensis : The soil bacteria which naturally produces the Bt protein toxic to European corn borer (ECB).

Backcross : To cross (mate) an offspring with one of its parents. OR A breeding method used to move one or a only a few desirable genes from an agronomically poor crop line to an elite line. This is done by crossing a donor parent to an elite line, and crossing offspring with the 'desired gene(s)' back to the elite parent.

Backcross breeding : After a plant has been transformed, the event must be moved into an elite genetic background. This is done by mating the transformed plant back to elite plants over several generations.

Backcrossing : A breeding method used to move one or only a few desirable genes from an agronomically poor crop line to an elite line. This is done by crossing a donor parent to an elite line, and crossing offspring with the 'desired gene(s)' back to the elite parent.

Bacterial streaming : Large populations of bacteria that exude from the cut surface of infected plant tissue when observed with a microscope.

Bactericide : A chemical or physical agent that kills bacteria.

Bacteriocin : A protein antibiotic, one or more types of which can be produced and excreted by certain strains of bacteria..

Bacteriophage : A virus that infects a bacterium.

Bacterium (pl. Bacteria) : A prokaryotic, microscopic, single-celled organism with a cell wall that increases by binary fission

Bar gene : Also called the pat gene; it is the bacterial gene that encodes the protein phosphinothricin (pat) which detoxifies the liberty herbicide molecule resulting in liberty resistance.

Bark : All tissues outside the vascular cambium of a woody plant, generally including the cork layers, cork cambium (phellogen), and phloem.

Basal knob (syn. Stylet knob) : Structure at the base of a nematode stylet.

Base : Having a ph of greater than 7.

Base : A chemical compound that accepts a proton.

Base pair : A pair of nucleotide bases in the double stranded DNA double helix structure. Base pairs form the 'rungs' on the DNA ladder. A single base pair is made when adenine from one DNA single strand binds with thymine in the other strand or when guanine from one strand binds with cytosine in the other.

Basic : Having a ph higher than 7, having a lower concentration of H+ ions.

Basidiocarp (syn. Basidioma) : Sexual fruiting body of a basidiomycetous fungus.

Basidioma (pl. Basidiomata; syn. Basidiocarp) : Sexual fruiting body of a basidiomycetous fungus.

Basidiomycetes (adj. Basidiomycetous) : A fungus group, also called the basidiomycota, characterized by the formation of external basidiospores on basidia.

Basidiospore : Haploid (1n) sexual spore produced on a basidium.

Basidium (pl. Basidia; adj. Basidial) : Specialized cell or organ, often club-shaped, in which karyogamy and meiosis occur, followed by production of externally-borne basidiospores (generally four) that are haploid. There are several types of basidia.

Basipetal : Downward from the apex toward the base of a shoot; referring to development in the direction of the base so that the apical part is oldest (see acropetal).

Basipetal : Unidirectional transport from the apex to the base.

Bc 1 generation : (backcross) 1 generation is the offspring resulting from the first cross back to the recurrent parent in the backcross breeding method.

Bearing strength : The average laod per unit area required to rupture a supporting soil mass. Measured in units of kg m2.

Beetle : Any insect of the order coleoptera characterized by elytra (thickened outer wings), chewing mouth parts, and complete metamorphosis.

Benzoic acids : Family of herbicides that have activity on synthetic auxins belonging to the growth regulators mode of action.

Bicellular : Two-celled.

Biennial : A plant that produces seed and dies at the end of its second year of growth (see annual, perennial).

Biflagellate : Having two flagella.

Biliabate : Divided into two upper lobes.

Binary fission : A type of asexual reproduction in which two cells, usually of similar size and shape, are formed by the growth and division of one cell.

Binomial, latin : The scientific name of an organism, composed of two names, the first designating the genus and second the specific epithet, together making the species name.

Binucleate : Having two nuclei.

Bioassay : Any test (assay) using a living organism.

Biocide : A compound toxic to all forms of life.

Biocontrol (syn. Biological control) : Use by humans of one species of organism to eliminate or control another species of organism.

Biocontrol (syn. Biological control) : Exploitation by humans of the natural competition, parasitism and/or antagonism of organisms for management of pests and pathogens.

Biolistic transformation (see gene gun) : A method used for genetic engineering in which plant cells are bombarded with metal particles coated with foreign genes using a "gene gun".

Biolistics : A method for introducing DNA into a target organism using highspeed particle bombardment.

Biological control (syn. Biocontrol) : Exploitation by humans of the natural competition, parasitism and/or antagonism of organisms for management of pests and pathogens.

Biomolecules : Molecules, usually organic molecules, found in cells and living organisms.

Biotechnology : The development of genetically modified organisms through the use of modern technology and processes, including genetic engineering.

Biotic : Relating to life, as disease caused by living organisms.

Biotic factors : Biological aspects of the environment, one of the conventionally accepted soil forming factors.

Biotransformation : Change in the herbicide's chemical structure within plant cells.

Biotroph (syn. Obligate parasite) : An organism that can live and multiply only on another living organism (see necrotroph).

Biotype : A subdivision of a species, subspecies, or race based on some identifiable physiological trait such as a specific virulence pattern.

Bitunicate : Having a double ascus wall (see unitunicate).

Blasting : A symptom of plant disease characterized by shedding of unopened buds; classically, the failure to produce fruit or seed.

Bleeding : Sap flow from a wound .

Blight : Sudden, severe, and extensive spotting, discoloration, wilting, or destruction of leaves, flowers, stems, or entire plants.

Blotch : Necrotic area of tissue irregular in form.

Bolting : Flower stalk elongation.

Borer : Insect or insect larva that forms tunnels or cavities in the bark or within the wood of trees.

Boundary layer : The calm layer of air on surfaces below the layer of more turbulent air. OR A region of still air near the surface of a leaf through which water must diffuse en route to the atmosphere.

Bract : A reduced leaf associated with a flower or inflorescence; modified leaf from the axil of which a flower arises.

Branchedchain amino acids : Amino acids whose side chain chemical structrue is branched. Examples include leucine, isoleucine and valine.

Breaking : Disease symptom, usually caused by a virus, involving addition or loss of flower color to create a variegated pattern. (see flower break).

Breeding line : Plant strain used in a plant breeding program and usually containing one or more desirable agronomic or breeding characteristics.

Broadcast application : Application by spreading or scattering on the soil surface.

Brooming : Profuse branching of woody stems from single stem position.

Brown rot (of wood) : Wood decay resulting from selective removal of cellulose and hemicellulose, leaving a brown amorphous residue that usually cracks into cubical blocks and consists largely of slightly modified lignin.

Bt corn : Corn that has been transformed with the bt gene and is resistant to european corn borer.

Bt corn hybrids : Corn hybrids that have been transformed with the bt gene and are resistant to european corn borer (ecb).

Bt gene : A gene originating from the bacillus thuringiensis soil bacteria that encodes a protein toxic to the european corn borer (ecb).

Bt protein : Another name for the cryla proteins. These proteins occur naturally in bacillus thuringiensis bacteria and are toxic to the european corn borer (ecb).

Bud : A terminal or axillary structure on a stem consisting of a small mass of meristematic tissue, generally covered wholly or in part by modified scale leaves.

Bud scale : A specialized protective leaf of a bud.

Bud wood : Wood consisting of strong, young shoots bearing buds suitable for use in budding.

Budding : A special type of plant grafting using a single bud on the rootstock of another plant; method of asexual reproduction in fungi, such as yeasts.

Buffer pH : A test used to estimate lime requirements in soil, which means the amount of agricultural lime required to neutralize the H+ from the soil exchange sites and in the soil solution.

Buffer strips : Vegetated areas designed to allow runoff to pass through the strips for filtration, sedimentation and adsorption of nutrients.

Bug : Any insect of the order hemiptera characterized in part by piercing-sucking mouth parts, a triangular scutellum, two pairs of wings, and gradual metamorphosis.

Bulb : A short, flattened, usually globose or disc-shaped, underground, perennial, storage organ composed of concentric layers of overlapping fleshy scale leaves attached to a stem plate at the base; essentially a subterranean bud.

Bulbils : Small bulb or bulblike structure found in the place of a flower or leaf axil of a plant; found on tigerlilies.

Bulbs : Fleshy, specialized underground stem modified for storage and surrounded by modified leaves called scales, i.e., tulips.

Bulk density : The dry weight of soil per unit volume of soil; most often expressed on a soil volume basis, rather than on a particle basis. Bulk density considers both the solids and the pore space; whereas, particle density considers only the mineral solids.

Bunt ball : Smut sorus filled, with teliospores, that replaces a cereal or grass kernel but is covered by plant tissue at maturity.

Burr knot : A rough outgrowth, often present on the trunks or roots of certain trees.

Bursa : Extension or flap of cuticle at the side of the male nematode sex organ, used for orienting during mating.

Bush-type : Plant with determinate growth.

C horizon : Horizons or layers, excluding hard bedrock, that are little affected by pedogenic processes and lack properties of o, a, e, or b horizons. (sssa)

C:n ratio : The amount of carbon in organic material divided by the amount of total nitrogen in the same organic material.

Calcium sulfate : A byproduct of manufacturing concentrated superphosphate; contains 1518 percent sulfur.

Calciumcalmodulin : A regulatory protein that has four highaffinity calcium binding sites. The calmodulincalcium complex can bind to and activate several enzymes.

Calibration (soil test) : The process of determining the crop nutrient requirement at different soil test values.

Callose : Amorphous, hardened carbohydrate constituent of plant cell walls, commonly developing upon injury.

Callus : Specialized tissues that form over a wound or cut in a plant; cork cambium may form and the cells produced will gradually seal the wound.

Callus : A mass of undifferentiated cells used in tissue culture.

Calyx : Outermost flower whorl; sepals, collectively. OR The sepals collectively, directly below the corolla.

Canker : A plant disease characterized (in woody plants) by the death of cambium tissue and loss and/or malformation of bark, or (in non-woody plants) by the formation of sharply delineated, dry, necrotic, localized lesions on the stem; "canker" may also be used to refer to the lesion itself, particularly in woody plants.

Canopy : The expanded leafy top of a plant or plants.

Capsid (syn. Coat protein) : The protective layer of protein surrounding the nucleic acid core of a virus; the protein molecules which make up this layer.

Capsule : Gel-like material surrounding a bacterial cell.

Carbohydrate : Any of various chemical compounds composed of carbon, hydrogen, and oxygen, such as sugars, starches, and cellulose.

Carbon cycle : The continuous circulation of carbon atoms from inorganic carbon dioxide (co2) to organic molecules and back to co2.

Carbon/nitrogen (c/n) ratio : The ratio of the mass of organic carbon to the mass of organic nitrogen in soil, organic material, plants, or microbial cells.

Carbonation : Reaction with carbonic acid.

Carbonic acid : A weak acid formed when water combines with carbon dioxide.

Carcinogen : A substance or agent that causes cancer.

Carotenoid : A plant pigment that protects chlorophyll from being photodegraded; enzymes that aid in the production of carotenoids are the target site of herbicides belonging to the pigment synthesis inhibitor mode of action.

Carpel : The ovule-bearing structure of a flower in angiosperms.

Casparian strip : A ribbonlike layer of waterrepellent suberin (waxy layer) that is wrapped around the tops, bottoms, and two side walls of endodermal cells in roots of plants.

Casting : Premature loss of abscised leaves or twigs.

Catabolic : Degradative processes that generate energy for the cell through oxidation of organic molecules.

Catalyst : A material that increases the rate of a chemical reaction without undergoing a chemical change.

Catalyzes : Speeds up.

Cation : A positivelycharged ion.

Cation exchange : The interchange between a cation in solution and another cation at the surface of a negatively charged material such as clay or organic matter.

Cation exchange capacity (cec) : The sum of exchangeable bases plus total soil acidity at a specific ph, values, usually 7.0 or 8.0. When acidity is expressed as salt extractable acidity, the cation exchange capacity is called the effective cation exchange capacity (ecec) because this is considered to be the cec of the exchanger at the native ph value. It is usually expressed in centimoles of charge per kilogram of exchanger (cmolckg1) or millimoles of charge per kilogram of exchanger.

Cation exchange capacity (cec) : The ability of a soil to hold certain elements which have a positive charge; mainly a function of clay content and organic matter. Some plant nutrients, such as potassium and caldium are cations, so a soil with higher cec will generally be more fertile because of its greater ability to hold these nutrients. Measured in milliequivalents/100 grams.

Cationic : Having positive charge.

Catkin : Conelike cluster of male or female flowers; conelike fruit of angiosperms.

Causal agent : Organism or agent that incites and governs disease or injury.

Cavitation : Spontaneous occurrence of a vapor phase in a liquid under tension; occurs in xylem of plants that are under moderate to severe waterdeficit stress.

Cavities : Tunnels left inside the corn stalk from European corn borer feeding.

Cell membrane : Structure that bounds a cell and helps control the movement of substances into and out of the cell.

Cell membrane disrupters : Mode of action of a group of herbicides that kill by destroying cell integrity.

Cell wall : Protective, resistant, but permeable structure secreted externally to the cell membrane in plants, bacteria, fungi, and certain other organisms. OR A thin, mechanically strong structure surrounding all plant cells consisting of a complex mixture of polysaccharides and other polymers that are secreted by the cell and are assembled into an organized network linked together by a mixture of covalent and noncovalent bonds. Cell walls regulate cell volume and determine cell shape.

Cellular path : Water moves from cell to cell using membranes, either across membranes or using plasmodesmata .

Cellulose : Carbohydrate polysaccharide composed of glucose units; major component of plant primary cell wall. OR A polysaccharide of glucose units that constitutes the chief part of the cell walls of plants.

Center of origin : Geographical area where a plant originated.

Centromere : The constricted portion of a chromosome to which, in mitosis, the chromosomal fiber is attached.

Certification : Seeds, propagative plant material, or nursery stock produced and sold under inspection to maintain genetic identity and purity, freedom from harmful pathogens, insect pests, and weed seeds. It is approved and certified by an official certifying agency.

Certified organic foods : "Organic" is a labeling term that denotes products produced under the authority of the organic foods production act. The term "organic" details the methods, practices and substances that can be used in producing and handling organic crops and livestock, as well as processed products. Organic agriculture practices cannot ensure that products are completely free of residues, however methods are used to minimize pollution from air, soil, and water. For more information contact the usda at www.usda.gov/news/release/2000/12/0425.htm.

Cfu (abbr. For colony forming unit) : The number of colonies formed per unit of volume or weight of a cell or spore suspension.

Channels : A transport protein in the membrane of cells that are selective pores, through which molecules can diffuse across the membrane.

Chemotaxis (syn. Chemotropism) : Movement or growth of an organism in response to changing concentration of a chemical stimulus, often in relation to food or for mating.

Chemotherapy : Treatment of plant disease with chemicals (e.g. Antibiotics or fungicides) absorbed and translocated internally.

Chimera (or chimaera) : Plant or organ consisting of two or more genetically different tissues.

Chimeric plant : A transformed plant that has the new gene in only some of its cells.

Chisel : A type of minimum tillage that breaks up the aggregates of the soil more than discing.

Chitin : Complex polysaccharide carbohydrate in fungal cell walls, animal exoskeletons, and nematode egg shells.

Chlamydospore : Thick-walled or double-walled asexual resting spore formed from hyphal cells (terminal or intercalary) or by transformation of conidial cells that can function as an overwintering stage.

Chloroacetamides : Family of herbicides that inhibit shoot inhibition and belong to the seedling growth inhibitor mode of action.

Chlorophyll (adj. Chlorophyllous) : One of a group of green pigments found in chloroplasts and important in photosynthesis.

Chloroplast : Disklike structure containing chlorophyll in which photosynthesis occurs in green plants.

Chloroplast : The organelle in plant cells that contains chlorophyll and is the site of photosynthesis.

Chlorosis (adj. Chlorotic) : Failure of chlorophyll development, caused by disease or a nutritional disturbance; fading of green plant color to light green, yellow, or white.

Chromatid : One of the two sidebyside replicas of a chromosome produced through DNA replication during mitosis/meiosis.

Chromosome : The structure that contains the genes of an organism; in eukaryotes, chromosomes are in the nucleus and can be visualized with an optical microscope as threads or rods during meiosis and mitosis; in bacteria, the chromosome is usually a single circle of DNA that cannot be visualized with an optical microscope. OR A unit of hereditary material composed of 1000's of genes. During cell division chromosomes duplicate and divide, maintaining the genetic integrity of the organism. OR A genetic structure in a cell composed of condensed DNA, which contains the genetic code for an organism.

Chromosome map : A genetic map showing the location of genes on a chromosome.

Chronic : Pertaining to slow-developing, persistent, or recurring symptoms (see acute).

Chronic toxicity : Poisoning due to low levels of exposure to a compound over a period of time (see acute toxicity).

Chytridiomycetes : A fungus group, also called the chytridiomycota, characterized by the production of motile zoospores and resting sporangia; the plant-pathogenic species are all biotrophs that are restricted to the cells of their host.

Circulative transmission (syn. Persistent transmission) : Virus transmission characterized by a long period of acquisition of the virus by a vector (typically an insect), a latent period of several hours before the vector is able to transmit the virus, and retention of the virus by the vector for a long period, usually several days; the virus circulates in the body of the vector.

Circulative-propagative transmission(syn. Propagative transmission) : Pathogen transmission characterized by a long period of acquisition of the pathogen (usually a mollicute, e.g. Phytoplasma or spiroplasma, and sometimes a virus) by a vector (typically an insect), a latent period before the vector is able to transmit the pathogen, and retention of the pathogen by the vector for a long period because the pathogen reproduces or replicates in the vector.

Cirrhus (pl. Cirrhi) : A curled, tendril-like mass of exuded spores, held together by a slimy matrix.

Citoplasma : Un compartimiento que rodea al núcleo de las células tanto vegetales como animales y en el cual se encuentran suspendidos muchos organelos y moléculas.

Citosol : La substancia líquida del protoplasma, excluyendo todos los organelos, tales como el núcleo, los plastidios y los ribosomas.

Clamp connection : A bridge- or buckle-hyphal protrusion in basidiomycetous fungi, formed at cell division and connecting the newly divided cells.

Clavate (or claviform) : Club-shaped.

Clay : A soil inorganic separate of less than 2 micrometers (< 0.002 mm.). This is smallest of the mineral soil separates.

Cleistothecium (pl. Cleistothecia) : A spherical ascocarp that is closed at maturity.

Climate : Precipitation, temperature and other aspects of climate affect how specific soils are formed.

Clone : N. A population of cells or organisms of identical genotype; population of recombinant DNA molecules all carrying the same inserted sequence; V. The vegetative propagation of an organism to produce a population of identical individuals; the use of in vitro recombination techniques to insert a particular DNA sequence into a vector. OR Genetically identical cells or organisms all derived from a single ancestor by asexual or parasexual methods.

Cloned gene : A copy of a gene that has been combined with a plasmid and placed into a bacteria cell.

Cloning : Asexually propagating an identical plant.

Coalesce : Grow together into one body or spot.

Coat protein (syn. Capsid) : The protective layer of protein surrounding the nucleic acid core of a virus; the protein molecules which make up this layer.

Coccus (pl. Cocci) : A spherical (or near-spherical) bacterial cell.

Coding region : Region of the DNA sequence between the promoter and the termination sequence. It contains the instructions about how to make a specific protein.

Codominance : The situation in which both alleles in a heterozygous organism are equally expressed resulting in a phenotype having characteristics from both alleles.

Codon : A particular sequence of three nucleotides in a nucleic acid that codes for a specific amino acid or termination of a polypeptide chain (see triplet codon). OR Three consecutive nucleotides in an RNA molecule that code for a single amino acid.

Codon bias : The frequency with which individual codons are used by an organism.

Coelomycetes : A group of fungi in the deuteromycetes (imperfect fungi) that produce pycnidia or acervuli (see hyphomycetes).

Coenocytic : Having multiple nuclei embedded in cytoplasm without cross walls; nonseptate

Cofactor : The nonprotein component of an enzyme. They often carry out activities for which amino acids are not suited.

Cohesion : A molecular attraction of molecules.

Cohesive : Able to stick together tightly.

Cohesivo : Capaz de permanecer fuertemente unido.

Colchicine : Alkaloid used to induce chromosome doubling in plant breeding; found in Colchicum autumnale (autumn crocus or meadow saffron).

Collar : The portion of the seedling or plant near the surface of the soil; in grafted woody plants, the scion portion of the plant near the soil surface.

Collenchyma : Supporting tissue in soft stems and other plant parts, composed of elongated living parenchyma cells with unevenly thickened primary walls, often bordering veins in dicot leaves; the "strings" in celery.

Colonization : Establishment and ramification of a pathogen within a host plant.

Colonize : To infect and ramify through plant tissue with the growth of a pathogen.

Colony : Growth of a microorganism in mass, especially as a pure culture.

Colony : A group of identical cells (or bacteria) that develop from a single cell when the cell divides multiple times. The cells are considered 'clones' of the original cell.

Colony forming unit (abbr. Cfu) : The number of colonies formed per unit of volume or weight of a cell or spore suspension.

Common name : The common name describes the chemical name of the active ingredient of the herbicide.

Community : (Plant community). An assemblage of plants occurring together at any point in time, while denoting no particular ecological status. A unit of vegetation.

Compaction : The process by which soil particles are rearranged to decrease void space and bring them into closer contact with one another, thereby increasing the bulk density.

Companion cell : Specialized cell adjacent to sieve tubes in phloem.

Compartmentalization : Isolation of a specific tissue area by host barrier tissues.

Compartmentalize : To separate with a barrier. This can mean a physical barrier such as a membrane, or two immiscible solvents such as oil and water

Competitive inhibitor : A molecule which has a similar structure as the normal substrate an enzyme acts upon. A competitive inhibitor inhibits an enzyme's activity by directly competing with the normal substrate in binding to the active site on the enzyme.

Complementary : Term used in science to refer to the normal nucleotidepairing in DNA replication, with cytosine pairing with guanine and adenine pairing with thymine.

Complete resistance : A resistant reaction in which some aspect of disease development, usually symptom expression or pathogen reproduction, is completely stopped (see partial resistance).

Composite : Member of the Asteraceae family.

Compost : A mixture of organic residues and soil that is allowed to decompose biologically.

Compressibility : The susceptibility of a soil to decrease in volume when subjected to load (pressure).

Concentrated superphosphate (0460) : A material manufactured by reacting phosphoric acid with finely ground rock phosphate; often used in bulk blended products.

Concentric : One circle within another with a common center.

Conduit : A channel through which something (as a fluid) is conveyed.

Conidiogenesis : The formation of asexual spores (conidia).

Conidiogenous : Producing and bearing conidia.

Conidioma (pl. Conidiomata) : Specialized conidia-bearing structure, e.g. Acervulus, pycnidium, sporodochium, synnema.

Conidiophore : Simple or branched hypha on which conidia are produced .

Conidium (pl. Conidia) : An asexual, nonmotile fungal spore that develops externally or is liberated from the cell that formed it.

Conjugated double bonds : Double bonds on alternating pairs of carbon molecules in a linear or cyclic organic molecule. An example is as follows c=cc=cc=c the electrons in conjugated double bonds form a resonance structure that is smeared over the double bond system. These electrons are responsible for light absorption.

Conjugation : The temporary contact of bacterial cells during which genetic material is transferred. Attachment of a molecule (herbicide) to another molecule like sugar, lipid, lingnin, cellulose, which inactivates the original molecule.

Conk : A shelf-like, typically hardened basidiocarp of a wood decaying fungus, usually a polypore.

Conservation tillage : Any tillage sequence, the object of which is to minimize or reduce loss of soil and water; operationally, a tillage or tillage and planting combination which leaves a 30% or greater cover of crop residue on the surface.

Consistency : The resistance of a soil to deformation or rupture.

Constitutive : Present or produced under all conditions (see induced).

Contact : Process resulting in two objects (i.e. Herbicide molecule and a plant) touching each other.

Contact fungicide (syn. Protectant fungicide) : A fungicide that remains on the surface where it is applied; no after-infection activity (see systemic fungicide).

Contact" herbicide : A herbicide that does not move from the point of initial uptake.

Continuous cropping : Growing the same crop in the same location repeatedly.

Coppice : Regrowth of woody plants by sprouts from stumps or roots.

Coremium (pl. Coremia; syn. Synnema) : Compact or fused, generally upright conidiophores, with branches and spores forming a headlike cluster.

Cork : External protective tissue of a stem or root, impermeable to water and gasses; the primary component of bark.

Corolla : Petals, collectively.

Cortex (adj. Cortical) : Region of parenchyma tissue between the epidermis and the phloem in stems and roots; region beneath the rind of a sclerotium.

Corymb : More or less flattopped, indeterminate inflorescence, the outer flowers open first.

Cosuppression : The silencing of a gene do to introduction of a second copy of the gene.

Cotransforming : Transforming a cell with two different genes at the same time.

Cotyledon : Seed leaf, one in moncots and two in dicots; primary embryonic leaf within the seed in which nutrients for the new plant are stored. The first leaf or leaves produced by the embryo of a flowering plant.

Cretaceous : The last period of the Mesozoic, approximately 144 to 65 million years ago.

Critical point : The point at which soil test values delineate responsive soils (those where fertilizer additions increase yields) and nonresponsive soils.

Crop rotation : The successive planting of different crop species; often used to improve soil fertility or to reduce disease and pest problems.

Cross : The deliberate mating of two parental types of organisms in genetic analysis.

Cross breeding : Breeding from a male and female of different lineage.

Cross-fertilization : A process in which sexual reproduction occurs as a result of the fusion of sex cells from different individuals (see self-fertilization).

Crossing : The deliberate mating of two parental types of organisms in genetic analysis.

Crossover events : A crossover event is the exchange between two DNA molecules. The resultant DNA contains sequences (hence genes) not found in either parental DNA molecule. Each exchange is an event. Because the exchange occurs between strands of DNA, the genes are physically linked. The maxiumum number of crossover events between two genes is 50% (which is identical to independent assortment).

Cross-pollination : Transfer of pollen from the anthers to the stigma of another flower (see self-pollination).

Cross-protection : The process whereby a normally susceptible host is infected with a less virulent pathogen (usually a virus) and thereby becomes resistant to infection by a second, usually related, more virulent pathogen.

Crossresistance : The development of resistance of an European Corn Borer to a Bt protein as a result of resistance development to a different Bt protein even though the European Corn Borer may have never been exposed to the second protein. OR Resistance within a plant to herbicides with the same site of action but from different herbicide families.

Crown : Upper dome of tree, bearing leaves, flowers, and fruits; junction of root and stem of a plant, usually at the soil line; in grafted woody plants, the rootstock portion of the plant near the soil surface.

Crozier : The hook of an ascogenous hypha before ascus development; the curved apical portion of a blighted stem.

Cruzamiento : El apareamiento deliberado de dos organismos progenitores en un estudio genético.

Cry proteins : Protein crystals produced naturally by bacillus thuringiensis bacteria that are toxic to some insects. The gene that produces this protein can be inserted into the corn genome giving it european corn borer (ecb) resistance.

Cryptobiosis (hidden life): : The ability to enter a state of suspended metabolic activity during unfavorable environmental conditions (drying, cold); survival strategy of some plant pathogenic nematodes.

Crystalline wax : Crystalline looking wax occurring on leaf surfaces that is difficult to wet.

Ctp : (chloroplast transit peptide): specific amino acid sequences added to proteins which direct them to go to the chloroplast of the plant cell.

Culm : Stem of grasses, cereals and bamboos.

Cultivar (abbr. Cv.; syn. Variety) : A plant type within a species, resulting from deliberate genetic manipulation, which has recognizable characteristics (color, shape of flowers, fruits, seeds and height or form).

Cultivo de Tejidos : Células vegetales cultivadas in vitro, que pueden ser manipuladas y luego inducidas para desarrollar plantas completas.

Cultural practices : The manner in which plants are grown, such as: application of nutrients, irrigation practices, type of cultivation; may be used for disease management.

Culture : Growth and propagation of microorganisms on nutrient media; growth and propagation of living plants.

Curl : Distoration, puffing, and crinkling of a leaf resulting from unequal growth.

Cuticle : Waxy layer of material on the surface of the plant epidermis to help plants retain water and resist microbe invasion.

Cuticle (adj. Cuticular) : Noncellular outer layer of an insect or a nematode; water-repellent, waxy layer of epidermal cells of plant parts, such as leaves, stems and fruit.

Cutícula : Capa cerosa que se encuentra sobre la epidermis de las plantas que ayuda a retener agua y resistir la invasión de microbios.

Cv. (abbr. For cultivar; syn. Variety) : A plant type within a species, resulting from deliberate manipulation, which has recognizable characteristics (color, shape of flowers, fruits, seeds and height or form).

Cyclohexanediones : Family of herbicides that belong to the lipid synthesis inhibitor mode of action (dims).

Cymes : More or less flattopped, determinate inflorescence, the outer flowers open last.

Cyst : In fungi, a resting structure in a protective membrane or shell-like enclosure;

In nematode females, the egg-laden carcass of a female nematode;

In bacteria, a specialized type of bacterial cell enclosed in a thick wall, often dormant and resistant to environmental conditions.

Cysteine : Nonessential amino acid (essential in infants and those with chronic diseases).

Cystenine : Nonessential amino acid (essential in infants and those with chronic diseases).

Cytochrome P450 : Also known as cytochrome P450 monooxygenases or cytp450s; multifunctional enzymes widely distributed among living organisms. They function as monooxygenases by binding and activating molecular oxygen, incorporating one of its atoms into an organic substrate, and reducing the second atom to form water.

Cytokinins : Plant hormones (growth regulators) that control cell division and are important for shoot stimulation of callus in tissue culture.

Cytopathology : The study of changes induced by disease at the cellular level.

Cytoplasm : Living protoplasm in a cell, except the nucleus. OR A compartment in a plant or animal cell surrounding the nucleus in which many cell organelles and molecules are suspended.

Cytoplasmic cell : The smaller, cytoplasmrich cell from the cell division of a progenitor cell. Also called embryogenic cell. It will develop into either proembryonal complex or a somatic embryo.

Cytoplasmic inheritance (also extrachromosomal or maternal inheritance) : Inheritance of genes not located in the nucleus, i.e. Those in mitochondria and chloroplasts.

Cytosol : The liquid substance of protoplasm, excluding all the organelles such as nuclei, plastids, ribosomes.

Damping-off : Death of a seedling before or shortly after emergence due to decomposition of the root and/or lower stem; it is common to distinguish between preemergence damping-off and postemergence damping-off.

Days to harvest : Term for the prescribed minimum number of days required by u.s. Government regulations between a pesticide application and the harvest of the crop.

Decay : The gradual decomposition of organic matter.

Deciduous : Describing a tree that sheds its leaves completely at the end of its annual growth period.

Decomposition : Breakdown into constituent elements through chemical change.

Decoy crop : Crop that stimulates germination of seeds of a parasitic plant such as witchweed (striga spp.), but is not susceptible to infection by the parasitic plant; helps reduce seed populations of the parasite in soil so a susceptible crop can be planted.

Dedifferentiation : The developmental process from a highly differentiated cell to an undifferentiated cell. It is a reverse to differentiation.

Deficiency correction : One of three approaches to making fertilizer recommendations which is based on the concept that a nutrient should be applied only if there is a reasonable expectation of a crop response. Under this approach, fertilizers should be applied only if they increase yields, and then only at optimum rates.

Defoliation : Loss of leaves from a plant, whether normal or premature.

Degradation : The breakdown of a chemical by either chemical or physical reactions that make it inactive.

Degree-day : The departure of the average daily temperature from a defined base (e.g. The minimum recognized temperature for the growth of a plant species). The number of degree-days may be totaled to assess the accumulated warmth of a particular year's growing season.

Dehiscent : Opening by breaking into parts.

Demicyclic : A rust fungus that lacks the urediniospore (repeating) stage (e.g. Many species of gymnosporangium) (see macrocyclic, microcyclic).

Denaturation step : The first temperature step in pcr. During this step, the DNA is made singlestranded, denatured.

Denature : A process which breaks the Hbonds between DNA bases, making the molecule become singlestranded.

Denitrification : NO3 > N2O and/or N2 Reduction of nitrogen oxides (usually nitrate and nitrite) to molecular nitrogen or nitrogen oxides with a lower oxidation state of nitrogen by bacterial activity (denitrification) or by chemical reactions involving nitrite (chemodenitrification). Nitrogen oxides are used by bacteria as terminal electron acceptors in place of oxygen in anaerobic or microaerophilic respiratory metabolism. OR The conversion of nitrate to various gaseous forms of nitrogen which can be lost to the atmosphere (nitric oxide, nitrous oxide, dinitrogen). Denitrification occurs under oxygenlimiting conditions, such as when soils are saturated.

Dentate : Toothed.

Deoxyribonucleicacid (abbr. DNA) : The double-stranded, helical molecule that contains genetic code information; each repeating unit, or nucleotide, is composed of deoxyribose (a sugar), a phosphate group, and a purine (adenine or guanine) or a pyrimidine (thymine or cytosine) base.

Deprotonated : Dissociated; a molecule that has lost its proton due to a change in ph.

Desiccate : To dry out.

Detachment : The removal of transportable fragments of soil material from a soil mass by an eroding agent, usually falling raindrops, running water, or wind; through detachment, soil particles or aggregates are made ready for transport.

Determinate : Ceasing vegetative growth when the first flower or reproductive structure forms. OR Refers to an inflorescence whose center flower opens first and axis prolongation is thereby arrested. Armitage, A; Herbaceous Perennial Plants

Detoxication : Altered so the compound is less toxic.

Deuteromycetes (syn. Fungi imperfecti) : Group of fungi without a sexual stage; the asexual stage of ascomycetes and basidiomycetes.

Dextranase : Catalyzes the endohydrolysis of α-d-1,6 glucosidic linkages in dextran.

Diacylglycerol (DAG) : A precursor in membrane lipid synthesis; diacylglycerol has two fatty acid groups attached to a glycerol back bone.

Diagnostic (n. Diagnosis) : A distinguishing characteristic important for the identification of a disease or other condition.

Diapause : A period of spontaneous dormancy, independent of environmental conditions, interrupting developmental activity in an embryo, larva or pupa. OR A type of hibernation that is a physiological condition resulting in suspended development of larva, such as ECB. It is controlled by day length, temperature, genetic composition of the insect and sometimes the nutritional quality of the host plant.

Dichotomous : Branching, often successively, into two more or less equal arms.

Dicot (syn. Dicotyledon) : Plant with two cotyledons or seed leaves (see monocot).

Dicots : A subclass of Angiosperms. They tend to have:broad leaves two cotyledons netlike veins in the leaves flower parts are usually in fours or fives a ring of primary vascular bundles in the stem taproot system ex. Most trees and shrubs, broadleafs.

Dieback (v. Die back) : Progressive death of shoots, leaves, or roots, beginning at the tips.

Differential host (syn. Differential cultivar) : A plant host that on the basis of disease symptoms serves to distinguish between various strains or races of a given plant pathogen.

Differential medium : Culture medium that is used to distinguish between organisms that can grow on it.

Differentiation : The physiological and morphological changes that occur in a cell, tissue, or organ during development from a juvenile state to a mature state. OR The developmental process that specializing cells' function and shape.

Diffusion : Process whereby molecules move from areas of higher concentration to areas of lower concentration.

Digitate : Having lobes radiating from a common center.

Dikaryon (adj. Dikaryotic) : Having two sexually compatible haploid nuclei per cell, that divide simultaneously; this phase is called the dikaryophase

Dilution plating : A method to obtain pure colonies of bacteria and fungi in which infected plant material or infested soil is diluted in sterilized water. Small samples of the water are spread on the medium surface of several petri plates to find which dilution will produce pure colonies.

Dilution streaking : Repeated streaking of bacteria on the surface of a nutrient medium with a sterile metal loop to allow pure colonies to grow.

Dimorphic : Having two distinct shapes or forms.

Dinitroanilines : Family of herbicides that inhibit microtubule assembly and belong to the seedling growth inhibitor mode of action (DNA).

Dioecious : Having male and female reproduction structures on separate individuals (see monoecious).

Diphenylether : Family of herbicides belonging to the cell membrane disrupter mode of action.

Diploid : Having two complete sets of chromosomes (2n chromosomes) (see haploid, polyploid). OR Containing two sets of chromosomes; 2n.

Dipole : A molecule or a functional group with a dipole moment.

Dipole moment : An electron asymmetry produces a local partial negative and partial charge. This separation of weak charge can be felt by the surrounding molecules. Fully charged functional groups (e.g. Phosphate) also contain a dipole moment.

Direct penetration : Penetration of plant tissues by a pathogen through barriers such as leaf cuticle by chemical and physical means (e.g. Penetration peg).

Disc : A type of minimum tillage that breaks up the organic matter and partially buries it in the soil. Depth of soil disturbance is seldom over 45 inches.

Discomycetes : A group of the ascomycetes in which the hymenium is exposed at maturity; one in which the fruiting body is an apothecium or discocarp.

Disease cycle : Succession of all of events and interactions among the host, parasite and environment that occur in a disease, from initial infection of the plant by a causal agent, through pathogenesis, to over-seasoning, until another infection occurs.

Disease incidence : Number of plants affected by a disease within a population.

Disease progress curve : Graph of some measure of disease (i.e. Severity) over time .

Disease pyramid : A memory aid similar to the disease triangle but including, in addition, the factor of time in the development of a disease.

Disease severity : The measure of damage done by a disease.

Disease triangle : A memory aid that diagrams the three important components necessary for disease: susceptible plant, virulent pathogen and favorable environment.

Disinfect : To eliminate a pathogen from infected plant tissues

Disinfest : To kill pathogens that have not yet initiated disease, or other contaminating microoganisms, that occur in or on inanimate objects as such soil or tools, or that occur on the surface of plant parts such as seed.

Dispersal (syn. Dissemination) : Spread of infectious material (inoculum) from diseased to healthy plants.

Disproportionating enzyme : Catalyzes the transfer of a segment of an α-d-1,4 glucan to a new position in an acceptor, which may be glucose or a 1,4-linked α-d-glucan.

Dissemination (syn. Dispersal) : Spread of infectious material (inoculum) from diseased plants to healthy plants.

Dissociate : (dissociation) – The splitting of a single chemical species into two ions, for example H2O to produce H+ and OH.

Dissociation : The release of a proton from a molecule due to a change in ph.

Distal : Far from the point of attachment or origin; in a direction away from main body (see proximal).

Division : Asexual propagation of a plant by breaking the crown or suckers (must contain a node) to producte an identical plant.

DNA (abbr. For deoxyribonucleic acid) : (deoxyribonucleic acid) the molecule that encodes genetic information. DNA is a doublestranded molecule held together by weak bonds between base pairs of nucleotides. It is the fundamental substance of which genes are composed. OR The double-stranded, helical molecule that contains genetic code information; each repeating unit, or nucleotide, is composed of deoxyribose (a sugar), a phosphate group, and a purine (adenine or guanine) or a pyrimidine (thymine or cytosine) base.

DNA extraction : The process of isolating the DNA from cells of an organism to allow scientists to work with and study it further.

DNA markers : DNA sequences that are found at specific loci in species that may vary in sequence among members of that species.

DNA polymerase : The enzyme which catalyzes the reaction of adding new DNA bases to a growing DNA strand.

DNA replication : The process in which a DNA molecule is copied. This takes place in pcr and naturally in cell division.

DNA sequence : The order of nucleotides, whether in a fragment of DNA, a gene, a chromosome, or an entire genome.

DNA template : That particular portion of a DNA molecule which is copied in pcr.

Dodder : Parasitic seed plant (cuscuta spp.) Without leaves; a yellow filamentous vine.

Dolipore septum : Cross wall found in basidiomycetes and characterized by special swellings and membranes in association with the septal pore.

Domain : A discrete portion of a protein with its own function. The combination of domains in a single protein determines its overall function.

Dominant : Describes a phenotypic trait that is expressed in hybrid progeny of diploid organisms even when contributed by only one of the parents (see recessive). OR The allele that is expressed in a heterozygous organism.

Donor parent : The parent from which one or a few genes are transferred to the recurrent parent in backcross breeding.

Dormancy (adj. Dormant) : A condition of suspended growth and reduced metabolism of an organism, generally induced by internal factors or environmental conditions as a mechanism of survival.

Double helix : The spiral shape that two linear strands of DNA assume when bonded together.

Downy mildew : A plant disease in which the fungus appears as a downy growth on the host surface; caused by a member of the oomycetes.

Drift (of pesticides) : Movement of airborne particles of a spray, dust, or vapor away from the target area during or shortly after an application.

Drought : Insufficient soil water for normal plant growth.

Durable resistance : Resistance that remains effective during prolonged and widespread use in an environment favorable to disease.

Dwarfing : Underdevelopment of a plant or plant organs, which may be caused by disease, inadequate nutrition, or unfavorable environmental conditions.

Dynamic : Something that changes or moves; not static or stationary.

Ear shank : The part of the ear that is attached to the stalk of the corn plant. The ear shank is a transport for water and nutrients to the ear from the plant.

Echinulate : Having small spines projecting from cell walls.

Ecology : The study of the interactions that occur between individual organisms, groups of organisms and organisms and their environment.

Economic threshold : The pathogen density at or above which the value of crop losses (in the absence of management efforts) would exceed the cost of management practices.

Ecosystem : Community of living things and their environment.

Ectomycorrhiza (pl. Ectomycorrhizae) : Symbiotic association between a nonpathogenic or weakly pathogenic fungus and the roots of plants with fungal hyphae between and external to root cells (see arbuscular mycorrhiza).

Ectoparasite : Parasite that feeds from the exterior of its host (see endoparasite)

Ectotrophic : Fungal development primarily over the root surface

Edema (also oedema; syn. Intumescence) : Blisters produced on leaves and other plant parts under conditions of high moisture and restricted transpiration.

Efflux : Movement from the cytoplasm to the cell wall region; out of the cell.

Egg mass : Group of eggs held together by a gelatinous matrix.

Electrochemical potential : Gradients of concentration and electric potential across cell membranes.

Electron asymmetry : The electrons that form a chemical bond between two atoms are not shared equally, resulting in one partner with a slightly negative charge (the tight sharer) and one partner with a slightly positive charge (the weaker sharer)

Electron orbitals : Electrons move about nuclei in spaces that have a characteristic shape. These orbiting shapes are called orbitals.

Electron transfer : The transfer of electrons between a series of components in a controlled fashion. Examples include the mitochondrial electron transfer chain involved in oxidative phosphorylation and the chloroplast electron transfer chain carrying out photophosphorylation. Electrons are transferred from molecules with a higher free energy to those with a lower free energy.

Electrophilic substrate : A functional group that likes electrons; is seeking a negative charge.

Electrophoresis : A technique which uses electricity to separate molecule fragments according to size so they can be studied.

Electrophoresis buffer : A solution of water and electrolytes that will conduct current for electrophoresis.

Electrophoresis gel : A gel matrix made from agarose, starch or acrylamide that is used to separate large molecules such as proteins or DNA based on the rate they move with an electric current.

Electroporation : A method of cell transformation using a pulse of electricity to open small temporary holes in the plant or bacteria cells. The foreign DNA mixed with these cells is able to pass into the cell through these holes.

Elemental sulfur : The most concentrated form of sulfur. Elemental sulfur must be oxidized to the sulfate form before plants can use it. It must be finely ground to particle sizes of 80100 mesh to be oxidized and effective during the same growing season.

Elevated meniscus : A meniscus is formed when water interacts with the sides of a vessel or tube and does not form a flat surface. In an elevated meniscus, the water forms a rounded surface toward the air.

Elicitor : A molecule produced by the host (or pathogen) that induces a response by the pathogen (or host).

Elisa (acronym for enzyme-linked immunosorbent assay) : A serological test in which the sensitivity of the reaction is increased by attaching an enzyme that produces a colored product to one of the reactants.

ELISA (enzymelinked immunosorbent Assay) : A test which can be used to detect the presence of proteins encoded by a specific gene.

Elite : A crop line that has many genes for good agronomic traits that result in high yields in a particular environment.

Elite experimental line : A line that is very close to release as a cultivar

Elongation step : The last temperatures step in pcr. During this step a DNA polymerase adds nucleotides to the growing DNA strand.

Embryo : An organism in the early stages of development, such as a young plant in the seed, or a nematode before hatching from the egg. OR The undeveloped plant in a seed.

Embryo proper : A cell mass that will develop into an embryo.

Emergence : Growth of the seedling shoot through the surface of the soil.

Emulsifier : A surfaceactive agent promoting the formation and stabilization of an emulsion; promotes the mixing of two normally unmixable liquids; an emulsifier is used to disperse an oil within water.

Emulsion : A system consisting of a liquid dispersed with an emulsifier in a second but immiscible liquid.

Enation : Abnormal outgrowth from the surface of a stem or leaf.

Encapsidate : To cover virus nucleic acid with a protein coat.

Encyst : To form a cyst or protective covering.

Endemic : Native to a particular place; pertaining to a low and steady level of natural disease occurrence.

Endocarp : Inner layer of a fruit wall.

Endoconidium (pl. Endoconidia) : A conidium produced inside a hypha or conidiophore.

Endodermis : Layer of cells within the root, between the vascular tissue and the cortex. OR The cells on the innermost cortex that form a barrier between the cortex and the vascular tissue in roots of plants.

Endogenous : Arising from the inside (see exogenous).

Endomycorrhiza (pl. Endomycorrhizae; syn. Arbuscular mycorrhiza) : Symbiotic association between a nonpathogenic or weakly pathogenic fungus and the roots of plants in which fungal hyphae invade cortical cells of the root (see ectomycorrhiza).

Endoparasite : Parasitic organism that lives and feeds from inside its host (see ectoparasite)

Endophyte : Plant developing inside another organism; also used for endoparasitic fungi found in grass species.

Endoplasmic reticulum : System of interconnected cytoplasmic membranes that transports materials within the cell.

Endosperm : Nutritive tissue formed within the embryo sac of seed plants.

Endospore : Resistant, thick-walled asexual spore formed within a bacterial or fungal cell .

Endotoxin : A protein that works internally as a toxin.

Energy of Activation : The amount of kinetic energy required for conversion of a substrate to a product.

Enthalpy : A measure of the total bondenergy (heat) in a molecule.

Entropy : A measure of the degree of disorder in a system.

Environment : The combinations of all the conditions external to the genome that potentially affect gene expression.

Enzyme : Protein that catalyzes a specific biochemical reaction. OR A protein that catalyzes, or speeds up, a specific biochemical reaction without changing the nature of the reaction and without being changed in the the chemical reaction.

Enzyme-linked immunosorbent assay (acronym elisa) : A serological test in which the sensitivity of the reaction is increased by attaching an enzyme that produces a colored product to one of the reactants.

Epa (acronym for environmental protection agency) : The federal agency responsible for controlling the various aspects of air, water, and soil pollution, including pesticide regulations.

Ephemeral gully erosion : See gully erosion.

Epicotyl : Portion of the stem of a plant embryo or seedling above the node where the cotyledons are attached (see hypocotyl).

Epidemic : An increase of disease in a population; a general and serious outbreak of disease (see epiphytotic).

Epidemiology (adj. Epidemiologic) : The study of factors influencing the initiation, development, and spread of infectious disease; the study of disease in populations of plants.

Epidermis (adj. Epidermal) : Surface layer of cells of leaves and other plant parts.

Epimers : A pair of isomers that differ from each other only in their configuration at a single asymetric center.

Epimers a : Pair of isomers that differ from each other only in their configuration at a single asymetric center.

Epinasty : Abnormal, downward curling of a leaf, leaf part, or stem. OR The downward curvature of leaves that occurs when the upper side of the petiole grows faster than the lower side.

Epiphyllous : On the upper surface of a leaf.

Epiphyseal cells : The meristem cells in epiphysis.

Epiphysis : The shoot apical meristem tissue.

Epiphyte : Organism growing on a plant surface, but not as a parasite.

Epiphytic : Living on the surface of plants, but not as a parasite.

Epiphytotic : An epidemic in a plant population.

Epistasis : Interaction between two loci, whereby the phenotypic effect of an allele at one locus depends on which allele is present at the other locus.

Epitope : An amino acid (or other) sequence that effects formation of an antibody.

Eps (extracellular polysaccharides) : Sugar polymers that contribute to the slimy appearance of bacteria probably promote colonization of plant tissues and disease development in bacteria.

Epsps : 5enolpyruvalshikimate3 phosphate syntheses; an enzyme essential for amino acid production that is the target site for glyphosate (roundup).

Eradication : Control of plant disease by eliminating the pathogen after it is established or by eliminating the plants that carry the pathogen.

Ergot : A disease of certain grasses and cereals, especially rye, caused by claviceps spp.; A sclerotium, or resting structure, produced by claviceps species and other closely related fungi in infected flowers of parasitized grain plants.

Ergotism : A disease caused by ingestion of grain contaminated with alkaloids of ergot fungi, often called the holy fire or st. Anthony's fire in the middle ages.

Erumpent : Bursting or erupting through the substrate surface.

Escape : Failure of inherently susceptible plants to become diseased, even though disease is prevalent.

Esr : Electron spin resonance spectroscopy.

Essential fatty acids : Fatty acids that are not synthesized by the human body, but are necessary for normal growth and life. These essential fatty acids are the linolenic and linoleic acids, and must be obtained from the diet.

Ethidium bromide : Dye molecule with a structure that binds to DNA and will fluoresce under ultraviolet light.

Ethylene : A plant hormone influencing various aspects of vegetative growth, fruit ripening, abscission of plant parts, and the senescence of flowers. OR A plant hormone produced by most plant tissues involved in fruit ripening, seed germination, senescence, abscission, and other aspects of plant development.

Etiolation : Elongation of stems caused by reduced light intensities.

Etiology : The study of the causes of diseases.

Eukaryote : An organism containing a membrane-bound nucleus and other organelles, i.e. All higher plants, animals, fungi, and protists (see prokaryote).

European corn borer (ecb) : Ostrinia nubilalis (hübner) is a major insect pest in corn that costs farmers millions of dollars annually in control expenditures and yield loss. The european corn borer (ecb) larva feed on the leaves and bore holes into the stalks where they tunnel.

Euthrophication : A condition in surface waters where high nutrient concentrations stimulate algal blooms which are sufficient to choke out some other forms of life. A more advanced state of nutrition pollution is a hypereutrophic condition where many clear water species cannot survive. Less contaminated states include oligiotrophic with clear waters and little biological activity and mesotrophic with an intermediate level of nutrient supply and biological activity.

Evaporation : To convert from a liquid to a vapor.

Evapo transpiration : The combined loss of water from a given area, and during a specific time, by evaporation from the soil surface and by transpiration from plants. (1997. Soil Science Society of America. P. 36).

Event : The insertion of a particular transgene into a specific location on a chromosome. Events are differentiated by two factors: 1) what transgene was inserted, and 2) where on the chromosome it inserted and how many gene copies inserted at that locus. OR The term used to differentiate different transgenic crops. Each event is the insertion of a particular transgene into a specific location on a chromosome. Events are differentiated by two factors: (1) what transgene was inserted, and (2) where on the chromosome it inserted and how many gene copies inserted at that locus.

Exclusion : Control of disease by excluding the pathogen or infected plant material from crop production areas (e.g. By quarantines and embargoes).

Exogenous : Originating from the outside (see endogenous). OR Applied externally; not from within the organism.

Exothermic : Characterized by or formed with evolution of heat.

Explants : A piece of plant tissue cut for tissue culture.

Exserted : Extends beyond the throat.

Extracellular : Outside a cell.

Extracellular polysaccharides (eps) : Sugar polymers that aid in movement, help prevent desiccation, and contribute to the slimy appearance of bacteria, fungi, and nematodes; and probably promote colonization of plant tissues and disease development in bacteria.

Extrachromosomal inheritance(also cytoplasmic or maternal inheritance) : Inheritance of genes not located in the nucleus, i.e. Those in mitochondria and chloroplasts.

Exudate : Liquid excreted or discharged from diseased tissues, from roots and leaves, or by fungi.

F. Sp. (abbr. For forma specialis) : A taxonomic group within a pathogenic species defined in terms of host range, i.e. Members of different formae speciales infect different groups of plants.

F1 generation : (first filial generation) the offspring produced from crossing two parental lines or individuals.

F2 generation : (second filial generation) the offspring produced from self crossing or intercrossing individuals in the f1 generation.

Facultative : Capable of changing life-style, e.g. From saprophytic to parasitic or the reverse.

Facultative parasite : Organism that is normally saprophytic but is capable of being parasitic.

Facultative saprophyte : Organism that is normally parasitic but is capable of being saprophytic.

Fallow : Cultivated land kept free from a crop or weeds during the normal growing season. .

Fallow : The practice of leaving land either uncropped and weedfree, or with volunteer vegetation during at least one period when a crop would normally be grown; objective may be to control weeds, accumulate water, and/or available plant nutrients.

Fasciation : Malformation in shoots or floral organs manifested as enlargements and flattening as if several parts were fused.

Fascicle (adj. Fasciculate) : Small group, bundle, or cluster.

Fastidious : In reference to prokaryotic organisms, having special growth and nutritional requirements.

Fatty acids : A class of compounds containing a long hydrocarbon chain and a terminal carboxyl group. Fatty acids can be saturated (containing no carboncarbon double bond), monounsaturated (containing one carboncarbon double bond), or polyunsaturated (containing multiple carboncarbon double bonds).

Fauna : Animal life.

Feedback inhibition : Type of enzymmatic regulation where the end product of a pathway regulates the enzyme's activity rate.

Feeder root : Fine root that absorbs water and dissolved nutrients.

Fermentation : Oxidation of certain organic substances in the absence of molecular oxygen.

Ferritin : Storage form of iron in the liver and other tissues.

Ferrous sulfate : The most commonly used inorganic iron source, which is 20 percent iron and 18.8 percent sulfur. It is not effective as a soilapplied material because it quickly reverts to an unavailable form.

Fertigation : Application of plant nutrients in irrigation water (1997. Soil Science Society of America. P. 38).

Fertilizer analysis : The percent composition of a fertilizer as determined in a laboratory and expressed as total n, available phosphoric acid (p2o5) equivalent, and watersoluble potash (k2o) equivalent.

Ffdca (acronym for federal food, drug, and cosmetic act) : The federal law which sets legal tolerances of pesticides on food products at harvest.

Fi hybrid : The first generation of offspring produced by crossing two parents of different genotypes

Fiber : Elongated plant cell with thick walls (in xylem and phloem), dead at maturity

Fibrous : Composed of profusely branched roots with many lateral rootlets but with no main or tap root development.

Field capacity : Soil water content after the soil has been saturated and allowed to drain freely for 24 to 48 hours. Free drainage occurs because of the force of gravity pulling on the water. When water stops draining, we know that the remaining water is held in the soil with a force greater than that of gravity.

Field water capacity (fc) : The content of water, on a mass or volume basis, remaining in a soil 2 or 3 days after having been wetted with water and after free drainage is negligible (1997. Soil science society of america. P. 40).

Fifra (acronym for federal insecticide, fungicide, and rodenticide act) : The federal law which legislates the conditions for use of pesticides.

Filamentous(syn. Filiform) : Threadlike.

Fission : Form of cell division in prokaryotes.

Fitness : The ability of an organism to survive and reproduce;

The ability of an organism to pass its genes to the next generation.

Fixation, nitrogen : Conversion of molecular nitrogen (n_2) to ammonia and subsequently to organic nitrogen utilizable in biological processes.

Flaccid : Wilted, lacking turgor. OR Body becomes soft and weak, lacking firmness.

Flag : Isolated, wilted or necrotic branch with dead leaves attached.

Flagellum (pl. Flagella) : Hairlike, whiplike, or tinsel-like appendage of a motile cell, bacterium or zoospore that provides locomotion.

Flanking : Flanking means sequences that border something. Flanking primers are located on either side of a DNA sequence of interest.

Flash grazing : A relatively short grazing period usually involving high grazing pressure.

Fleck : Minute, discolored spot in green tissue.

Flexuous : Having turns or windings; capable of bending.

Flora : Plant life.

Florets : Small or reduced flowers; often found in grasses and the Asteraceae family.

Flower break : Break or stripe in flower color (see breaking).

Fluorescein : A fluorescent dye.

Fluorescence : Light emitted by an excited singlet molecule as part of the deexcitation process. Because a small amount of the energy in the original exciting photon is lost by molecular vibration before the emission of the fluorescence, the wavelength of fluorescence is greater (i.e. Less energy) than the wavelength of the exciting photon.

Fluorophore : A molecule which fluoresces, give off light energy, when excited by an external light source.

Focus (pl. Foci) : Small area of diseased plants within a population.

Foliar : Pertaining to leaves.

Foraging : The search and consumption of forage.

Forest decline : Combination of biotic factors (e.g. Insect pests, pathogens) and abiotic factors (e.g. Drought, air pollution) that contributes to poor growth and death of populations of trees.

Form genus : A genus based on morphology and not on evolutionary relationships, such as used for imperfect fungi.

Forma specialis (abbr. F.sp.; pl. Formae speciales) : A taxonomic group within a pathogenic species defined in terms of host range, i.e. Members of different formae speciales infect different groups of plants.

Fqpa (acronym for food quality protection act) : The federal law which sets a health-based standard governing pesticide tolerance levels in food.

Fragment banding pattern : Bands made when many copies of the same DNA fragment move together through an electrophoresis gel. The pattern of bands reveals the number of different size fragments in a sample.

Free radical : Any molecule with one unshared electron giving the molecule a partial negative charge; highly reactive and can cause permanent damage to cellular membranes.

Free water : Unbound water; often used to describe a film of water on a plant surface.

Frequency : A measure of the number of times per second (units of sec^1) that the electric field in a photon vibrates.

Fret : Fluorescence resonance energy transfer a real time pcr DNA detection system, using flanking primers.

Fructification : General term for spore-bearing organs in both macro- and microfungi.

Fruiting body : Any of various complex, spore-bearing fungal structures.

Fumigant (v. Fumigate) : A gas or volatile substance used to kill or inhibit the growth of microorganisms or other pests.

Fungi imperfecti (syn. Deuteromycetes) : Group of fungi without a sexual stage; also, the asexual stage of ascomycetes and basidiomycetes.

Fungicide (adj. Fungicidal) : Chemical or physical agent that kills or inhibits the growth of fungi.

Fungistasis : Inhibition of fungal growth, sporulation, or spore germination but not death; used to describe the nonspecific phenomenon in natural soils where spore germination .is inhibited and often overcome by rhizosphere nutrients.

Fungistat (adj. Fungistatic) : A chemical or physical agent that inhibits fungal growth, sporulation, or spore geminiation, but does not cause death.

Fungus (pl. Fungi) : A eukaryotic organism that is usually filamentous (forming a mycelium) and heterotrophic, has cell walls composed of chitin, and reproduces by sexual and/or asexual spores.

Fusiform : Spindle-shaped; tapering at each end.

Gall (syn. Tumor) :

Gallery : Insect tunnel in bark and wood.

Gametangium (pl. Gametangia) : Cell containing gametes or nuclei that act as gametes.

Gamete : Sex cell.

Gamete : A mature male or female reproductive cell (sperm or ovum) containing half of the total number of chromosomes in a cell (i.e., humans have 46 chromosomes per cell; gametes would have 23).

Gametophyte : The haploid stage of a plant (see sporophyte).

Gc/ms : Gasliquid chromatograph coupled to a mass spectrometer .

Gene : Unit within an organism controlling heritable characteristics; genes are organized on chromosomes. OR The fundamental unit of heredity that carries genetic information from one generation to the next. A gene is an ordered sequence of nucleotides located on a particular position on a particular chromosome that encodes a specific functional protein.

Gene action : The phenotypic response observed when the dosage of an allele changes. For example, in the allelic series aa, ab, and bb, the dosage of the b allele changes from 0 to 1 to 2. Gene action may be additive (if aa<ab<bb), dominant (if ab=bb), or overdominant (if ab>bb).

Gene cloning : Finding and making copies of a specific gene. This is the first part of the genetic engineering process.

Gene expression : The production of a protein encoded by a gene. Gene expression is controlled by the promoter region of the gene.

Gene flow : Movement of genes among and between species or populations.

Gene gun : Also known as microprojectile bombardment and particle acceleration. The method used to transform cells using small gold or tungsten particles which are coated with DNA and literally shot into the cell.

Gene gun (see biolistic transformation) : A device used to bombard plant cells with metal particles coated with foreign genes to accomplish genetic engineering.

Gene library : A collection of cloned DNA molecules representing all or part of an individual's genome.

Gene mapping : Determination of the relative positions of genes on a DNA molecule or chromosome and the distance between them.

Gene silencing : The inactivation of a gene by an organism to prevent the gene from expressing.

Gene stacking : Creating a plant with a desired combination of two or more unique genes. This can be done when the genes are first transferred into the cells by cotransformation, or during the breeding process by crossing two lines that each contain a different gene resulting in progeny with both genes.

Gene-for-gene hypothesis : The hypothesis that corresponding genes for resistance and virulence exist in the host and pathogen, respectively.

General resistance (syn. Horizontal resistance, race non-specific resistance) : Resistance that is effective against all biotypes of the pathogen (see durable resistance, specific resistance, vertical resistance).

Genetic : Relating to heredity; referring to heritable characteristics.

Genetic code : The system of triplet codons composed of nucleotides of DNA or rna that determine the amino acid sequence of a protein. OR The sequence of nucleotides, read three at a time, that determines the order of amino acids in a protein.

Genetic engineering : The transfer of specific genes between organisms using enzymes and laboratory techniques rather than biological hybridization.

Genetic engineering : The process of adding foreign DNA to the genome of an organism.

Genetic fingerprinting : Determining the genetic makeup of an organism with the use of DNA markers.

Genetically modified organism; (gmo) (syn. Transgenic) : Organism possessing a gene from another species; used to describe the organisms that have been the subject of genetic engineering.

Genome : The complete genetic information of an organism or virus. OR All the genetic material in the haploid set of chromosomes for a particular organism.

Genotype : Genetic constitution of an individual or group; class or group of individuals sharing a specific genetic makeup (see phenotype). OR The allelic composition of a cell or organism.

Genus (pl. Genera) : A taxonomic category that includes a group of closely related (structurally or phylogenetically) species; the genus or generic name is the first name in a latin binomial.

Geocarp : Fruit that ripens beneath the ground (e.g. Peanut pod).

Germ theory : The theory that infectious or contagious diseases are caused by microorganisms (germs) .

Germ tube : Hypha resulting from an outgrowth of the spore wall and cytoplasm after germination.

Germinate (n. Germination) : Begin growth (as of a seed, spore, sclerotium, or other reproductive body).

Germline cells : Cells that give rise to gametes, the reproductive cells of an organism, (such as an egg or sperm), that have only one set of chromosomes. When the two unite, they form a living embryo.

Germplasm : Bearer of heredity material, often loosely applied to cultivars and breeding lines.

Germplasm : The total of the inherited materials, or genes, in a species.

Giant cell : Enlarged, multinucleate cell formed in roots by repeated nuclear division without cell division induced by secretions of certain sedentary plant-parasitic nematodes

Gibberellins : Plant hormones (growth regulators) that affect stem elongation.

Gill : Thin, radial membrane producing basidiospores in the cap of certain basidiomycetous fungi (e.g. Mushrooms).

Girdle : To circle and cut through a stem or the bark and outer few rings of wood, disrupting the phloem and xylem.

Glabrous : Smooth; without hairs.

Globular stage : The developmental stage of embryogenesis that the embryonic tissues organized as a ball.

Glucoside : A substance that, on decomposition, yields glucose and certain other compounds; some glucosides are defense compounds produced by plants, e.g. Cyanogenic or phenolic glucosides

Glutathione transferase : An enzyme that catalyzes the binding of xenobiotics to glutathione.

Glycosidic bond : A bond between the c-1 of glucose and the oxygen atom of another hexose moeity sugars can be linked to each other by o-glycosidic bonds to form oligo- and polysaccharides.

Glyphosate : A herbicide which inhibits the production of amino acids.

Gonad : Sexual structure in animals that produces gametes.

Graft : Transfer of aerial parts of one plant (e.g. Buds or twigs - the scion) into close cambial contact with the root or trunk (the rootstock) of a different plant; a method of plant propagation; the joining of cut surfaces or growing roots of two plants to form a living union.

Graft transmission : Transmission of a pathogen from one host plant to another through fusion of living tissue from the diseased host with living tissue of a healthy host.

Gram stain : Procedure used for identification of bacteria in which crystal violet stain, gram's iodine, ethyl alcohol and safranin stain are applied in succession to cells of the bacteria.

Graminicide : Herbicides that target the grass family (Graminae); refers to the lipid synthesis inhibitor herbicides that or only effective on grasses.

Gram-negative : Bacteria staining red or pink in the gram staining procedure after treatment with gram's stain.

Gram-positive : Bacteria staining violet or purple in the gram staining procedure after treatment with gram's stain.

Grandular hair : A hair with a minute bead of clear liquid at its apex.

Grassed waterway : A natural or constructed waterway, usually broad and shallow, covered with grasses, used to conduct surface water from or through cropland.

Grassland : Land on which the vegetation is dominated by grasses, grasslike plants, and/or forbs. Herbaceous vegetation provides at least 80% of the canopy cover.

Gravid : Containing an egg or eggs; capable of depositing eggs

Gravitational water : The amount of water held by the soil between saturation and field capacity.

Grazing capacity : The optimal stocking rate that will achieve a target level of animal performance (e.g., average daily gain) or other specific objective, while preventing deterioration of the ecosystem.

Grazing distribution : Dispersion of animals during grazing within a management unit or area.

Grazing efficiency : The percentage of the total standing crop by weight that is ingested.

Grazing pressure : Animal demand per unit weight of forage dry matter either at an instant in time (au/t) or over a period of time (aum/t).

Green manure : A crop plowed under while still green and growing to improve the soil; sometimes used to enhance populations of antagonistic microorganisms for biological control.

Ground meristem : Mode of action of herbicides that mimic plant hormones causing unregulated growth.

Growth regulator (syn. Hormone) : A chemical substance produced in one part of an organism and transported in minute quantities to induce a growth response in another part, e.g. In plants, auxins, cytokinins, and gibberellins

Guard cells : Paired, specialized, epidermal cells that contain chloroplasts and surround a stoma.

Gully : A channel resulting from erosion and caused by the concentrated but intermittent flow of water usually during and immediately following heavy rains. Deep enough (usually >0.5 m) to interfere with, and not to be obliterated by, normal tillage operations.

Gully erosion : The erosion process whereby water accumulates and often recurs in narrow channels and, over short periods, removes soil from this narrow area to considerable depths, often defined for agricultural land in terms of channels too deep to easily ameliorate with ordinary farm tillage equipment, typically ranging from 0.5m to as much as 25 to 30 m.

Gum : Gelatinous, sugary aggregate that is synthesized and exuded by plant tissues.

Gummosis (pl. Gummoses) : Pathologic condition characterized by excessive formation of gums; the products of cell degeneration.

Guttation : Exudation of watery, sticky liquid from hydathodes, especially along leaf margins.

H^+ ATPase : A membrane protein located in the plasma membrane and tonoplast, used to carry out primary active transport of H+ out of the cytosol by coupling directly to the energy released during ATP hydrolysis.

Habitat : A place with a particular kind of environment suitable for the growth of an organism.

Halflife : The time it takes for a compound to degrade to half of its original concentration.

Halophytic : Having characteristics of salt tolerance.

Haploid : Having a single complete set of chromosomes (see diploid, polyploid)

Hardiness : Ability to withstand stress.

Haulm : Stem or stalks collectively.

Haustorium (pl. Haustoria) : Specialized branch of a parasite formed inside host cells to absorb nutrients.

Head : The inflorescence (composite), consisting of ray and disc flowers: Gaillardia, sunflower, coneflower, plants of the Asteraceae (composite) family.

Heart stage : A developmental stage in embryogenesis that the young embryo looks like a heart. At this stage, most all organs in a young plant have been formed.

Heartwood : Central cylinder of nonfunctional xylem in a woody stem (see sapwood).

Heat shock : A method of cell transformation that opens small temporary holes in the plant cell wall by alternating temperatures between hot and cold. Foreign DNA mixed with these cells is able to pass into the cell through these holes.

Hemiparasite : Parasitic flowering plant that contains chlorophyll when mature (photosynthetic) and obtains water with dissolved nutrients by connecting to the host xylem via the haustorium, e.g. True mistletoe (phoradendron spp. Or viscum spp.) Or witchweed (striga spp.).

Hemizygous An organism that has a gene copy in only one chromosome in a chromosome pair (i.e., during the transformation process, the transgene will insert into only one chromosome of a pair).

Henry's law : The moles of a particular gas that can be dissolved into a solvent depends on the partial pressure of that gas in the atmosphere.

Herbaceous : Describing primary, soft, nonwoody tissue, as a plant or plant part; having the characteristics of an herb. A characteristic that distinguishes a herb from a woody; no persistant, woody tissue above ground.

Herbicide : A chemical used for killing plants or inhibiting plant growth, e.g. A weed or grass killer. OR A pesticide used to kill plants.

Herbicide family : A group of herbicides that have similar chemical structure and kill plants by the same mechanism of action.

Herbicide resistance : The inherited ability of a plant to survive and reproduce following exposure to a dose of herbicide normally lethal to the wild type. In a plant, resistance may be naturally occurring or induced by such techniques as genetic engineering or selection of variants produced by tissue culture or mutagenesis.

Herbicide tolerance : The inherent ability of a species to survive and reproduce after herbicide treatment. This implies that there was no selection or genetic manipulation to make the plant tolerant; it is naturally tolerant.

Heritability : A measure of the ability of a characteristic to be passed from one generation to the next. OR The proportion of phenotypic variation that is due to genetics, as opposed to environment or genotype x environment interaction.

Hermaphrodite (adj. Hermaphroditic) : Having both male and female reproductive organs.

Heteroecious : Pertaining to a rust fungus requiring two unrelated host plants for completion of its life cycle (see autoecious).

Heterokaryon (adj. Heterokaryotic) : Cell with genetically different nuclei.

Heterologous genes : A gene transferred from another organism which expresses the same trait as an endogenous gene.

Heterosis (syn. Hybrid vigor) : Increased vigor of hybrid offspring when compared to either parent.

Heterothallism (adj. Heterothallic) : Condition in which sexual reproduction can occur only in the presence of genetically different mycelia (see homothallism).

Heterotroph : An organism that obtains nourishment from outside sources and must obtain its carbon from organic carbon compounds (see autotroph).

Heterozygote : An organism having two different alleles at one or more locations.

Heterozygous : Having alternate forms (allele) of a gene on homologous chromosomes (see homozygous). OR An organism that has two different alleles at one or more locations on a chromosome.

Hexane : A organic molecule that forms a nonpolar solvent [CH3(CH2)4CH3].

Holomorph : The whole fungus in all its states (see anamorph, teleomorph)

Holoparasite : Parasitic flowering plant that lacks chlorophyll (nonphotosynthetic) and must rely totally on the contents of the xylem and the phloem of the host plant, e.g. Broomrape (orobanche spp.)

Homeothermy : Refers to body temperature that is constant and largely independent of the temperature of the animal's surroundings.

Homologous : Having the same function, structure, position or value.

Homothallism (adj. Homothallic) : Condition in which sexual reproduction occurs with a single thallus; self-fertile (see heterothallism).

Homozygosity : Having like alleles at corresponding loci on homologous chromosomes. An organism can be homozygous at on, several, or all loci.

Homozygote : An organism having identical alleles at one or more locations.

Homozygous : Having the same form (allele) of a gene on homologous chromosomes (see heterozygous).

Homozygous : An organism that has two identical alleles at one or more locations on a chromosome. Inbred lines are homozygous at many of their gene loci.

Honeydew : Sugary ooze or exudate, often from aphids, and a characteristic symptom of ergot .

Horizons, soil : The layers of soil, usually three, which make up the soil profile. Soil horizons differ in color, texture, structure and organic matter content. The a horizon is the upper surface or topsoil and usually has the highest organic matter content; the b horizon is the subsoil; and the c horizon is the parent material. A given soil may have one or all three horizons.

Horizontal resistance (syn.) General resistance, race non-specific resistance : Resistance which is effective against all biotypes of the pathogen (see durable resistance, specific resistance, vertical resistance).

Hormone (syn. Growth regulator) : A chemical substance produced in one part of a an organism and transported in minute quantities to induce a growth response in another part, e.g. In plants, auxins, cytokinins, and gibberellins.

Host plant : Living plant attacked by or harboring a parasite or pathogen and from which the invader obtains part or all of its nourishment.

Host range : The range of plants on which an organism, particularly a parasite, feeds.

HPLC/MS : High performance liquid chromatograph coupled to a mass spectrometer.

Humus : An organic material in soil which is a product of plant and animal remains that have decomposed and then synthesized into something new.

Hyaline : Transparent or nearly so; translucent; often used in the sense of colorless.

Hybrid : An individual produced by crossing two parents of different genotypes.

Hybrid (v. Hybridize) : Offspring of two individuals of different genotypes.

Hybrid vigor (syn. Heterosis) : Increased vigor of hybrid offspring when compared to either parent.

Hybridoma : A cell produced by the fusion of an antibody-producing cell and a lymphoma (cancer) cell for production of monoclonal antibodies.

Hybrids : Individuals produced by crossing two parents of different genotypes.

Hydathode : Epidermal leaf structure specialized for secretion or exudation of water; leaf opening at terminus of vein.

Hydrogen bonding : An electrostatic attraction between a hydrogen atom in one polar molecule of water and a small electronegative atom (oxygen) in another molecule of water.

Hydrogen fluoride (hf) : A common pollutant from volcanoes and several industries such as brick factories, aluminum smelters, and phosphate fertilizer factories.

Hydrolysis : (Hydrolyze) – A chemical reaction in which some reactant combines with water and splits the water into hydrogen and hydroxyl ions.

Hydrolyze : A chemical reaction in which some reactant combines with water and splits the water into hydrogen and hydroxyl ions.

Hydrophilic : Having chemical properties relating to water (waterlike). Compounds are polar or ionizable and can be soluble in water; lipophobic.

Hydrophobic : Water "fearing"; compounds that are relatively insoluble in water and other polar solvents and soluble in lipids and other nonpolar solvents.

Hydroseeding : A process of spray grass seeding with a mixture or blend of cellulose fiber mulch, grass seed and water. This forms a mat or blanket on the soil to enhance seed germination, vigorous growth, and good rooting system. Also known as "hydromulching."

Hydrostatic pressure : The positive pressure built up when water is compressed.

Hymenium : Continuous, spore-bearing layer of a fungus fruiting body.

Hyperhydricity : A network formed by hydroxyl connection.

Hyperparasite : Organism that is parasitic upon another parasite.

Hyperplasia (adj. Hyperplastic) : Abnormal increase in the number of cells, often resulting in the formation of galls or tumors.

Hypersensitive : Extremely or excessively sensitive; often refers to an extreme reaction to a pathogen.

Hypersensitive reaction and pathogenicity (hrp) gene : Gene required for elicitation of the hypersensitive (hr) response in resistant plants and causation of disease in susceptible plants; a type of avirulence gene.

Hypersensitive response (hr) : Rapid and localized cell death at the site of infection in resistant interactions between plants and pathogens.

Hypertrophy : Excessive development of an organ or part; exaggerated growth or complexity.

Hypertrophy (adj. Hypertrophic) : Abnormal increase in the size of cells in a tissue or organ, often resulting in the formation of galls or tumors.

Hypha (pl. Hyphae; adj. Hyphal) : Single, tubular filament of a fungal thallus or mycelium; the basic structural unit of a fungus.

Hyphal sheath (syn. Mantle) : Dense hyphal mass of ectomycorrhizal fungus enclosing the short feeder roots of plants.

Hyphomycetes : Imperfect fungi that produce conidia freely or on a sporodochium or synnema (see coelomycetes).

Hyphopodium : Stalked, thick-walled, lobed cells that stick to plant surfaces; sometimes used to describe the infection structures produced by ectotrophic hyphae of certain root-infecting fungi such as gaeumannomyces (take-all pathogen).

Hypocotyl : Portion of the stem below the cotyledons and above the root (see epicotyl).

Hypophyseal region : The tissue of a developing embryo that is derived from the topmost cell of the suspensor and is later incorporated into the root apex of the embryo, giving rise to part of the root cap, its initial cells and the ground meristem initial cells.

Hypovirulence : Reduced ability to cause disease.

Hypoxia : Oxygen deficiency in bodies of water created by pollution from nutrients and sediments.

Icosahedral (n. Icosahedron) : Having 20 faces, as a polyhedral-shaped virus particle (see isometric).

Imidazolinones : A family of herbicides that are similar in chemical structure (imidazolinone ring) and have the same mechanism of action amino acid synthesis inhibition.

Immiscible : Two solvents that are not able to dissolve into each other when mixed.

Immobilization : The conversion of an element from the inorganic to the organic form in microbial tissues, so that the element is not readily available to other organisms.

Immune : Cannot be infected by a given pathogen.

Immunoassay : A detection method based on antibodies specifically selected to react with the substance to be detected (the antigen).

Immunosupressant : Chemical or activity that suppresses the natural immune responses in animals, including humans.

Imperfect fungi (syn. Fungi imperfecti, deuteromycetes) : Group of fungi without a sexual stage; also, the asexual stage of ascomycetes and basidiomycetes.

Imperfect state (syn. Anamorph) : The asexual form in the life cycle of a fungus, when asexual spores (such as conidia) or no spores are produced.

In planta : In a plant.

In situ : In its original place or environment.

In vitro : In glass, on artificial media, or in an artificial environment; outside the host. OR In an experimental situation outside the organism (literally, "in glass").

In vivo : Within a living organism. OR In a living cell or organism.

Inactivate : The process of making a herbicide nonherbicidal by degradation or bonding to another molecule.

Inbred : An organism that has been selffertilized for several generations until it is homozygous at all important gene loci, "genetically pure."

Inbreeding : Mating between closely related organisms.

Inclusion body : Structure developed within a plant cell as a result of infection by a virus, often useful in identifying the virus.

Incomplete dominance : The interaction of alleles of a gene that produces an intermediate phenotype, as in the production of pink flowers when red and white alleles are present in a heterozygous individual.

Incomplete dominance : Expression of a heterozygous phenotype which is different from, and often intermediate to, that of either homozygous parent.

Incubation period : The time between penetration of a host by a pathogen and the first appearance of disease symptoms; the time during which microorganisms inoculated onto a medium are allowed to grow.

Indehiscent : Pertaining to fruit that does not split open at maturity.

Indeterminate : Continuing to grow vegetatively while producing flowers or reproductive structures. OR Center flowers open last in a inflorescence; the growth of the plant's main axis is not stopped with the opening of the first buds.

Indexing : Testing of a plant for infection, often by mechanical transmission or by grafting tissue from it to an indicator plant.

Indicator plant : Plant that reacts to a pathogen or an environmental factor with specific symptoms, used to detect or identify the pathogen or determine the effects of the environmental factor.

Indirect penetration : Penetration of plant tissues by a pathogen through natural openings (e.g. Stomata) or wounds.

Induced : Produced in response to a stimulus (see constitutive).

Induced systemic resistance (isr) : Reduced disease symptoms on a portion of a plant distant from the area where the inducing agent is active, caused by the triggering of active plant defenses against a variety of pathogens; used to describe increased resistance in plants induced by certain rhizobacteria (see systemic acquired resistance- sar).

Inducing medium : The tissue culture media for helping the start of organ or embryonic cell development.

Infect : Process in which an organism enters, invades, or penetrates and establishes a parasitic relationship with a host plant.

Infection court : Site in or on a host plant where infection can occur.

Infection cushion : Organized mass of hyphae formed on the surface of a plant from which numerous infective hyphae develop.

Infection focus : Initial site of infection, generally with reference to a population of plants.

Infection peg (syn. Penetration peg) : The specialized, narrow, hyphal strand on the underside of an appressorium that penetrates host cells.

Infection period : The time required for infection to occur under conducive environmental conditions, usually hours of leaf wetness and temperature.

Infectious : Pertaining to disease, capable of spreading from plant to plant.

Infective : Referring to an organism able to attack a host and cause infection; referring to a vector carrying or containing a pathogen and able to transfer it to a host plant.

Infest (n. Infestation) : To attack as a pest (used especially of insects and nematodes); to contaminate, as with microorganisms; to be present in large numbers.

Infiltration : Penetration of water into and through soil pore spaces. How much water penetrates depends on several factors, including the rate of precipitation, soil moisture condition and soil texture.

Inflorescence : Flower or flower cluster. OR The arrangement of flowers on a stem; inflorescence types include corymb, cyme, head, panicle, solitary, raceme, spike thyrse, and umbel.

Influx : Movement from the cell wall region into the cytoplasm; into of the cell.

Initial -inoculum (syn. Primary inoculum) : Inoculum, usually from an overwintering source, that initiates disease in the field, as opposed to inoculum that spreads disease during the season.

Injury : Damage caused by transitory interaction with an agent such as an insect, chemical, or unfavorable environmental condition.

Inner bark : Interior, living portion of the bark; secondary phloem.

Inoculate (n. Inoculation) : To place inoculum in an infection court; to insert a pathogen into healthy tissue.

Inoculum (pl. Inocula) : Pathogen or its parts, capable of causing infection when transferred to a favorable location.

Inoculum density : A measure of the number of propagules of a pathogenic organism per unit area or volume.

Insect : Member of the class hexapoda (phylum arthropoda) possessing three sets of limbs attached to a central body segment.

Insoluble : This occurs when molecules in a solid material are put into a solvent and are unable to leave the solid and enter the solvent.

Instar : One of the stages of growth between molts, in the life history of certain insects.

Integrated pest management (abbr. IPM) : A combination of strategies to reduce losses due to pests and pathogens based on environmental and economic considerations.

Intense selection : Selection for the trait(s) of interest with little or no retention of lines not having the trait(s) of interest.

Intercalating dye : A dye molecule which can insert between bases of DNA. Used in laboratories to detect DNA sequences.

Intercalation : When a molecule, such as Ethidium Bromide or SYBR Green inserts in between DNA nucleotides of a doublestranded DNA moledule. Ethidium Bromide and SYBR Green are therefore, used in detecting the presence of DNA.

Intercellular : Between or among cells.

Intercrop : To grow two or more crops simultaneously on the same area of land.

Internal transcribed spacers : Regions of a DNA sequence that separate genes for certain ribosomal rnas (rrnas). These genes appear to be nearly identical among a wide variety of species, but the DNA spacers between them are quite variable and may be species-specific. This makes them good potential targets for "primers" for polymerase chain reaction (pcr) amplification that is used to identify a wide variety of organisms.

Internode (adj. Internodal) : The portion of a stem between two successive nodes.

Interphase : The stage of cell division when DNA is replicated; a copy of each chromosome is made.

Interveinal : Between (leaf) veins.

Intracellular : Through or within cells.

Intron : A portion of DNA located between coding regions in a gene which is transcribed, but spliced out of the mrna before translation. This region of nucleotides does not actually code for any proteins.

Intumescence (syn. Edema or oedema) : Blisters produced on leaves and other plant parts under conditions of high moisture and restricted transpiration .

Involucral bracts : A series of bracts beneath or around a flower or cluster of flowers.

Ions : Molecules with a positive or negative charge.

IPM (abbr. For integrated pest management) : A combination of strategies to reduce losses due to pests and pathogens based on environmental and economic considerations.

Iron oxides : Group name for the oxides and hydroxides of iron. Includes the minerals goethite, hematite, lepidocrocite, ferrihydrite, maghemite, and magnetite. Sometimes referred to as "sesquioxides," or "iron hydrous oxides."

Isoamylase : Catalyzes the hydrolysis of α-d-1,6 glucosidic branch linkages in glycogen, amylopectin, and their β-limit dextrins.

Isolate : (n.) A culture or subpopulation of a microorganism separated from its parent population and maintained in some sort of controlled circumstance; (v.) To remove from soil or host material and grow in pure culture.

Isometric : Usually used for virus particles to describe those that are icosahedral in structure and appear approximately round.

Isoprenoids : Also known as terpenes, this is a diverse class of biomolecules. They have in common a five carbon branched molecule as the basic building block.

Isozyme : An enzyme that is similar in structure and function to another enzyme.

ITS (acronym for internal transcribed spacers) : Regions of a DNA sequence that separate genes for certain ribosomal rnas (rrnas). These genes appear to be nearly identical among a wide variety of species, but the DNA spacers between them are quite variable and may be species-specific. This makes them good potential targets for "primers" for polymerase chain reaction (PCR) amplification that is used to identify a wide variety of organisms.

Juvenile : An immature form that appears similar to but usually smaller than the adult and is not sexually mature (e.g. Insects with gradual metamorphosis, nematodes)

Karyogamy : The fusion of nuclei.

Karyotype : A photograph of an individual's chromosomes arranged in order showing the number, size, and shape of each individual chromosome.

Kcat : The number of reactions per second that the enzyme can catalyze under optimum conditions.

Ketose : Sugar with a terminal ketone group.

Key management species : Plant species on which management of a specific unit is based.

Km : The Michaelis constant. It is an indication of the affinity of the enzyme for the substrate.

Knockout genes : Transgenes designed to silence an existing gene.

Knot : Gall; localized abnormal swelling.

Koch's postulates : The procedure used to prove the pathogenicity of an organism, i.e. Its role as the causal agent of a disease.

Labile phosphorus : Intermediate form of phosphorus that is rather weakly adsorbed or bound to various compounds and clay in the soil.

Lamina : Expanded part of a leaf (see petiole).

Land race : Plant stocks selected by farmers on a local basis over many years.

Larva : Newly hatched, wingless, wormlike stage of an insect.

Larva (pl. Larvae) : Immature stage of certain animals (especially insects) that undergo complete metamorphosis; the wormlike or caterpillar stage of the life cycle of such insects

Latent : Present but not manifested or visible, as a symptomless infection by a pathogen.

Latent infection : Infection unaccompanied by visible symptoms.

Latent period : The time between infection and the production of new inoculum; the time after a vector has acquired a pathogen and before it can be transmitted.

Lateral bud (syn. Axillary bud) : Bud that develops in the axil of a leaf.

Lateral flow strip : A variation on the elisa tests, using strips as opposed to microtiter wells to detect the presence of a protein produced.

Lateral moraines : An accumulation of soil, pebbles, rocks, and boulders distributed and marking the lateral edges of a glacier or ice sheet.

Layering : Asexual propagation of a plant that has roots develop on a shoot before it is detached from the parent plant.

Ld50 (abbr. For lethal dose 50%) : The dose of chemical that can kill 50% of the test subjects. OR A measure of relative acute toxicity; the lethal dose of a compound that causes death in 50% of the test animals treated.

Leaching : The removal of soluble materials from one zone in soil to another via water movement in the profile. (Soil Science Society of America. 1997, p. 62).

Leaf dip : Inspection of a sample of sap from a suspected virus-infected leaf using an electron microscope.

Leaf sheath : Lower, tubular part of a grass leaf that clasps the culm.

Leaf spot : A plant disease lesion typically restricted in development in the leaf after reaching a characteristic size.

Leafhopper : Mobile insects with sucking mouthparts (order homoptera).

Leaflet : One of the separate blades or divisions of a compound leaf.

Leafroll : Disease symptom in which the edges of the leaf roll or turn up; often a symptom of virus infection.

Legume : Simple, dry, dehiscent fruit developed from a simple pistil and splitting at maturity along two seams; A plant of the family fabaceae (formerly leguminosae).

Lenticel : A natural opening in the surface of a stem or tuber, fruit or root for gas exchange.

Lesion : Localized diseased area or wound.

Life cycle : The cyclical stages in the growth and development of an organism.

Ligase : An enzyme that can rejoin cut strand of DNA.

Lignification : Hardening of tissue through the deposition of lignin in the cell wall.

Lignin : The complex phenolic polymer that endows the xylem and other tissues in plants with compression and decay resistance.

Lime : A naturally occurring material composed of carbonates of calcium and magnesium.

Limit dextrinase : Catalyzes the hydrolysis of α-d-1,6 glucosidic linkages in α- and β-limit dextrins of amylopectin and glycogen, and in amylopectin and pullulan.

Line : Plants within a species that have the same genetic composition and are genetically pure, (i.e., inbred line). Lines are typically not agronomically competitive and are used only in plant breeding.

Linkage : The association of genes on the same chromosome. The shorter the distance between two genes, the greater the probability they will be inherited together.

Linkage disequilibrium : The condition whereby alleles at two loci occur together in a population at a greater frequency than that predicted by the product of the individual allele frequencies.

Linkage map : An ordered representation of genes on a chromosome derived from observed percentages of recombination. Units on a linkage map are centimorgans (cm), with 1 cm approximately equal to 1% recombination.

Linkage drag : The inheritance of undesirable genes along with a beneficial gene due to their close linkage.

Lipid bilayer : Made up of compounds which have both hydrophobic and hydrophilic moieties, this is the structure of membranes such as the cellular membrane.

Lipid synthesis inhibitors : Mode of action of herbicides that target the production of lipids which are essential in plant growth and function.

Lipophilic : Having chemical properties relating to lipids (lipidlike), nonpolar compounds that are highly soluble in organic solvents, but not water; hydrophobic.

Local lesion : Small, restricted lesion, often the characteristic reaction of differential cultivars to specific pathogens, especially in response to mechanical inoculation with a virus.

Locus : The site on a chromosome where a particular gene (or other sequence) is located (plural=loci).

Lod : Abbreviation for logarithm of odds, a significance measure used in maximum likelihood statistics. In interval mapping, the lod score is $\log_{10}$ of the ratio of the probability that a qtl is present to the probability that a qtl is absent. For example, if the presence of a qtl is 1000 times more probable than its absence, then the lod score = 3, because $\log_{10}$ of 1000 = 3.

Lodge : To fall over.

Lodged : Occurs when plant stalks (such as corn plants) have been damaged by insect, disease, or climate which results in the plant falling over.

Loess : Material transported and deposited by wind and consisting of predominantly siltsized particles. (SSSA).

Log k_o w : Octanol/water partition coefficient, relative measurement of how water soluble a chemical is. A compound with a low log kow would be more water soluble than a compound with a high log k_o w.

Lumen (pl. Lumina) : Central cavity of a cell or other structure.

Lysine : Nonessential amino acid found in cheese, eggs, milk, meat, legumes, whole grains, and nuts.

Lysis : The dissolution or destruction of cells by the disruption of the cell membrane.

Macerate : To cause disintegration of tissues by separation of cells; to soften by soaking.

Macroconidium (pl. Macroconidia) : The larger of two kinds of conidia formed by certain fungi (see microconidium).

Macrocyclic : A rust fungus that typically exhibits all five stages of the rust life cycle (see demicyclic, microcyclic).

Macromolecules : Large molecules which are composed of smaller individual molecules.

Macronutrient : An element needed in relatively large quantities for plant growth, e.g. Nitrogen (n), phosphorus (p), and potassium (k); (see micronutrient).

Macronutrients : Elements needed in relatively large amounts for plant growth. Macronutrients include carbon, hydrogen, sulfur, nitrogen, phosphorus, potassium, calcium, magnesium, and oxygen.

Maintenance approach : One of three approaches to making fertilizer recommendations which sets soil test level goals for specific nutrients and then recommends fertilizer to build the soil to those nutrient goals. The maintenance approach emphasizes maintaining the soil fertility level at or above the point of the economic maximum yield.

Malonylation : Conjugation of a glucose conjugate of xenobiotic to malonyl in the presence of malonylcoa and malonylcoatransferase; considered a Phase III reaction or secondary conjugation.

Malto-oligosaccharides (maltose, maltotriose, maltotetraose, maltopentaose, maltohexaose) : The series of linear oligosaccharides composed of two, three, four, five and six, respectively, units of glucose all linked via an α-d-1,4 bond.

Maltose phosphorylase : Catalyzes the phosphorolysis of maltose yielding d-glucose and β-d-glucose 1-phosphate.

Mantle (syn. Hyphal sheath) : Dense hyphal mass of ectomycorrhizal fungus enclosing the short feeder roots of plants.

Marker : An easy to detect trait controlled by a known gene. Markers, such as antibiotic or herbicide resistance, are often used to determine if an organism is transgenic.

Marker gene : A gene controlling an easily detectable trait used to select transgenic cells or plants from nontransgenic ones, (i.e., herbicide and antibiotic resistance).

Marker-assisted selection : A procedure used in plant breeding in which genetic markers that are (a) easy to identify and (b) linked to desirable genetic traits that are difficult to identify (such as disease resistance) are used to aid in selection from a population.

Mass flow : Process by which a dissolved or suspended molecule is moved with movement of the solvent, i.e., a floating piece of wood moving with the current in a stream.

Mass wasting : Movement of disintegrated rock materials by sliding or falling to the base of a slope by the force of gravity.

Master horizon : A principal horizon, such as the a, e, b, c, or r.

Maternal inheritance (also cytoplasmic or extrachromosomal inheritance) : Inheritance of non-nuclear genes, i.e. Those in mitochondria and chloroplasts.

Mating types : Compatible strains, usually designated + and - or a and b, necessary for sexual reproduction in heterothallic fungi.

Maximum contaminant limit (mcl) : The highest level of a contaminant that is allowed in drinking water.

Mechanical analysis : The laboratory procedure used to identify soil separates.

Mechanical injury : Injury of a plant part by abrasion, mutilation, or wounding.

Mechanical transmission : Spread r introduction of inoculum to an infection court (wounding) by human manipulation, accompanied by physical disruption of host tissues.

Mechanism of Action : The specific process inhibited by a herbicide.

Medium (pl. Media) : A mixture of organic and/or inorganic chemical compounds and water that provides the nutrients needed for the growth of a microorganism in vitro; for higher plants, a mixture of fertilizers and other components in which a plant is growing.

Meiosis : Process of nuclear division in which the number of chromosomes per nucleus is halved, i.e. Converting the diploid state to the haploid state (see mitosis). OR A type of cell division which results in the formation of gametes, cells with half the normal number of chromosomes.

Melanin : Brown-black pigment; common in sclerotia and other survival structures.

Meristem (adj. Meristematic) : Plant tissue characterized by frequent cell division, producing cells that become differentiated into specialized tissues.

Meristem culture : Aseptic in vitro culture of a plant or plant part from a portion of the meristem; a method used to produce pathogen-free plants.

Meristems : Area of cell division.

Messenger rna (abbr. mrna) : A form of rna that carries information to direct the synthesis of protein.

Metabolism : A general term referring to the change of a herbicide from an active to an inactive state. OR The sum of all chemcial processess occurring within a cell or an organism.

Metabolite : Any chemical participating in metabolism; a nutrient.

Methionine : An essential amino acid found in cereal, whole grains, sesame and sunflower seeds, and yeast.

Microarray : A small glass, or filter, square containing probes for thousands of gene products.

Microbial : Pertaining to or relating to microbes, or microorganisms.

Microclimate : Weather conditions on a small scale, e.g. At the surface of the plant or within a crop.

Microconidium (pl. Microconidia) : The smaller of two kinds of conidia formed by certain fungi (see macroconidium).

Microcyclic : Describing a rust fungus that produces only teliospores and basidiospores (see demicyclic, macrocyclic).

Microfibers : A method used to transform tissue culture cells using tiny fibers coated with DNA. These fibers are combined in solution with the cells and shaken vigorously causing the fibers to stab the plant cells delivering the DNA to them.

Microfibrils : In guard cells, microfibrils are made of cellulose and are oriented transversely to the long axis of the cell resulting in cell expansion in the direction of its long axis because the cellulose reinforcement offers the least resistance at right angles to its orientation.

Microfiltration : Removes particles, such as microorganisms and macromolecules, from solution by passing it through a microfilter.

Microflora : The combination of all microorganisms in a particular environment.

Micron (syn. Micrometer) : One millionth of a meter.

Micronutrient : An essential element needed by plants for growth and required in very small amounts (see macronutrient). OR Elements needed in very small amounts for plant growth. Micronutrients include zinc, iron, chlorine, copper, manganese, boron and molybdenum.

Microorganism (syn. Microbe) : An organism of such small size that it can only be seen as an individual organism with the aid of microscope.

Micropropagation : Tissue culture technique that asexually propagates a clone of a plant from tissue.

Microsclerotium : Microscopic, dense aggregate of darkly pigmented, thick-walled hyphal cells.

Middle lamella : The layer, consisting largely of pectic substances, between the walls of adjacent plant cells.

Midrib : Central, thickened vein of a leaf.

Midvein : Runs down the center of the leaf blade; is part of the framework of the leaf and gives support.

Migratory : Migrating from place to place on a plant or from plant to plant when feeding (see sedentary).

Mildew : Thin coating of mycelial growth and spores on the surfaces of infected plant parts.

Mineralization : The conversion of an element from an organic form to an inorganic state as a result of microbial activity.

Mineralized : The conversion of an element from an organic form to an inorganic state as a result of microbial activity.

Mineralizes : The conversion of an element from an organic form to an inorganic state as a result of microbial activity.

Miscible : Two solvents that are able to dissolve into each other when mixed.

Mitigation : Correction of a problem, often due to the presence of a pollutant.

Mitochondrion (pl. Mitochondria) : Cellular organelle outside the nucleus that functions in respiration.

Mitosis : Nuclear division in which the chromosome number remains the same (see meiosis).

Mixedfunction oxidase : Enzymes that catalyze oxidation of several substrates.

Mlo (syn. Mycoplasmalike organsim) : Archaic term for phytoplasma; plant-parasitic pleomorphic mollicute (prokaryote with no cell wall) found in phloem tissue; cannot yet be grown on artificial nutrient media.

Modes of Action The method by which an organism is affected, damaged or killed. The mode of action usually involves a chemical reaction instigated by another source. (herbicide, etc.).

Mold : Any microfungus with conspicuous, profuse, or woolly superficial growth (mycelium and/or spore masses) on various substrates; especially an economically important saprobe; molds commonly grow on damp or decaying matter and on the surface of plant tissues.

Molecular marker : An identifiable DNA sequence on a chromosome. A marker can be a gene, part of a gene, or a sequence in a nongene region. Ssr, rflp, rapd, and aflp are acronyms for commonly used marker techniques.

Mollicute : One of a group of prokaryotic organisms bounded by flexuous membranes and lacking cell walls (see phytoplasma and spiroplasma).

Molt : To shed a cuticle or body encasement during a phase of growth.

Monoclonal antibody : Antibody produced from clones of a single antibody-producing cell (see polyclonal antibody).

Monocot (syn. Monocotyledon) : Plant with one cotyledon, such as grasses, grain crops, corn (see dicot).

Monoculture : The growth of the same plant species in close proximity, with few or no other types of plant present.

Monocyclic : Having one disease or life cycle per growing season (see polycyclic).

Monoecious : Having male and female reproductive organs on a single individual (see dioecious).

Monogenic : Determined by a single gene (see polygenic).

Monogenic resistance (syn. Single gene resistance) : Resistance conferred by a single gene (see oligogenic resistance, multigenic resistance, polygenic resistance).

Monopolar : Develop or move toward one direction.

Monoxenic culture : A culture containing one species of organism growing in the presence of one other species of organism.

Morphogenesis : The process of morphology development.

Morphology : The study of the form and structure of an organism.

Morphology (adj. Morphologic) : The study of the form of organisms; form and structure of organisms.

Mosaic : Disease symptom characterized by non-uniform coloration, with intermingled normal, light green and yellowish patches, usually caused by a virus; often used interchangeably with mottle.

Most limiting nutrient : Justus Von Liebig formulated the law of the minimum: plant growth is limited by the most deficient nutrient, even if the other elements are abundant.

Motile : Capable of self-propulsion by means of flagella, cilia, or amoeboid movement. OR Disease symptom comprising light and dark areas in an irregular pattern, usually caused by a virus; often used interchangeably with mosaic.

Movement protein : For viruses, a protein encoded by the pathogen that is required for spread within an infected plant.

mRNA : (messenger ribonucleic acid) The message made during transcription by reading the DNA sequence to build a particular protein. A singlestranded nucleic acid similar to DNA but having a ribose sugar rather than deoxyribose sugar and a uracil rather than thymine as one of the bases.

MS medium : A popular plant tissue culture medium developed by t. Murashige and f. Skoog.

Mucilaginous : Viscous, slimy.

Mulch : Layer of material, such as organic matter or plastic, applied to the surface of the soil for purposes such as retention of water and inhibition of weeds.

Multicellular : Composed of many cells (see unicellular).

Multigenic resistance (syn. Polygenic resistance) : Resistance conferred by several genes (see monogenic resistance, oligogenic resistance, single gene resistance).

Multiline : Mixture of cultivars that differ only in single, major genes for resistance to a pathogen, giving the opportunity to modify the mixture from year to year as needed.

Multinucleate : Having more than one nucleus per cell (see uninucleate).

Multiparticulate virus : Plant virus in which the genome is divided into more than one piece of nucleic acid, which are encapsidated in more than one particle.

Multipartite virus : Plant virus in which the genome is divided into more than one piece of nucleic acid.

Multiple resistance : Evolution of populations resistant to herbicides not chemically related and having different sites of action, such as a weed resistant to both sulfonylurea and aryl.

Multiseptate : Having many septa or crosswalls.

Mummy : A dried shriveled fruit; plant part or organ partially or completely replaced with fungal structures.

Mushroom : A fleshy fruiting body of a fungus, especially of a basidiomycete of the family agaricaceae.

Mutagen : A substance that causes a mutation.

Mutation (n.mutant) : An abrupt heritable or genetic change in a gene or an individual as a result of an alternation in a gene or chromosome, or of an increase in chromosome number.

Mycelia sterilia : A group in the fungi imperfecti that does not produce spores, but may produce chlamydospores or sclerotia, e.g. Rhizoctonia and sclerotium; subsequent research has defined sexual reproduction for some species.

Mycelium (pl. Mycelia, adj. Mycelial) : Mass of hyphae constituting the body (thallus) of a fungus.

Mycology : The study of fungi.

Mycoparasite : A fungus that attacks another fungus.

Mycoplasmalike organism (syn. Mlo) : Archaic term for phytoplasma; plant-parasitic pleomorphic mollicute (prokaryote with no cell wall) found in phloem tissue; cannot yet be grown on artificial nutrient media.

Mycorrhiza (pl. Mycorrhizae; adj. Mycorrhizal) : Symbiotic association between a nonpathogenic or weakly pathogenic fungus and the roots of plants (see ectomycorrhiza and endomycorrhiza).

Mycotoxin : Poisonous compound produced by a fungus.

Mycovirus : A virus that infects fungi.

Myxomycetes (syn. Slime molds) : Saprophytic organisms that form vegetative amoeboid plasmodia and spores.

NADPH : Reduced nicotinic adenine dinucleotide phosphate.

Necrosis (adj. Necrotic) : Death of cells or tissue, usually accompanied by black or brown darkening.

Necrotroph : A parasite that typically kills and obtains its energy from dead host cells (see biotroph).

Nectary : A nectar-secreting gland in a flower.

Needle cast (of conifers) : Disease symptom caused by fungi resulting in premature drop of needles.

Negative cross resistance : Occurs when a resistant biotype is more susceptible to other classes of herbicides than the susceptible biotype.

Negative sense rna : Ribonucleic acid complimentary to the positive or plus sense; not translated into protein (see positive sense rna).

Nematicide : Agent, usually a chemical, that kills nematodes.

Nematode : Nonsegmented roundworm (animal), parasitic on plants or animals, or free living in soil or water.

Nitrate : One of the mineral forms of nitrogen (NO_3).

Nitrification : The conversion of ammonium to nitrate by bacteria in soil (NH4 > NO3).

Nitrification inhibitor : A chemical which, when applied to the soil, slows or temporarily prevents the biological oxidation of ammonium to nitrite and nitrate.

Nitrogen cycle : The cycling of the element nitrogen from gaseous forms to various inorganic forms, such as ammonium, nitrates and nitrites, and organic forms, such as nucleic acids and proteins and back to elemental nitrogen.

OR

Nitrogen cycle : The transitional process taking nitrogen from its simplest stable form (dinitrogen) and following it through fixation, mineralization, nitrification, leaching, plant assimilation, ammonia volatilization, denitrification and immobilization.

Nitrogen fixation : The process of converting dinitrogen gas to chemically reactive forms, where nitrogen combines with other elements, such as oxygen, hydrogen and carbon.

Nitrogen gas : The most abundant form of nitrogen existing as a gas, making up the bulk of the gas in the air (~78%). N_2.

Nitrogen oxides : Primary air pollutants produced from internal combustion engines and other industrial processes.

Nitrogenphosphoruspotassium fertilizers : Fertilizers manufactured using ammonium phosphate, to which potassium (usually potassium chloride) and water have been added; additional nitrogen and other nutrients also may be added.

Nitrous oxide : One of the gaseous form of mineral nitrogen. N_2o .

NMR : Nuclear magnetic resonance spectrometry.

Node (adj. Nodal) : Enlarged portion of a shoot at which leaves or buds arise

Nodes : A joint on a stem where shoots, leaves and flowers arise.

Nodule (v. Nodulation) : Small knot or irregular, rounded lump; on leguminous plants, structures on roots that contain nitrogen-fixing bacteria.

Nomenclature : The scheme by which names are attached to objects, including organisms.

Noninfectious disease : A disease that is caused by an abiotic agent that cannot be transmitted from one plant to another.

Nonlabile phosphorus : A slowly available form of phosphorus.

Nonlimiting water range : The region bounded by the upper and lower soil water content over which water oxygen and mechanical resistance are not limiting to plant growth (1997. Soil science society of america. P. 71).

Nonpathogenic (syn. Avirulent) : Unable to cause disease.

Nonpersistent transmission (syn. Stylet-borne transmission) : A type of virus transmission in which the virus is acquired and transmitted by the vector after short feeding times, and is retained by the vector for only a short period of time.

Nonpolar : A molecule with functional groups lacking dipole moments (e.g. Alkyls and aryls).

Nonpolar functional groups : Functional groups generally containing c and h atoms that do not form dipole moments.

Nonpolar solvent : A solvent with functional groups lacking dipole moments (e.g. Alkyls and aryls).

Nonreducing end : The extreme of the polysaccharide chain that holds the aldol c-1 of the hexose moeity (oxidized with alkaline copper).

Nonseptate : Without cross-walls; coenocytic.

Nontarget genes : Genes in the recurrent parent that are not at the locus of the gene of interest. Basically, nontarget genes are all those genes in the recurrent parent that the backcross breeder would like to retain in the backcrossing program.

Normal : Conforming with, adhering to, or constituting a norm, standard, pattern, level, or type; typical.

Northern Blot : A Method used to transfer RNA molecules separated during electrophoresis to membrane filters for detection of specific nucleotide sequences using probes.

No-till : A cultural system most often used with annual crops, in which the new crop is seeded or planted directly in a field on which the preceding crop plants were cut down, had the tops harvested, or were destroyed by a nonselective herbicide (the old crop is not removed or incorporated into the soil as is common in preparing a plant bed).

Nucleic acid : Genetic material of all living organisms, including DNA (deoxyribonucleic acid) and rna (ribonucleic acid).

Nucleoid : Region in prokaryotes, chloroplasts, and mitochondria where the DNA is concentrated. Unlike a nucleus, it is not bound by a membrane.

Nucleophile : Seeks positive charge; may be negatively charged or neutral.

Nucleotide : A subunit of DNA or RNA consisting of a nitrogenous base (adenine, guanine, thymine, or cytosine in DNA; adenine, guanine, uracil, or cytosine in RNA), a phosphate molecule, and a sugar molecule. Thousands of nucleotides are linked together to form a DNA strand.

Nucleus : Dense aggregation of proteinaceous matter and nucleic acid in cells, surrounded by a membrane; contains chromosomes and controls heredity.

Nutrient removal : One of three approaches to making fertilizer recommendations, nutrient removal is the oldest method and was developed before reliable soil testing was readily available. Under this approach, fertilizer recommendations are based on how much nutrient a specific crop is likely to remove from the soil and how much nutrient will be needed to replace it. It does not account for soils' ability to supply many essential nutrients.

O line : A structure in young embryo which is the lower boundary of shoot apical tissues.

Obligate parasite (syn. Biotroph) : Organism that can grow only as a parasite in association with its host plant and cannot be grown in artificial culture media.

Occlusion : Block or plug that stops flow of liquids (as in vessels).

Oedema (also edema; syn. Intumescence) : Blisters produced on leaves and other plant parts under conditions of high moisture and restricted transpiration.

Offset : A young plant produced at the base of a parent plant.

Oil additives : An additive made either synthetically or extracted from seeds that enhances herbicide uptake into the plant; crop oils, methylated seed oils, crop oil concentrates.

Oligogenic resistance : Resistance conferred by a few genes (see monogenic resistance, multigeneic resistance, polygenic resistance, single gene resistance).

Oligonucleotide : Small fragments of nucleotides hooked together; primers used in PCR.

Oogonium (pl. Oogonia) : Female gametangium of oomycetes, containing one or more gametes.

Oomycetes (adj. Oomycetous) : Class of fungal-like organisms typically with nonseptate mycelium, asexual sporangia and zoospores, and sexual oospores.

Oospore : Thick-walled, sexually-derived resting spore of oomycetes.

Ooze : Mass of bacterial cells mixed with host fluids.

Open reading frame : Portion of a nucleic acid molecule that is translated into a protein.

Order : A group of soils in the broadest category. In the current system of soil classification in the U.S., there are 12 orders, differentiated by the presence or absence of diagnostic horizons: Alfisols, Andisols, Aridisols, Entisols, Gelisols, Histosols, Inceptisols, Mollisols, Oxisols, Spodosols, Ultisols, Vertisols. (adapted from SSSA).

Ordinary superphosphate (0200 + 12s) : A material produced by reacting sulfuric acid with finely ground rock phosphate; about 8590 percent water soluble. An excellent source of phosphorus and sulfur.

Organelle : A membrane-bound structure within a cell having a specialized function, e.g. Mitochondria and chloroplasts.

Organic : Describing a molecule containing carbon atoms; pertaining to living organisms.

Organic chemical : A chemical that is created naturally by using an organism(s) from the environment.

Organic matter : The total of the organic compounds in soil, exclusive of undecayed plant and animal tissues, their "partial decomposition" products, and the soil biomass. The term is often used synonymously with humus. (glossary of soil science terms. Soil sci. Soc. Am. 1998).

Organic molecule : A molecule containing carbon atoms.

Osmoticum : Particles that cause osmosis.

Ostiole (adj. Ostiolate) : Pore; opening in the papilla or neck of a perithecium, pseudothecium, or pycnidium through which spores are released.

Outerbark : Exterior, nonliving portion of bark in woody plants.

Ovary : The female reproductive structure of organisms; in plants, enlarged basal portion of a pistil, containing the ovules and developing into the fruit.

Overseason : To survive or persist from one planting season to the next.

Overwinter : To survive or persist through the winter period.

Oviposit : To deposit or lay eggs with an ovipositor.

Ovule : Enclosed structure that, after fertilization, becomes a seed; egg contained within an ovary.

Oxidation : Removal or exchange of electorns usually by combining with oxygen.

Ozone : A secondary air pollutant that is a highly reactive form of oxygen (o3) formed when the exhaust of automobiles and other internal combustion engines is in the presence of sunlight.

Ozone layer : A protective layer of ozone in the upper atmosphere that reduces ultraviolet radiation.

P680 : The reaction center of photosystem ii which best absorbs light at a 680 nm wavelength. P680 is a chlorophyll a molecule.

P700 : The reaction center of photosystem i which best absorbs light at a 700 nm wavelength. P700 is a chlorophyll a molecule.

Palatability : Summation of plant characteristics that determine the relish with which a particular plant species or plant part is consumed by an animal.

Palisade parenchyma : Tissue found beneath the upper epidermis of leaves, composed of elongate, tubular cells arranged upright in the manner of posts in a palisade fortification.

Pan (abbr. For peroxyacetyl nitrates) : A secondary air pollutant formed when the exhaust of automobiles and other internal combustion engines is in the presence of sunlight.

Pandemic : A widespread and destructive outbreak of disease simultaneously in several countries.

Panicle : An indeterminate inflorescence whose primary axis bears branches of pedicelled flowers.

Paniculate : An indeterminate inflorescence whose primary axis bears branches of pedicelled flowers.

Papilla : Nipple-like projection; used to describe the tip of some sporangia and the localized wall thickenings on the inner surface of plant cell walls at sites penetrated by fungi.

Paragynous : Having the antheridium contact the oogonium on the side, as in many pythium sp. (see amphigynous).

Parasexualism : A type of recombination, found in certain heterokaryotic fungi, that is based on mitosis rather than meiosis. Genetically distinct haploid nuclei fuse in the heterokaryon. The resulting diploid nuclei multiply by mitotic division, with some crossing-over, and a diploid homokaryon develops. During repeated mitotic divisions, chromosomes may be shed by the diploid nucleus to produce a haploid nucleus.

Parasite (adj. Parasitic) : Organism that lives in intimate association with another organism on which it depends for its nutrition; not necessarily a pathogen.

Parenchyma (adj. Parenchymatous) : Soft tissue of living plant cells with undifferentiated, thin, cellulose walls.

Parent material : Rock or minerals which are weathered to form smaller particles of a soil. Parent material is one of the five factors contributing to formation of a specific soil. In the great plains, much parent material is associated with ancient seas or glacier deposits.

Parthenogenesis (adj. Parthenogenetic) : Reproduction by the development of an unfertilized egg.

Partial resistance : Resistance in which disease develops more slowly on the host or to a lesser extent, but the host does become diseased (see complete resistance)

Particle density : The weight of an individual soil particle per unit volume. Particle density is usually expressed in units of grams per cubic centimeter (g/cm^2). Bulk density considers both the solids and the pore space; whereas, particle density considers only the mineral solids.

Pasteurization : To free a material, usually a liquid, of selected harmful microorganisms using heat.

Pat : The enzyme that controls resistance of liberty herbicide by detoxifying the herbicide molecule.

Patchy distribution : Close and often repeated grazing of small patches or even individual plants while adjacent similar patches or individual plants are lightly grazed or not grazed.

Pathogen (adj. Pathogenic) : A disease-producing organism or agent.

Pathogenesis : Production and development of disease.

Pathogenesis-related (pr) proteins : Proteins, such as antimicrobial proteins and hydrolytic enzymes, that are synthesized in the early events of the plant defense response.

Pathogenicity : Ability to cause disease.

Pathogens : An agent that causes disease, especially a living microorganism such as a bacterium or fungus.

Pathology : The study of diseases.

Pathotype : A subdivision of a pathogen species characterized by its pattern of virulence or avirulence to a series of differential host varieties.

Pathovar (abbr. Pv.) : A subdivision of a plant pathogenic bacterial species defined by host range; pathovar for bacteria is equivalent to forma specialis for fungi.

Pathway : A sequence of interconnecting enzyme reactions.

PCR (acronym for polymerase chain reaction) : A technique used to amplify the number of copies of a specific region of DNA in order to produce enough of the DNA for use in various applications such as identification and cloning.

Pectin : A methylated polymer of galacturonic acid found in the middle lamella and the primary cell wall of plants; jelly-forming substance found in fruit.

Ped : A unit of soil structure, formed together by joining fine sand, silt, or clay particles.

Pedicel : Small slender stalk; stalk bearing an individual flower, inflorescence, or spore

Pedicles : Flowerbearing stems.

Peduncle : Stalk or main stem of an inflorescence; part of an inflorescence, or a fructification.

Penetration : Initial invasion of a host by a pathogen.

Penetration peg (syn. Infection peg) : The specialized, narrow, hyphal strand on the underside of an appressorium that penetrates host cells.

Penicillin : Antibiotic produced by certain species of penicillium and effective against gram-positive bacteria.

PEP carboxylase promoter : Phosphenol pyruvate (PEP) carboxylase is an enzyme used in photosynthesis. The promoter of this gene induces protein production in actively photosynthesizing tissues of plants.

Percolation : The downward movement of water through the soil, made possible by pore space in the soil.

Perennial : Something that occurs year after year; plant that survives for several to many years (see annual, biennial).

Perfect flower : Flower possessing both stamens and pistils.

Perfect state (syn. Teleomorph) : The sexual form in the life cycle of a fungus (see anamorph).

Pericarp : Fruit wall.

Pericycle : Layer or layers of cells between the phloem and the endodermis of roots, giving rise to branch roots.

Perithecium (pl. Perithecia) : Flask-shaped or subglobose, thin-walled fungus fruiting body (ascocarp) containing asci and ascospores; spores are expelled or released through a pore (ostiole) at the apex.

Peritrichous : Having hairs or flagella distributed over the whole surface.

Permanent wilting point (pWP) : The soil water content when a plant has extracted all the water it can. At this point, a plant will wilt and not recover.

Peroxyacetal nitrates (PAN) : A secondary air pollutant formed when the exhaust of automobiles and other internal combustion engines is in the presence of sunlight.

Persistent transmission (syn. Circulative transmission) : A type of virus transmission in which the virus is acquired and transmitted by the vector after relatively long feeding times and remains transmissible for a prolonged period while in association with its vector.

Pest : Any organism that damages plants or plant products.

Pesticide : Any substance or mixture of substances used for controlling, preventing, destroying, repelling, or mitigating any pest. A pesticide also includes any substance or mixture of substances intended for use as a plant regulator, defoliant, or dessicant.

Petiole : Stalk portion of a leaf (see lamina).

pH : Negative logarithm of the effective hydrogen ion concentration; a measure of acidity (ph 7 is neutral; values less than ph 7, acidic; greater than ph 7, alkaline).

Phellem : Cork; a protective tissue composed of nonliving cells with suberized walls produced by the phellogen and accumulating on the surface of stems and roots.

Phelloderm : Tissue formed by and internal to the phellogen; resembles the cortical parenchyma in morphology.

Phellogen : Cork cambium; lateral meristem forming the periderm, a protective tissue in stems and roots; phellem (cork) is produced toward the surface, phelloderm toward the inside.

Phenological synchrony : The coordinated development of parasites and their hosts based on degree-days and other environmental factors so that parasites can successfully infect their hosts, such as the production and release of ascospores of venturia inaequalis, the apple scab pathogen, from fallen leaves from the previous season, when new apple leaves are expanding.

Phenotype : External, visible physical characteristics of an organism determined by the interaction of its genotype with the environment (see genotype).

Phenotypic trait : An observable physical characteristic of an organism. See phenotype.

Phenoxyacetic acids : A family of herbicides with similar chemical structure belonging to the growth regulating mode of action. An example would be 2,4d.

Pheromone : Chemical substance that attracts members of the same species or one sex of that species, esp. Insects and nematodes. OR A chemical secreted by an insect or animal that influences the behavior of others within the same species; often functions as an attractant of the opposite sex.

Phialide : End cell of a conidiophore with one or more open ends through which a basipetal succession of conidia develops.

Phloem : Food-conducting, food-storing tissue in the vascular system of roots, stems, and leaves. OR Tissue used for the movement of sugars from source tissue to sink tissue; Composed of living cells called sieve elements that have no nuclei, vacuoles, or ribosomes. The ends of the sieve elements contain connective pores, which join them together to form tubelike structures.

Phloem necrosis : Death of phloem cells, often visible, caused by infection by systemic phloem pathogens such as phytoplasmas, spiroplasmas, and viruses.

Phopt : The optimum pH for a reaction.

Phosphatase : An enzyme produced by some algae species under conditions of low P availability. Phosphatase converts organic P to inorganic P, which the algae can, in turn, use for growth.

Phosphate : Used, instead of phosphorus, in fertilizer technology to refer to available phosphoric acid (P2O5).

Phosphoinositide : Phospholipids containing inositol involved in linking extracellular signals to intracellular responses, or signal transduction.

Phospholipid molecules : A lipid that is mostly nonpolar, but contains a charged phosphate group and possibly other polar functional groups at one end. This type of lipid is a primary component of biological membranes.

Phosphorescence : Light emitted by an excited triplet molecule as part of the deexcitation process. Because a significant amount of the energy in the original exciting photon is lost in the conversion from the excited singlet to the excited triplet state, the wavelength of phosphorescence is greater (i.e. Less energy) than the wavelength of the exciting photon and greater than the wavelength of any fluorescence from the excited singlet state in that molecule.

Phosphoric acid (0540) : Material produced by treating finely ground rock phosphate with sulfuric acid or by smelting (burning) rock phosphate with coke and silica in an electric furnace and reacting with water. Agricultural grade phosphoric acid is seldom used for direct application because of its corrosive properties; it is usually used to make other forms of phosphoric fertilizers.

Phosphorolysis : The cleavage of a glycosidic linkage by the addition of a phosphoryl group to one of the products.

Phosphorus : A key element in the complex nucleic acid structure of plants which regulates protein synthesis; important in cell division and development of new tissues. Next to nitrogen, the most limiting nutrient in Nebraska crop production; naturally found in sufficient amounts in many Nebraska soils.

Phosphorus, bioavailable : This water p fraction, also refer to algalavailable p, is a combination of dissolved p plus the estimated portion of p to be released from sediment p within a short time, e.g. 45 hours, after the particles enter surface water.

Phosphorus, dissolved : This water p fraction, a near equivalent with dissolved reactive p, is readily available to aquatic plants and animals.

Phosphorus, labile : This soil p fraction is typically accounts for less than 1% to less than 5% of the p in soils, and is less tightly bond than stable p. Some labile p, as well as solution p, is measured with agronomic soil tests.

Phosphorus, sediment : This water p fraction, also referred to as particulate p, often is the greatest fraction in water bodies. It consists of organic and inorganic sediments. Its not available to most aquatic plants and animals, but phosphatase producing algae can cause organic sediment p to become available.

Phosphorus, solution : This soil p fraction is typically accounts for less than 1% of the p in soils and is readily available to plants.

Phosphorus, stable : This soil p fraction is also called nonlabile p. It accounts for most of the p in soils, typically more than 95% of the total, and includes tightly bond p in secondary and primary minerals and in organic forms.

Phosphorylation : The attaching of a phosphate group to a substrate.

Phosphotidylinositol : A a membrane lipid called a phosphoinositide. Phosphoinositides are phospholipids containing inositol involved in linking extracellular signals to intracellular responses, or signal transduction.

Photochemical : Chemical reactions that require sunlight.

Photomorphogenesis : The development of a plant cell or organ that is triggered by exposure to light. For example, if a corn plant is germinated in the dark, it will produce an etiolated shoot, lacking chlorophyll or the ability to carry out photosynthesis. Exposing an etiolated shoot to light will initiate development of pigment biosynthesis and the production of cellular components necessary for photosynthesis.

Photon : The basic unit of light. Although light has some properties consistent with a wave, light from the sun is in individual packages. Each photon is a particle of electromagnetic radiation traveling with the speed of light (3 X 10^8 m sec^1). The energy in each photon is dependent on the frequency with which the electromagnetic field vibrates.

Photophosphorylation : The coupling of photosystems I and II with an electron transfer chain that moves electrons from water (which is oxidized to form O_2) to $NADP^+$ (which is reduced to form NADPH). The transfer of electrons between photosystem II and photosystem I releases energy, which is conserved in the form of a transmembrane proton gradient and used to synthesize ATP.

Photosensitizers : Molecules which have efficient conversions to the triplet state after absorption of a photon. These triplet state molecules either reduce other molecules to generate free radicals, or they excite O_2, normally a triplet, to the excited singlet state; singlet oxygen is very reactive and can cause significant cellular damage.

Photosynthate : Product of photosynthesis.

Photosynthesis : Manufacture of carbohydrates from carbon dioxide and water in the presence of chlorophyll(s), using light energy and releasing oxygen. The process in which plants use light energy to make sugars and other organic food molecules from carbon dioxide and water.

Photosynthetic inhibitors : The mode of action of a group of herbicides that directly influence the light reaction of photosynthesis.

Photosystem : An array of pigmentprotein complexes and electron transfer components that function together to harvest light energy, transfer the energy to photochemical reaction centers, and move the excited electrons in a controlled fashion to produce usable biochemical energy. Each photosystem contains hundreds of chlorophyll and carotenoid molecules functioning as antennae, while only a few chlorophyll molecules are employed in the reaction centers.

Phycomycete : Archaic term for a member of a group of fungi without cross walls (septa) in their mycelium.

Phyllody : Change of floral organs to leaflike structures.

Phylloplane : Leaf surface.

Phylloplane-competent : Used to describe epiphytic microorganisms adapted to living on the leaf surface.

Phyllotaxy : Arrangement of leaves on a stem in relation to one another.

Physical structure : The combination or arrangement of soil particles into secondary units or peds. (glossary of soil science terms. Soil sci. Soc. Am. 1998).

Phytoalexin : Substance produced in higher plants in response to a number of chemical, physical, and biological stimuli that inhibits the growth of certain micro-organisms.

Phytoglycogen : A minor constituent of starch having a branched structure in which an α-d-1,6 bond occurs every 10–15 glucose units linked via α-d-1,4 bonds.

Phytohormones : Compounds produced by the plant to regulate growth and development of the plant.

Phytopathogenic : Able to cause disease in plants.

Phytopathology (syn. Plant pathology) : The study of plant diseases.

Phytoplasma (syn. Mycoplasmalike organism, mlo) : Plant-parasitic pleomorphic mollicute (prokaryote with no cell wall) found in phloem tissue; cannot yet be grown on artificial nutrient media.

Phytosanitary certificate : Official document that indicates that plant material has been inspected and found to be free of certain pathogens.

Phytotoxic : Harmful to plants (usually used to describe chemicals). OR A substance which is toxic to plants.

Pigment : A colored compound, such as chlorophyll, in the cells of plants or fungi

Pigment synthesis inhibitors : The mode of action of a group of herbicides that stops pigment synthesis (not chlorophyll)i these pigments are present to protect chlorophyll from breakdown by sunlight. Without the pigments present, chlorophyll is easily degraded showing bleaching symptoms on plant tissue.

Pipette : A tool used to accurately dispense specific amounts of a liquid.

Pistil : The female reproductive structure in plants; the stigma, style and ovary collectively.

Pith : Parenchymatous tissue occupying the center of the stem.

Pka : The dissociation constant of an acid; the ph at which 50% of the molecules are associated (have the H+ ion) 50% are dissociated (lost the H' ion).

Plan : Shape of the slope viewed from the (map view, bird'seye view).

Plant available water : That portion of soil water that can be absorbed by a plant. Generally, plant available water is considered to be 50 percent of the water holding capacity.

Plant pathology (syn. Phytopathology) : The study of plant diseases.

Plasma membrane : The semipermeable membrane that surrounds the protoplasm of a cell.

Plasmid : A small circular piece of DNA from bacteria that often contains antibiotic resistance genes. Some types have the capability to be replicated in the bacteria and integrate itself into the genome of a plant.

Plasmodesma (pl. Plasmodesmata) : Cytoplasmic strands that connect living cells.

Plasmodiophoromycetes : A fungus-like group of organisms, sometimes called the plasmodiophoromycota or the endoparasitic slime molds, characterized by the production of zoospores and plasmodia that are restricted to the cells of their host.

Plasmodium (pl. Plasmodia) : Naked multinucleate mass of protoplasm moving and feeding in amoeboid fashion.

Plasmogamy : The fusion of two sex cells.

Plasmolysis : Shrinking of a protoplast due to water loss from the vacuole of a plant cell.

Pleiotropy : The control of more than one trait by a single gene.

Pleomorphic : Able to assume various shapes (and perhaps sizes); occurrence of more than one distinct form in the life cycle of an organism.

Ploidy : The number of (complete) sets of chromosomes in a cell, e.g. Haploid, diploid, polyploidy. OR The number of sets in a cell of basic chromosomes; often affects plant vigor, flower size and number, and ability to hybridize; an important fact in daylilies and orchid species.

Plow : Operation that completely turns the soil over to a depth of about 8 inches; uses a moldboard plow.

Polar : More hydrophilic or water loving.

Polar Axis of Plant : The longitudinal directions a plant grows.

Polar functional groups : Functional groups with atoms such as o, n, s or p which do not equally share the electrons in the bonds, resulting in dipole moments

Polar molecules : Molecules with functional groups containing dipole moments (e.g., alcohols, carbonyls, amides, amines).

Polar solvent : A solvent with molecules containing functional groups with dipole moments.

Polarity : Property of a covalent bond or an entire molecule that results in a negative pole and a positive pole; A bond becomes polar if the electrons involved in the bond are not shared equally. Polarity results when the center of the negative charge does not coincide with the center of the positive charge.

Pollen : Male sex cells produced by anthers of flowering plants or the cones of seed plants. OR Dustlike material produced by the anther (male reproductive structure) that is composed of pollen grains (microgametophytes) which carry the male gametes of seed plants; important in fertilization.

Pollination : Transfer of pollen from anther to stigma or from a staminate cone to an ovulate cone.

Polyclonal antibody : A preparation containing antibodies against more than one epitope of an antigen (see monoclonal antibody).

Polycyclic : Having several to many disease cycles in a growing season (see monocyclic).

Polyetic : Requiring more than one year to complete one life or disease cycle.

Polygenic : Pertaining to, or governed by, many genes (see monogenic).

Polygenic resistance (syn. Multigenic resistance) : Resistance conferred by several genes (see monogenic resistance, oligogenic resistance, single gene resistance).

Polymerase chain reaction (PCR) : A technique used to amplify the number of copies of a specific region of DNA in order to produce enough of the DNA for use in various applications such as identification and cloning.

Polymorphic : Existing in multiple forms. When a marker locus is polymorphic, there are detectable differences in DNA sequence at that locus.

Polyploid : Having three or more complete sets of chromosomes (see haploid, diploid). OR An organism or cell having more than two complete sets of chromosomes.

Polyploidy : Having more than 2n (diploid) number of basic chromosomes such as triploid (3n) or tetraploid (4n); very common in angiosperms.

Polyprotein : A protein translated from an entire viral genome which is then cleaved by proteases (protein-degrading enzymes) into the active protein products.

Pome (syn. Pome fruit) : Simple, fleshy, indehiscent fruit derived from several carpels, e.g. Apple, pear, quince.

Pore space : Spaces in soil, between the mineral and organic matter, that are filled with water or air.

Porosity : The volume of soil voids that can be filled by water and/or air; inversely related to bulk density. Porosity is also known as "pore space."

Positive sense RNA : Rna that can serve directly as messenger rna (see negative sense rna)

Postemergence : Application of a herbicide after the plant has emerged.

Posterior : Toward the back or rear; (see anterior).

Postzygotic mechanisms : Barriers to gene flow between to two populations *after fertilization*, causing reproductive isolation that leads to two different species (genotype) of an organism; but they may not be distinguishable by appearance (phenotype). Barriers include death of zygote, hybrid inviability, hybrid sterility, and hybrid breakdown (reduced viability and fertility of second generation).

Potassium : An essential plant nutrient needed in large amounts. Postassium is vital to plant nutrient absorption, respiration, transpiration and enzyme activity. The major portion of potassium is contained in minerals such as feldspar and mica, and clays such as montmorillonite, vermiculite and illite.

Potassium chloride : The most commonly used and most economical potassium fertilizer. It is mined as sylvinite from underground mines or by solution mining in new mexico, utah, california and michigan.

Potassium sulfate : Material used as a potassium fertilizer for crops where the chloride content of potassium chloride is undesirable.

Potassiummagnesium sulfate (sul po mag)(0022 + 22s) : A double salt of potassium and magnesium sulfate; commonly used as a dry granular source of potassium, magnesium and sulfur.

Powdery mildew : A white, powdery, superficial ascomycetous fungus that is an obligate parasite and generally is found only on one or a few closely related species of plants.

PPO : Protoporphyrinogen oxidase (protox); enzyme utilized in photosynthesis and is the target site of many herbicides in the cell membrane disrupter mode of action.

Predispose (n. Predisposition) : To make prone to infection and disease.

Preference : Selection of certain plants, or plant parts, over others by grazing animals.

Prezygotic mechanisms : Barriers to gene flow between two populations *berfore* *fertilization,* causing reproductive isolation leading to two different species (genotype) of an organism; but they may not be distinguishable by appearance (phenotype). Barriers include geographical isolation, egolocial isolation, seasonal isolation, behavioural isolation, mechanical isolation, and gametic incompatibility.

Primary inoculum (syn. Initial inoculum) : Inoculum, usually from an overwintering source, that initiates disease in the field, as opposed to inoculum that spreads disease during the season.

Primary leaf : First true leaf that emerges on a plant following the cotyledons.

Primary pollutant : Air pollutant that is released directly into the atmosphere and is harmful to plants, e.g. So2 or nox.

Primary root : Root that develops directly from the radicle of an embryo rather than from a crown or node

Primer : Small fragment of nucleic acid with a free 3'-hydroxyl group necessary for initiation of DNA, and, sometimes, rna synthesis; often specific fragments chosen for use in polymerase chain reaction (pcr) for rapid identification of pathogens. OR Small section of DNA nucleotides which bind to the singlestranded DNA template during PCR. These can be sequences specific for a gene or totally random, depending upon the experiment's objective.

Primer annealing step : The second step of pcr. In this step the primers will bind to their complementary DNA sequences on the singlestranded DNA template.

Primerdimers : Primerdimers occur when singlestranded primer oligonucleotides bind to each other rather than the DNA template.

Probe : A macromolecule such as DNA or RNA that has been labeled with either radioactivity or antibodies and can be detected by an assay. Probes are used to identify target molecules, genes, or gene products.

Procambium : The tissue in the central portion of an embryo.

Proembryonal complex : A mass of embryogenic cells.

Profile : The shape of a slope profile. Profile is the slope viewed in a vertical crosssection.

Progenitor cell : A cell that is able to produce embryogenic cells by cell division.

Progeny : The offspring of an organism.

Progeny testing : A test of the value of a genotype based on the performance of its offspring produced in some definite system of mating.

Proherbicide : A molecule that doesn't become herbicidally active until a reaction within a plant makes it a herbicide.

Proinhibitin : A plant defense compound that is already present before invasion by a pathogen

Prokaryote : An organism without internal membrane-bound organelles, lacking a distinct nucleus, such as bacteria and mollicutes (see eukaryote).

Promoter : A specific DNA sequence to which RNA polymerase binds and initiates transcription. This region contains information which regulates when and how often the gene is transcribed and ultimately the amount of protein it produces.

Promycelium (pl. Promycelia) : In rust and smut fungi, a germ tube issuing from the teliospore and bearing the basidiospores.

Propagative transmission (syn. Circulative propagative transmission) : Pathogen transmission characterized by a long period of acquisition of the pathogen (usually a mollicute, e.g. Phytoplasma or spiroplasma, and sometimes a virus) by a vector (typically an insect), a latent period before the vector is able to transmit the pathogen, and retention of the pathogen by the vector for a long period because the pathogen reproduces or replicates in the vector.

Propagule : Any part of an organism capable of independent growth.

Propanoic herbicides : A family of herbicides that have similar chemical structure and belong to the lipid synthesis inhibiting mode of action (fops).

Protectant : Agent, usually a chemical, applied to a plant surface in advance of a pathogen to prevent infection.

Protectant fungicide (syn. Contact fungicide) : A fungicide that remains on the surface where it is applied; no after-infection activity (see systemic fungicide).

Protection : A principle of plant disease control in which a barrier is placed between the susceptible plant and pathogen (e.g. The use of protective chemical dusts or sprays).

Protein : A large molecule composed of one or more chains of amino acids in a specific order. Proteins are necessary for the structure, function, and regulation of an organism's cells, tissues, and organs. Each protein has a unique function determined by its shape.

Protein kinases : Enzymes that phosphorylate proteins using atp.

Protein synthesis : The production of proteins in a cell. Proteins are chains of amino acids linked in the order determined by the genetic code.

Proteins 14-3-3 : Phosphoserine-binding proteins that regulate the activities of a wide array of targets via direct protein–protein interactions. The 14-3-3

proteins were first identified as abundant brain proteins and named after their elution position on ion exchange chromatography and mobility in starch gel electrophoresis.

Proteomics : A system for identifying proteins within a cell or tissue sample.

Protoderm : A primary embryonic tissue that is located on the surface of young embryos.

Proton : A positivelycharged atomic particle; a hydrogen ion H+.

Proton motive force (pmf) : Created when protons are extruded from the cytosol by electrogenic h^+atpases, both at the plasma membrane and the tonoplast, a membrane potential and a ph gradient are created at the expense of atp hydrolysis. This gradient of electrochemical potential for h^+ represents stored free energy in the form of the h^+ gradient.

Protonated : Having a proton attached, usually it is a H+ when considering herbicides.

Protoplasm : Living contents of a cell.

Protoplasm : All the substance of a cell, usually considered not to include the cell wall. The protoplasm of a single cell is a protoplast.

Protoplast : Living cell exclusive of a wall. OR A plant cell that has had the cell wall removed.

Protoporphyrin : An intermediate in the biosynthetic pathways for chlorophyll and heme. This molecule is a strong photosensitizer and causes much cellular damage if it accumulates in tissues exposed to light.

Proximal : Nearest to the point of attachment (see distal).

Pseudothecium (pl. Pseudothecia) : Perithecium-like fruiting body containing asci and ascospores dispersed rather than in an organized hymenium; an ascostroma with a single locule or cavity and containing bitunicate asci.

Pubescent : Covered with hairs; ex. Stachys, Pulsatilla.

Puccinia pathway : The region through which rust urediniospores move from southern areas through all grain-producing areas of the united states to canada each season.

Pullulanase : Catalyzes the hydrolysis of α-d-1,6 glucosidic linkages in pullulan [a linear polymer of α-1,6-linked maltotriose units] and in amylopectin and glycogen, and the α- and β-limit dextrins of amylopectin and glycogen.

Pupa : The nonfeeding stage between the larva and adult. The organism undergoes a complete transformation within a protective cocoon or hardened case.

Pupa (pl. Pupae; v. Pupate) : Quiescent stage between the larva and the adult of certain insects.

Pustule : Small blisterlike elevation of epidermis formed as spores emerge.

Pv. (abbr. For pathovar) : A subdivision of a plant pathogenic bacterial species defined by host range; pathovar for bacteria is equivalent to forma specialis for fungi.

Pycnidiospore : Spore (conidium) produced in a pycnidium.

Pycnidium (pl. Pycnidia) : Asexual, globose or flask-shaped fruiting body of certain imperfect fungi producing conidia.

Pycniospore (syn. Spermatium) : Haploid, sexually derived spore formed in a pycnium of rust fungi.

Pycnium (pl. Pycnia; syn. Spermagonium) : Globose or flask-shaped haploid fruiting body of rust fungi bearing receptive hyphae and pycniospores.

Pyramiding : The addition, through plant breeding or genetic engineering, of several resistance genes into a single plant cultivar.

QTL : Quantitative trait locus. (1) a locus that influences the expression of a quantitative trait. (2) a chromosome region detected by statistical analysis that is significantly associated with variation for a quantitative trait.

Qualitative characters : A character in which variation is discontinuous.

Quantitative characters : A character in which variation is continuous so that classification into discrete categories is not possible.

Quantitative resistance : Resistance reactions that have no distinct classes but vary continuously from resistant to susceptible, the result of few to many genes the individual effects of which may be small and difficult to detect. (see qualitative resistance).

Quarantine : Legislative control of the transport of plants or plant parts to prevent the spread of pests or pathogens.

Quenching : The process of causing the deexcitation of a molecule from an excited singlet or triplet state back down to the ground state. Quenching of chlorophyll is an important protective role of carotenoids. If a triplet molecule is not rapidly quenched, it can react with oxygen to generate singlet oxygen. Singlet oxygen can cause much cellular damage.

Quiescent : Dormant or inactive.

Quorum sensing : The ability of bacteria to interact with each other through a variety of mechanisms; allows a population of bacteria to behave more like a multicellular organism.

Race : Subgroup or biotype within a species or variety, distinguished from other races by virulence, symptom expression, or host range, but not by morphology.

Race non-specific resistance (syn. General resistance, horizontal resistance) : Resistance which is effective against all biotypes of the pathogen. (see durable resistance, specific resistance, vertical resistance).

Raceme : A simple, indeterminate inflorescence with pedicelled flowers.

Rachis : Elongated main axis of an inflorescence.

Radicle : Part of the plant embryo that develops into the primary root.

Range condition : The present state of vegetation of a range site in relation to the natural potential plant community for that site.

Rangeland : Land on which the native vegetation is predominantly grasses, grasslike plants, forbs, or shrubs.

RAPD : Random amplified polymorphic DNA. These markers involve pcr amplification of DNA fragments using primers that are random 10base oligonucleotides. A single rapd reaction may amplify several to many DNA fragments. Rapds are dominant markers and, therefore, less informative than codominant markers because heterozygous individuals cannot be identified. A major advantage is that no prior knowledge of an organism's genome is necessary for their use. A common problem with rapds has been lack of reproducibility among labs. The technique is described in more detail in s. Grandillo and t.m. Fulton (2002), approaches to gene mapping. Pp. 101136. In p.m. Gilmartin and c. Bowler (ed.) Molecular plant biology vol. 1. Oxford univ. Press, oxford and new york.

Ray : Tissue that extends radially in the secondary xylem and phloem of a woody plant

Reaction center : A specialized component of the photosystem that actually carries out the photochemistry with the excitation energy harvested by antenna pigments.

Receptacle : The structure of a flower that bears the reproductive organs.

Receptacle : The expanded end of a pedicel (stalk) that holds the flowers or floral organs.

Receptive hypha : The part of a rust fungus pycnium (spermogonium) that receives the nucleus of a pycniospore (spermatium).

Receptor : A site that recognizes and binds an elicitor; any organ or molecular site that is sensitive to a distinct (specific) signal molecule. OR Special proteins in the cell membrane that bind to a signal or chemical messenger from the environment and transmit signals to the cell to initiate a response.

Recessive : Describes a phenotypic trait that is expressed in diploid organisms only if both parents contribute the trait to the progeny (see dominant). OR A trait that will not be expressed unless there are two copies of the gene allele present in an organism.

Recessive allele : An allele whose phenotype is not expressed in a heterozygote.

Recombinant DNA : DNA molecules in which sequences, not normally contiguous, have been placed next to each other by in vitro methods. OR DNA that has been altered and is different from the original sequence.

Recombinant plasmid : A plasmid that has been cut with special enzymes and "recombined" in a new order or with DNA from another source.

Recombination frequencies/map Distance : Recombination frequency is the frequency of crossovers between two genes. The map distance (in this case a genetic or recombinational map) is the recombination frequency between two genes times 100. For example, if the frequency of recombination is 0.10, the map distance is 10 map units.

Recurrent parent : The parent to which successive backcrosses are made in backcross breeding.

Reduction : The opposite of oxidation. When a molecule acquires one or two electrons it becomes more reduced.

Regiospecificity : The ability of an enzyme to target a particular bond with in a molecule.

Regulatory protein : A protein that binds to the promoter of a gene to regulate transcription.

Relative fitness : The fitness of one genotype when compared with another.

Relative humidity : The ratio of the amount of water vapor actually present in the air to the greatest amount possible at the same temperature.

Reniform : Kidney-shaped.

Replicase : An enzyme involved in the replication (copying) of nucleic acid.

Replicate : The copying of a DNA molecule in a cell.

Replicated : Copied DNA. To make copies of DNA.

Replication : The process by which a DNA or rna molecule makes an exact copy of itself;

Repetition of an experiment or procedure at the same time and place (one of several identical experiments, procedures, or samples);

The process by which a virus particle induces the host cell to reproduce the virus. OR The copying of a DNA molecule.

Resin : Sticky to brittle plant product derived from essential oils; often a defense compound of conifers.

Resinosis : A disease symptom in which there is a copious flow of resin from a wound or infection site of a conifer.

Resistant (n. Resistance) : The ability of an organism to survive and thrive in the presence of something that would normally cause damage or death, i.e., herbicideresistant corn, Roundupready corn. OR The inherited ability of a population to survive and reproduce following repeated exposure to a dose of herbicide normally lethal to the wild type. Resistance may also be induced by such techniques as genetic engineering or selection of variants produced by tissue culture or mutagenesis. OR Possessing properties that prevent or impede disease development (see susceptible).

Respiration : Series of chemical reactions that make energy available through oxidation of carbohydrates and fat. OR The chemical process by which simple sugars such as glucose are oxidized to carbon dioxide and water with the liberation of energy in the form of ATP.

Resting spore : Spore, often thick-walled, that can remain alive in a dormant state for some time, later germinating and capable of initiating infection.

Restriction endonuclease : Enzyme that cleaves DNA at a particular base sequence; sometimes informally referred to as a restriction enzyme.

Restriction enzymes : Enzymes that are able to detect a particular nucleotide sequence and cut the DNA, rna, or protein at that location.

Reticulate : Having netlike markings.

Reverse transcriptase : Enzyme used to make complementary DNA from a piece of rna, such as a plant virus.

RFLP : Restriction fragment length polymorphism. DNA is digested with a restriction enzyme and the fragments are separated by electrophoresis. The DNA fragments are then transferred to a membrane and hybridized with a labeled DNA probe. Polymorphisms are detected as labeled DNA fragments of different sizes. Rflps are very informative due to their codominant nature (they distinguish between homozygotes and heterozygotes). Large numbers of rflp markers are available for many species. Disadvantages are that they are relatively expensive and time consuming, and they require a large quantity of DNA. Most commonly, radioactive labeled probes are used, but nonradioactive methods are also available.

RFLP mapping : (restriction fragment length polymorphism) a technique which uses restriction enzymes to cut DNA and then compare the different sizes of specific fragments created with those from another individual to determine genetic difference.

RH : Relative humidity.

Rhizobium : Genus of bacteria that lives symbiotically with roots of leguminous plants; during the symbiosis, atmospheric nitrogen gas is converted into a form useable by the plant.

Rhizobium nodules : Galls on roots caused by rhizobium spp.

Rhizomatous : Having or looking like a rhizome.

Rhizome : A mostly horizontal, jointed, fleshy, often elongated, usually underground stem. OR A specialized horizontal stem that gives rise to shoots and roots at its nodes; usually found underground, i.e., iris, rhubarb.

Rhizomorph : Macroscopic rope-like strand of compacted tissue formed by certain fungi.

Rhizosphere : Microenvironment in the soil, immediately around plant roots.

Rhizosphere-competent : Used to describe microorganisms adapted to living in the rhizosphere of a plant.

Ribonucleic acid (abbr. RNA) : Several nucleic acids composed of repeating units of ribose (a sugar), a phosphate group, and a purine (adenine or guanine) or a pyrimidine (uracil or cytosine) base; transcribed from DNA and involved in translation to proteins.

Ribosomal RNA (abbr. rRNA) : RNA molecules forming part of the ribosomal structure.

Ribosome : A large molecule that catalyzes the translation of mrna codons into an amino acid sequence.

Rill erosion : The erosion process which causes the formation of numerous small channels of less than to two to three inches deep.

Ringspot : Disease symptom characterized by yellowish or necrotic rings enclosing green tissue, as in some plant diseases caused by viruses.

RNA (abbr. For ribonucleic acid) : Several nucleic acids composed of repeating units of ribose (a sugar), a phosphate group, and a purine (adenine or guanine) or a pyrimidine (uracil or cytosine) base; transcribed from DNA and involved in translation to proteins.

RNA polymerase : An enzyme that catalyzes the synthesis of rna by copying the nucleotide sequence of the DNA.

Rogue : To remove and destroy individual plants that are diseased, infested by insects, or otherwise undesirable.

Root cap : A group of cells on a root that protects the growing tip.

Root exudates : The various compounds that leak from growing and expanding sections of roots as well as from broken cells at exit points of lateral roots.

Root graft : The fusion of roots of two adjacent plants so that their water and food conducting (vascular) systems become joined.

Root hair : Threadlike, single-celled outgrowths from a root epidermal cell.

Rootstock : Portion of the stem (trunk) and associated root system into which a bud or scion is inserted in grafting; fleshy overwintering part of a herbaceous perennial plant with buds and eyes (see scion).

Rosette : Disease symptom characterized by short, bunchy growth habit due to shortened internodes and no comparable reduction in leaf size.

Rot : Softening, discoloration, and often disintegration of plant tissue as a result of fungal or bacterial infection.

Rotation : Growth of different kinds of crops in succession in the same field.

Roundup : A herbicide that provides nonselective control of several annual and perennial weeds. Roundup will also damage crops, such as corn and soybeans that are not Roundupresistant.

Roundup ready : A trade name given to certain varieties of corn or soybean which are resistant to the herbicide, Roundup, whose active ingredient is glyphosate.

Roundup resistant : A plant that is not harmed by the application of Roundup herbicide, i.e., Roundup Ready Corn, Roundup Ready Soybeans.

Roundworm : Nematode.

rRNA (abbr. For ribosomal rna) : Rna molecules forming part of the ribosomal structure.

RTPCR : Realtime PCR. This is a term used to describe PCR methods which detect DNA amplification in "real time" as the experiment is taking place, rather than at the end of the experiment.

Rugose : Wrinkled, roughened.

Runner (syn. Stolon) : A slender horizontal stem that grows close to the soil surface.

Runner plants : New plants produced asexually on a runner or stolon.

Runoff : The portion of precipitation or irrigation on an area that does not infiltrate but, instead, is discharged from the area. That which is lost without entering the soil is called "surface runoff."

Russet : Brownish, roughened areas resulting from cork formation.

Rust : A disease caused by a specialized group of basidiomycetes that often produces spores of a rusty color.

Safener : A compound used in conjunction with a herbicide to provide extra protection to the crop without reducing weed control. A safener usually acts by increasing GSH and GST levels and activity.

Salinity : The amount of soluble salts in a soil. The conventional measure of soil salinity is the electrical conductivity of a saturation extract, usually in units of decisiemens per meter (ds m1). (SSSA and Dennis mccallister).

Saltation : The detachment of soil particles from aggregates due to the impact of windblown soil particles.

Sand : Individual rock or mineral fragments in a soil that range from 0.05 to 2.0 millimeters in diameter. Most sand grains consist of quartz, but they can be of any mineral composition. Sand is also the textural class name of any soil that contains 85 percent or more sand and no more than 10 percent clay.

Sanitation : Destruction or removal of infected and infested plants or plant parts; decontamination of tools, equipment, containers, work space, hands, etc.

Sap transmission : Transmission, usually of viruses, by rubbing sap from an infected plant onto a healthy plant to cause infection.

Saprobe (syn. Saprophyte) : Organism that obtains nourishment from non-living organic matter.

Saprophyte (adj. Saprophytic; syn. Saprobe) : Organism that obtains nourishment from non-living organic matter.

Sapwood : Physiologically active zone of wood contiguous to cambium (see heartwood).

Saturation : Soil conditions where all the pores are completely filled with water. The soil tension exhibited on water under this condition is zero, and gravitational is the dominant force moving water downward.

Scab : Roughened, crustlike diseased area on the surface of a plant organ.

Scald : A necrotic condition in which tissue is usually bleached and has the appearance of having been exposed to high temperatures.

Scar : Sequence characterized amplified region. To improve their specificity and repeatability, a rapd or aflp marker can be converted into a scar marker.

Scars are developed by obtaining the DNA sequence of a rapd or aflp band and designing 1825 base pcr primers to amplify the same DNA segment. Scars are easier to employ in markerassisted selection than rapds or aflps.

Scion : Portion of a shoot used for grafting onto the root stock (see rootstock).

Sclerenchyma (adj. Sclerenchymatous) : Tissue made up of thick-walled plant cells.

Sclerotium (pl. Sclerotia) : A vegetative resting body of a fungus, composed of a compact mass of hyphae with or without host tissue, usually with a darkened rind.

Sccrch : Any symptom that suggests the action of flame or fire on the affected part, often seen at the margins of leaves.

Secondary embryogenesis : The development of embryos from young embryos.

Secondary infection : Infection resulting from the spread of infectious material produced after a primary infection or from secondary infections without an intervening inactive period.

Secondary inoculum : Inoculum produced by infections that took place during the same growing season.

Secondary messengers : Transient secondary signals inside the cell that greatly amplify the original signal.

Secondary metabolite : A compound produced in microbes (e.g., mycotoxins, syringomycins) or plants (e.g., caffeine or nicotine) that is not necessary for normal growth and development.

Secondary organism : Organism that multiplies in already diseased tissue but is not the primary pathogen.

Secondary pollutant : Air pollutant that must be chemically produced from other air pollutants, e.g. Ozone (o3) that is a product of a photochemical reaction of exhaust products from combustion engines in the atmosphere.

Secondary root : Branch from a primary root.

Sedentary : Remaining in a fixed location (see migratory).

Sediment : Eroded particles that are no longer part of the soil. These sediments, sand silt, or clay size, are carried by wind or water and deposited to a different area, like surface water bodies. Sedimentation occurs when eroded soil is deposited and/or settles in water. Erosion, itself, is the process of detaching and transporting soil particles.

Sedimentation : Occurs when eroded soil deposits and/or settles out in water.

Seed : Ripened ovule consisting of an embryo and stored food enclosed by a seed coat.

Seed treatment : Application of a biological agent, chemical substance, or physical treatment to seed, to protect the seed or plant from pathogens or to stimulate germination or plant growth.

Seedborne : Carried on or in a seed.

Seedling growth inhibitors : Mode of action of a group of herbicides that generally disrupt early growth of plants.

Segregating population : A collection of lines from a cross or population that is undergoing Mendelian or Hardyweinberg gene segregation.

Segregation distortion : Deviation of the expected segregation ratio from that expected by mendelian inheritance. Distorted segregation means that either the maternal or paternal allele was transmitted to the progeny at a higher frequency than expected.

Selectable marker gene : A gene encoding an easily detectable trait transferred into a cell with another gene encoding a less detectable trait. The selectable marker gene allows genetic engineers to select transgenic cells and plants from those that are nontransgenic.

Selection media : Agarose media containing the substance for which the selectable marker gene encodes resistance. When tissue culture cells are place on this media, cells that have not been transformed and do not contain the gene conferring resistance to the substance will die leaving only those cells that are transgenic.

Selection pressure : The process that changes the relative frequency of one or more genes within a population.

Selective : Herbicides which cause one species of plants to die, but another species will survive the same treatment. Selectivity allows for acceptable weed control without injury to crop plants.

Selective grazing : Taking certain plant species, individual plants, or plant parts to the exclusion of others.

Selective medium : A culture medium containing substances that specifically inhibit or prevent the growth of some species of microorganisms

Selectivity : Differential effect of a herbicide when applied to a mixed population of plants; sensitive plants die, tolerant plants are unharmed. Implies that different plant species do not respond in the same way to a particular herbicide. Herbicides which cause one species of plants to die, but not another, are selective. Selectivity allows for acceptable weed control without injury to crop plants.

Self-fertilization : A process in which sexual reproduction occurs as a result of the fusion of sex cells produced by the same individual (see cross-fertilization)

Senesce (adj. Senescent, n. Senescence) : To decline, as with maturation, age, or disease stress. OR The end of a plant's life cycle when it reaches maturity and leaves dry up.

Sepal : One of the modified leaves comprising a calyx.

Sepals : A modified leaf which is usually green and part of the caylx; can protect the flower when in bud.

Septate : With cross walls; having septa

Septum (pl. Septa; adj. Septate) : Dividing wall; in fungi, cross wall

Sequencing : Determination of the order of nucleotides in a DNA or RNA molecule or the order of amino acids in a protein.

Serology (adj. Serologic) : A method using the specificity of the antigen-antibody reaction for the detection and identification of antigenic substances and the organisms that carry them

Serrate : Edges with teeth, like a saw

Sessile : Used in reference to a leaf, leaflet, flower, floret, fruit, ascocarp, basidiocarp, etc., without a stalk, petiole, pedicel, stipe or stem; (of nematodes) permanently attached; not capable of moving about.

Sessile : Without a petiole (leaf) or pedicel (stalk of a flower).

Seta (pl. Setae) : Bristle or hair-like structure, usually deep yellow or brown and thick-walled.

Severity values : A means to quantify accumulating opportunities for pathogen infection (e.g. Infection periods) to a pre-determined threshold that requires a disease management activity (e.g. A fungicide application); used in disease prediction or forecasting.

Sexual reproduction : Reproduction involving fusion of two haploid nuclei (karyogamy) to form a diploid nucleus followed by meiosis (reduction division) back to haploid nuclei at some point in the life cycle, resulting in genetic recombination.

Sexual spore : Spore produced during the sexual cycle.

Sexually compatible : Able to be cross-mated or cross-fertile.

Shattercane : *Sorghum bicolor.* A summer annual that resembles grain sorghum and corn. It is derived from wild and/or cultivated sorghum varieties.

Shattercane is a major weed problem in corn, sorghum, and soybean fields in the Central United States. (California, 2001).

Shear strength : The ability of a soil or rock to resist shearing (tangential) forces and/or stresses.

Sheet erosion : The removal of a relatively uniform thin layer of soil for the land surface by unchanneled runoff, or sheet flow.

Shikimate pathway Important chemical pathway in plants. It is responsible for producing amino acids and other aromatic products.

Shoot : Portion of the developing seedling above the root system.

Shot-hole : Symptom in which small lesions fall out of leaves, giving the leaf the appearance of being hit by buckshot.

Shrinkswell potential : The susceptibility of a soil to a change in volume due to a loss or gain in moisture, especially those soils with a relatively high clay content.

Sieve element (syn. Sieve tube element) : A tube-shaped living cell in the phloem functioning in the transport of dissolved organic substances in the plant.

Sign : Indication of disease from direct observation of a pathogen or its parts (see symptom).

Signal transduction pathway : The process whereby a receptor initiates one or more sequences of biochemical reactions that connect the stimulus to a cellular response.

Silt : A soil inorganic separate in the range of 2 to 50 micrometers (or 0.002 to 0.05 mm.). Silt is smaller than sand but larger than clay.

Single gene resistance (syn. Monogenic resistance) : Resistance conferred by a single gene.

Singleseed descent : Breeding procedure in which progenies of a segregating population are selfpollinated and advanced to the next generation by a single seed in each progeny line.

Singlet states : An orbital to which an electron may move after being excited by a photon. Singlet state orbitals require that the excited electron have a spin opposite to that of the spin on the electron remaining in the ground state.

Site of Absorption : Site of Absorption is the location where the herbicide is taken up by the plant. This should not be confused with Site of Action which is the biochemical pathway within the plant where the herbicide acts.

Site of action : In a biochemical pathway, the point at which a herbicide functions. OR The Site of Action is the biochemical pathway a particular herbicide acts upon in a plant. Currently there are over 20 sites of action .

Site of Uptake : Site of Uptake is the location where the herbicide is taken up by the plant. This should not be confused with Site of Action which is the biochemical pathway within the plant where the herbicide acts.

Slime molds (syn. Myxomycetes) : Saprophytic organisms that form vegetative amoeboid plasmodia and spores.

Smectite : A group of silicate (i.e., rocks or minerals with their primary unit as silicon tetroxide, sio_4) clay minerals that have shrinkswell property.

Smut : A disease caused by a smut fungus (ustilaginales) in the basidiomycota or the fungus itself; it is characterized by masses of dark brown or black, dusty to greasy masses of teliospores that generally accumulate in black, powdery sori.

SNP : Single nucleotide polymorphism. A single position in a given DNA sequence that contains allelic variations within a population at relatively high frequencies. For example, replacement of a guanine (g) with a thymine (t) nucleotide is an example of a snp. Snps occur often in many genomes, thus facilitating the development of highdensity marker maps.

Snrk1 : The plant protein snrk1 is homologous to the product of the SNF1 (sucrose non-fermenting-1) gene that was identified genetically in screenings of budding yeast mutants. These variants fail to express the invertase gene, SUC2, in response to glucose deprivation. Snrk1 phosphorylates and, in so doing, inactivates in vitro sucrose 6F-phosphate phosphatase, 3-hydroxy-3-methylglutaryl-Coenzyme A reductase (linked to sterol/isoprenoid synthesis), and nitrate reductase (associated to nitrogen assimilation).

Soft rot : Softening, discoloration, and often disintegration of plant tissue as a result of fungal or bacterial infection.

Soil : The layer of the Earth's surface, consisting of four major components: air, water, organic matter and mineral matter. There are three categories of soil particlessand, silt and claywhich are called "soil separates."

Soil age : Determined by the amount of weathering that has occurred; to what extent the parent material has been converted to distinct horizons or soil layers. Usually described as young, mature or old.

Soil aggregation/aggregates : The process of soil particles binding together. The mineral components of soil are clay, silt and sand. Clay particles are too small to be seen with a light microscope, and sand particles can be seen with the naked eye. Silt particles are of intermediate size. Without soil aggregation, soils would not have pores large enough for movement of air and water, and roots would not be able to penetrate the soil.

Soil air : The soil atmosphere; the gaseous phase of the soil, being that volume not occupied by solid or liquid. (1997. Soil science society of america. P. 94).

Soil classification : A specific soil is classified according to the number of horizons in its soil profile and the soil properties of each horizon. Well developed soils are old and contain all three master horizons (a, b, and c), as well as several subdivisions of the master horizons.

Soil development : Five factors influence the development of a specific soil: parent material, climate, living organisms, topography and time.

Soil drench : Application of a solution or suspension of a chemical to the soil, especially pesticides to control soilborne pathogens.

Soil forming factors : The variables, usually interrelated natural agencies, that are active in and responsible for the formation of soil. The factors are usually grouped into five major categories as follows: parent material, climate, organisms, topography, and time. (sssa).

Soil horizons : A layer of soil or soil material approximately parallel to the land surface and differing from adjacent genetically related layers in physical, chemical, and biological properties or characteristics such as color, structure, texture, consistency, kinds and number of organisms present, degree of acidity or alkalinity, etc. (sssa).

Soil inhabitant : An organism that maintains its population in soil over a period of time

Soil invader : An organism whose population in soil diminishes in several months to years.

Soil organic matter : The organic fraction of the soil, exclusive of undecayed plant and animal residues. See also "humus."

Soil pasteurization : Process used to free soil of selected harmful microorganisms using heat.

Soil sample : A collection of individual cores from a known area.

Soil series : A unit of soil classification determined by studying horizon characteristics, such as: number of horizons, color, thickness, texture, erosion phase, slope, organic content and depth to hardpan. All soils given the same soil series name possess the same characteristics across the landscape.

Soil sterilization : Process used to free soil of all microorganisms.

Soil structure : The arrangement of soil separates into units called soil aggregates.

Soil taxonomy : The classification and naming of soils. See table 1.1 for a list of taxonomic levels and the number of classes for each level. Table 1.2 provides the taxonomy of the nebraska state soil, the holdrege silt loam.

Soil test : Chemical analysis of soil samples to assess soil nutrient levels and determine how fertilizer use can be improved.

Soil test calibration : A means of establishing a relationship between a given soil test value and the yield response from adding a nutrient to soil as fertilizer.

Soil test correlation : The relationship between the amount of nutrient extracted from soil by a laboratory test and nutrient uptake by plants in the greenhouse or field and/or crop yield.

Soil texture : This refers to the relative proportions of sand, silt and clay in a soil.

Soil water content : The water lost from the soil upon drying to a constant mass at 105 degrees c., expressed either as the mass of water per unit mass of dry soil or as the volume of water per unit bulk volume of soil.

Soilborne : Carried on or beneath the soil surface.

Solarization : Disease control practice in which soil is covered with polyethylene sheeting and exposed to sunlight, thereby heating the soil and controlling soilborne plant pathogens.

Solubility : The ability of a solute to dissolve in a solvent; the ability of a herbicide to dissolve in water.

Soluble : A state change when a solid material is put into a solvent. The molecules leave the solid phase and go into the liquid phase of the solvent .

Solute molecules : Molecules that are dissolved in a solvent.

Solute potential : The effect of dissolved solutes on water potential.

Solution : A liquid matrix where a solute has been dissolved in a solvent.

Solvent : A liquid that is capable of dissolving other molecules called solutes. In biological systems, water is usually the solvent.

Somatic cells : Body cells.

Sooty mold : Black, nonparasitic, superficial fungal growth on honeydew produced by aphids and other phloem-feeding insects.

Sorption : The removal of an ion or molecule from solution. Commonly used in regards to phosphorus fixation.

Sorus (pl. Sori) : Compact fruiting compact fruiting structure, especially the erumpent spore mass in the rust fungi (uredinales) and smut fungi (ustilaginales); occasionally a group of fruiting bodies as in synchytriaceae; a cluster of sporangia on a fern sporophyte.

Southern blot : A method used to transfer DNA fragments separated during electrophoresis to membrane filters for detection of specific nucleotide sequences using probes.

Sp. (abbr. For species; pl. Spp.) : A genus name followed by sp. Means that the particular species is undetermined; spp. After a genus name means that several species are being referred to.

Species : Any one kind of life subordinate to a genus but above a race; a group of closely related individuals of the same ancestry, resembling one another in certain inherited characteristics of structure and behavior and relative stability in nature; the individuals of a species ordinarily interbreed freely and maintain themselves and their characteristics in nature.

Specific epithet : The second word in a latin binomial.

Specific resistance (syn. Vertical resistance) : Resistance which is effective against some biotypes or races of the pathogen, but not others, usually inherited monogenically and expressed qualitatively. (see general resistance, horizontal resistance, race-nonspecific resistance).

Spermagonium (pl. Spermagonia; syn. Pycnium for rust fungi) : Structure in which male reproductive cells are produced; in rust fungi, globose or flask-shaped haploid fruiting body composed of receptive hyphae and spermatia (pycniospores).

Spermatium (pl. Spermatia; syn. Pycniospore for rust fungi) : A male sex cell; a nonmotile male gamete; a haploid male gamete.

Spicule : Copulatory organ of male nematode.

Spike : An unbranched, indeterminate iflorescence with sessile flowers.

Spikelet : Spike-like appendage comprised of one or more reduced flowers and associated bracts; unit of inflorescence in grasses; a small spike.

Spiroplasma : Spiral-shaped plant pathogenic mollicute (prokaryote without cell wall).

Spontaneous generation : The theory, now known to be invalid, that plants, animals and microorganisms arose suddenly from non-living materials under certain environmental conditions.

Sporangiophore : Sporangium-bearing body of a fungus.

Sporangiospore : Non-motile, asexual spore that is borne in a sporangium.

Sporangium (pl. Sporangia) : Saclike fungal structure in which the entire contents are converted into an indefinite number of asexual spores.

Spore : Reproductive structure of fungi and some other organisms, containing one or more cells; a bacterial cell modified to survive an adverse environment.

Sporidium (pl.sporidia) : Basidiospore of rust fungi, smut fungi, and other basidiomycota.

Sporocarp : Spore-bearing fruiting body.

Sporodochium (pl.sporodochia) : Superficial, cushion-shaped asexual fruiting body consisting of a cluster of conidiophores.

Sporophore : A spore-producing or spore-bearing structure such as a conidiophore, ascocarp or basidiocarp.

Sporophyte : The diploid stage of a plant (see gametophyte).

Sporulate : To produce spores.

Spot : A symptom of disease characterized by a limited necrotic area, as on leaves, flowers, and stems.

Springwood : Early part of the yearly xylem growth ring in woody plants consisting typically of cells that are larger than those formed later in the season (summerwood).

SSR : Simple sequence repeat, also known as a microsatellite. These DNA markers are based on differences in length of repeated di and trinucleotide sequences. By identifying the conserved flanking sequences of these SSRs, primers can be designed to selectively amplify them via PCR. SSRs are usually codominant and a potentially large number can be identified. Only small amounts of DNA are required, and analyses can be run quickly and relatively inexpensively. However, initial identification of SSRs followed by design, synthesis, and evaluation of primers is a largescale effort.

Stabilizing selection : The theorized competitive disadvantage of unnecessary virulence genes; races with excess genes would have decreased fitness relative to races with fewer virulence genes, so a "super-race" would be less likely to appear in multiline crops.

Staghead : Defoliated, dead or dying major branches in the crown of a tree, usually resulting from inadequate water uptake or translocation.

Stalk rot : One type of stalk rot, anthracnose stalk rot, is caused by a fungus (colletotrichum graminicola) that enters the corn plant above ground. The disease causes the inside of the stalk to degrade from bright white to gray. This disease weakens the stalk making it more likely for plants to lodge.

Stamen (adj. Staminal) : Male structure of a flower, composed of a pollen-bearing anther and a filament, or stalk.

Staminode : A stamen without anthers.

Starch : Plant storage form of carbohydrate.

Starch phosphorylase : Catalyzes the phosphorolysis of (α-d-1,4 glucosyl)n yielding (α-d-1,4 glucosyl)n–1 and α-d-glucose 1-phosphate.

Starch phosphorylase : Catalyzes the phosphorolysis of (α-d-1,4 glucosyl)n yielding (α-d-1,4 glucosyl)n–1 and α-d-glucose 1-phosphate.

Starch synthase : Catalyzes the reaction: adp-glucose + (α-d-1,4 glucosyl)n > adp + (α-d-1,4 glucosyl)n+1.

Starch synthase : Catalyzes the reaction: adp-glucose + (α-d-1,4 glucosyl)n > adp + (α-d-1,4 glucosyl)n+1.

Start/scodons : Codons used to start and stranslation during protein synthesis.

Stele : Central cylinder of vascular tissue (especially in roots).

Stem pitting : A viral disease symptom characterized by depressions on the stem.

Stems : Bears the buds, leaves and flowers of a plant.

Stereochemically : The specific 3 dimensional arrangement of atoms aroundd an asymmetric carbon.

Sterigma (pl. Sterigmata) : Small, usually pointed projection that supports a spore.

Sterile : Unable to reproduce sexually; To be free of living microorganisms.

Sterile fungus : A fungus that is not known to produce any kind of spores.

Sterility : Unable to produce seed, pollen or spores capable of fertilization.

Sterilization (adj. Sterilized) : The total destruction of living organisms by various means, including heat, chemicals or irradiation.

Stigma : Portion of a flower that receives pollen and on which the pollen germinates.

Stipe : Stalk.

Stippling : Series of small dots or speckles in which chlorophyll is absent.

Stipule : Small leaflike appendage at the base of a leaf petiole, usually occurring in pairs.

Stocking density : Animal demand per unitarea (au/acre) at an instant in time.

Stolon (syn. Runner) : A slender, horizontal stem that grows close to the soil surface; in fungi, a hypha that grows horizontally along the surface.

Stoma (pl. Stomata; adj. Stomatal; also stomate) : Multiple stomata. Minute pores in leaf surfaces defined by two guard cells that enable pores to open and close in response to changes in environmental conditions. OR Structure composed of two guard cells and the opening between them in the epidermis of a leaf or stem, functioning in gas exchange.

Stomata : An opening in the outer layer cells in a leaf bordered by two guard cells and serving in gas exchange.

Stone fruit : Fruit with a stony endocarp, e.g. Cherry, peach, plum.

Strain : A distinct form of an organism or virus within a species, differing from other forms of the species biologically, physically, or chemically.

Stratification : 1. Exposing seeds to a cold period to encourange germination or break seed dormancy;

Stratified : Exposed to a cold period to break seed dormancy.

Striate (n. Striations) : Marked with delicate lines, grooves, or ridges.

Stroma (pl. Stromata) : Compact mass of mycelium (with or without host tissue) that supports fruiting bodies or in which fruiting bodies are embedded.

Structure : The combination or arrangement of soil particles that forms peds or aggregates.

STS : Sequence tagged site. A DNA sequence of known location that occurs just once in the genome. STS loci are detectable by PCR, making them useful for a variety of molecular marker applications.

Stunting : Reduction in height of a vertical axis resulting from a progressive reduction in the length of successive internodes or a decrease in their number.

Style : Slender part of many pistils located between the stigma and the ovary and through which the pollen tube grows. OR Part of the pistil (female reproductive structure) that transports the sperm collected by the stigma to the ovary for fertilization and developement of the egg; situated between the ovary and stigma.

Stylet : Stiff, slender, hollow feeding organ of plant-parasitic nematodes or sap-sucking insects, such as aphids or leafhoppers.

Stylet knob (syn. Basal knob) : Structure at the base of a nematode stylet.

Stylet-borne transmission (syn. Nonpersistent transmission) : A type of virus transmission in which the virus is acquired and transmitted by the vector after short feeding times, and is retained by the vector for only a short period of time.

Suberize : To convert into cork tissue.

Subgenomic rna : A piece of viral rna, shorter than the entire genome of the virus, found in cells infected by the virus and sometimes encapsidated.

Subspecies : A subpopulation of a species, defined on the basis of more than one character (morphologic for many organisms) that distinguishes the members of the subpopulation from other members of that species.

Substrate : The substance on which an organism lives or from which it obtains nutrients; chemical substance acted upon, often by an enzyme. OR The substance, base, or nutrient on which an organism grows, (iii) Compounds or substances that are acted upon by enzymes or catalysts and changed to other compounds in the .chemical reaction. (1997. Soil Science Society of America. P. 107)

Substrate analogs Molecules that differ in structure, at one or more positions, from the endogenous substrate.

Suicide substrate : A substrate which irreversibly binds to the active site of an enzyme, causing it to be unable to catalyze further reactions.

Sulfonylureas : A herbicide family with similar chemical structure belonging to the amino acid synthesis inhibitors mode of action; these herbicides specifically target the ALSAHAS enzyme.

Sulfur dioxide (SO_2) : A primary air pollutant produced in industrial processes and coal burning that causes interveinal necrosis on broadleaf plants and tip necrosis on conifers.

Summerwood : Part of the yearly xylem growth ring in woody plants formed late in the growing season and consisting of cells smaller than those of springwood.

Sunscald or sunburn : Injury of plant tissues burned or scorched by direct sun or or sunburn.

Superphosphoric acid : A mixture of polyphosphate and orthophosphate made by dehydrating water from phosphoric acid; used to produce ammonium polyphosphate fertilizers.

Suppressive soil : Soil in which various diseases are naturally at lower levels than expected due to biological factors in the soil; an example of natural biological control.

Surface creep : The creeping of the surface.

Surfactant : An additive that reduces the surface tension ofa herbicide to allow for better contact between water droplet and plant surface; cationic surfactants, ionic surfactants, nonionic surfactants; surface active agent.

Suscept : An abbreviated term for a susceptible plant.

Susceptibility : A plant's inability to withstand a herbicide at normal use rates.

Susceptible : Sensitive to a stimulus.

Susceptible (n. Susceptibility) : Prone to develop disease when infected by a pathogen (see resistance).

Suspension : A solid dispersed in another solid, liquid or gas that does not dissolve but remains as a solid in the mixture. OR The containment or support in fluid media (usually air or water) of soil particles or aggregates, allowing their transport in the fluid when it is flowing. In fluids at rest, suspension follows Stoke's Law. In wind this usually refers to particles or aggregates <0.1 mm diameter through the air, usually at a height of >15 cm above the soil surface, for relatively long distances.

Suspensor : The organ anchoring and supporting a developing embryo. Suspensor connects embryo with its mother body.

Symbiosis : The living together of two or more biological entities.

Symbiosis (adj. Symbiotic; n. Symbiont) : Mutually beneficial association of two different kinds of organisms.

Symplastic path : Water travels from one cell to the next via plasmodesmata. Symplastic means "living" tissue.

Sympodial : Pertaining to proliferation of axes, in which each successive spore or branch develops behind and to one side of the previous apex where growth has ceased.

Symptom : Indication of disease by reaction of the host, e.g. Canker, leaf spot, wilt (see sign).

Symptomless carrier : A plant that, although infected with a pathogen (usually a virus), produces no obvious symptoms.

Syncytium (pl. Syncytia) : A multinucleate structure in root tissue formed by dissolution of common cell walls induced by secretions of certain sedentary plant-parasitic nematodes, e.g. Cyst nematodes.

Synergism : Increased phytotoxicity occurs when two or more chemicals are applied together, as compared to the individual chemicals applied alone.

Synergism (adj. Synergistic) : Greater than additive effect of interacting factors.

Synergist : A nonphytotoxic chemical which increases the phytotoxicity of a herbicide, usually by inhibiting metabolism of the herbicide through enzyme inhibition.

Synnema (pl. Synnemata; syn. Coremium) : Compact or fused, generally upright conidiophores, with branches and spores forming a headlike cluster.

Systematics : The study of the kinds of organisms and the relationships between them.

Systemic : Pertaining to a disease in which the pathogen (or a single infection) spreads generally throughout the plant; pertaining to chemicals that spread internally through the plant.

Systemic : Having the ability to move throughout the plant by either the xylem or phloem.

Systemic acquired resistance (SAR) : Reduced disease symptoms on a portion of a plant distant from the area where a hypersensitive response occurred or other stimulus was applied; a rapid and coordinated defense response against a variety of pathogens as a signal travels throughout the plant (see induced systemic resistance).

Systemic fungicide : A fungicide that is absorbed into plant tissue and may offer some curative or after-infection activity; includes fungicides that are locally systemic, xylem-mobile (upward moving), and amphimobile (move in phloem upward as well as downward in the plant) (see contact or protectant fungicide).

Systemic herbicide : A herbicide that can move throughout the plant; the activity is not limited to the point of application.

Taproot : Primary root that grows vertically downward and from which smaller lateral roots branch.

TAQ : A DNA polymerase often used in PCR which can withstand high temperatures.

TAQ polymerase : A DNA polymerase enzyme which is highly heat stable, used in pcr procedures.

Taqman : A real time PCR probe detection system, using two labelled probes.

Target DNA : The particular portion of the template DNA which is copied during PCR.

Taxonomy (adj. Taxonomic) : The science dealing with naming and classifying organisms.

Teleomorph (syn. Perfect state) : The sexual form in the life cycle of a fungus (see anamorph and holomorph).

Teliospore (sometimes called teleutospore, teleutosporodesm) : Thick-walled resting or overwintering spore produced by the rust fungi (uredinales) and smut fungi (ustilaginales) in which karyogamy occurs; it germinates to form a promycelium (basidium) in which meiosis occurs.

Telium (pl. Telia) : Fruiting body (sorus) of a rust fungus that produces teliospores.

Temporary wilt : Wilt due to insufficient soil water from which a plant can recover when water is supplied.

Teratogen : A chemical that causes malformations in the fetus.

Terminal moraines : An accumulation of soil, pebbles, rocks, and boulders carried forward, deposited and marking the farthest advance of a glacier or ice sheet.

Terminal tip : On the end of a branch or stem.

Termination sequence : The sequence of DNA which signals the transcription to stop.

Terminator genes : A set of transgenes designed to arrest the final stages of seed development in a crop plant. The resulting trait is the production of seeds that will not germinate. Also called the technology protection system.

Terraces : Physical earthen barriers to runoff and can be very effective in reducing P losses to erosion and runoff.

Testa (pl.testae) : Seed coat.

Tetrapyrrole : A class of molecules characterized by a series of four, fivemembered rings containing four carbon atoms and one nitrogen. Tetrapyrrole molecules include hemes, chlorophylls, vitamin B_{12} and phytochrome.

Textural class : Classification of a soil based upon the relative proportions of sand, silt, and clay present. There are 12 textural classes listed here in order from coarsest to finest: sand, loamy sand, sandy loam, loam, silt loam, silt, sandy clay loam, clay loam, silty clay loam, sandy clay, silty clay, and clay.

Texture : The relative proportions of the various soil separates (i.e., sand, silt, clay) in a soil.

Thallus : Vegetative body of a fungus.

Thatch : Tightly intertwined layer of plant litter from accumulations of undecomposed or partially decomposed plant residues.

Thermal cycler : The machine used in pcr which can be programmed for the heating and cooling cycles.

Thermophilic eubacterium : Type of bacterium found in a hot environment.

Thermotherapy : Use of heat to reduce or eliminate pathogens in plant tissue; often used on plants prior to meristem culture to produce pathogen-free plants.

Thorax : Insect body part between the head and abdomen.

Threonine : Nonessential amino acid found in cheese, eggs, milk, meat, legumes, nuts and seeds.

Threshold velocity : Threshold velocity is the wind speed at which particle movement is initiated.

Thylakoid membranes : A series of sealed membrane compartments found inside the chloroplast of plant cells. The membranes contain the photosynthetic apparatus and all of the chlorophyll in the leaf. Their structure as closed 'sacks' permits functional 'inside' and 'outside' spaces within the chloroplast across which photophosphorylation can generate the proton gradient necessary for atp synthesis.

Thyrse : A type of inflorescence having 3 pairs of cymes in 3 verticillasters.

Ti plasmid : A circle of DNA found in agrobacterium tumifaciens that encodes genes for moving the plasmid into a plant cell and inserting a portion of the plasmid into the plant chromosome.

Tillage : The process of turning or stirring the soil.

Tiller : A lateral shoot, culm, or stalk arising from a crown bud; common in grasses.

Tilth : The physical condition of soil as related to its ease of tillage, fitness as a seedbed, and its impedance to seedling emergence and root penetration.

Time : One of the conventionally accepted soil forming factors.

Tissue : Group of cells, usually of similar structure, that perform the same or related functions.

Tissue culture : *In vitro* method of propagating healthy cells from plant tissues.

Tissue culture : Plant cells are grown in culture which allows them to be manipulated and then induced to develop into whole plants.

Titer : Concentration of a virus.

Tolerance : The inherent ability of a species to survive and reproduce after herbicide treatment; this implies that there was no selection or genetic manipulation to make the plant tolerant; it is naturally tolerant.

Tolerance (adj. Tolerant) : Ability of a plant to endure an infectious or noninfectious disease, adverse conditions, or chemical injury without serious damage or yield loss; (of pesticides) the amount of chemical reside legally permitted on an agricultural product entering commercial channels and usually measured in parts per million (ppm).

Tolerant : The inherent ability of a plant to survive and reproduce after herbicide treatment. This implies that there was no selection or genetic manipulation to make the plant tolerant; it is naturally tolerant.

Tonoplast : The vacuolar membrane in plant cells.

Topography : The combination of land slopes and the way those slopes interrelate. Topography is one of the conventionally accepted soil forming factors.

Toposequence : A sequence of related soils that differ, one from the other, primarily because of topography as a soilformation factor. (SSSA)

Topt : The optimum temperature for an enzymatic reaction.

Totipotency : The concept that even specialized cells contain all of the genetic information for an organism and, therefore, any cell should be able to regenerate into any tissue or into an entire plant.

Totipotent : The ability of a single plant cell to grow, divide, and differentiate into an entire plant. Mammalian cells do not have this ability.

Toxicity : Capacity of a substance to interfere with the vital processes of an organism.

Toxin : Poisonous substance of biological origin.

Tracheids : One of the two major types of water conducting cells found in plants (vessel elements are the others). The elongate cells are dead at maturity, have tapered ends and are arranged end to end and have walls that are partially coated with lignin.

Tracking dye : A dye molecule visible in room light that is mixed with a sample. This allows the scientist to track the loading of the sample and indicate it's movement through an electrophoresis gel.

Trade name : The name under which a product is marketed.

Trait : The characteristic that results from an expressing gene(s). Ex. Upright leaves, drought tolerance, Bt resistance. A trait can be influenced by the environment.

Transcription : The process by which the nucleotide sequence of DNA is copied into a single stranded molecule of RNA. The nucleotide sequence of the RNA created is complementary to the DNA sequence except all thymine molecules are replaced with uracil molecules.

Transduce : The process whereby a receptor initiates one or more sequences of biochemical reactions that connect the stimulus to a cellular response.

Transduction : The transfer of genes from one organism to another by viruses, especially in bacteria.

Transfer rna (abbr. Trna) : The rna that moves amino acids to the ribosome to be placed in the order prescribed by the messenger rna.

Transformation : The transfer of genetic materials from one organism to another by humans (genetic engineering); a means of genetic variation in bacteria by absorption and incorporation of DNA from another bacterial cell.

Transforming : The process of introducing foreign DNA into an organism which is then passed on to the organism's progeny.

Transgene : A gene that has been genetically altered. They are usually used to transform organisms. A foreign gene that has been inserted into the host's genomic DNA.

Transgenic : An organism that has a new genetically engineered DNA sequence found in every one of its cells. Genetically engineered organisms are transgenic. These two terms are used interchangeably.

Transgenic plant : A plant that has a new genetically engineered DNA sequence present in every one of its cells. Genetically engineered plants are transgenic plants. These two terms are used interchangeably.

Transgressive segregation : Segregation of progeny phenotypes that are more extreme than either parent.

Transit peptide : Special sequences found at the end of a protein to direct its transportation to the needed cellular location.

Translation : The assembling of amino acids into a protein using messenger rna, ribosomes and transfer RNA. OR The process following transcription during which the nucleotide sequence of mrna is read and 'translated' into a chain of amino acids (protein). The mrna sequence is read three nucleotides (codon) at a time, and each codon codes for a specific amino acid.

Translocation : Movement of water, nutrients, chemicals, or food materials within a plant. OR Moving from one point to another within a plant by either the xylem or phloem. OR The transport of herbicide short or long distances. Herbicides movement from site of application to site of action might involve cell to cell transport or long distance transport in the xylem or phloem.

Translucent : So clear that light may pass through.

Transmembrane path : The route followed by water that sequentially enters a cell on one side, exits the cell on the other side, enters the next in the series, and so on. In this path, water crosses at least two membranes for each cell in its path.

Transmit (n. Transmission) : To spread or transfer, as in spreading an infectious pathogen from plant to plant or from one plant generation to another.

Transpiration : Water loss by evaporation from leaf surfaces and through stomata.

Transpiration : The loss of water as vapor from plants at their surfaces, primarily through stomata.

Trap crop : Crop planted around a field to protect the inner crop from diseases transmitted by aerial vectors; host crop of a parasitic plant, such as witchweed (striga spp.), that is planted to stimulate seed germination, and later sacrified by plowing under before the parasitic plant produces new seeds.

Trenching : Physical separation of soil in a vertical plane to sever grafted roots between trees.

Triazines : A family of herbicides with similar chemical structure and belonging to the photosynthetic inhibitors mode of action.

Trichogyne : A female receptive hypha.

Trichome : A plant epidermal hair, of which several types exist.

Triglyceride : Sn1,2,3triacylglycerol.

Triplet codon : A set of three nucleotide bases in DNA or rna that code for an amino acid.

Triplet state : A normally forbidden excited electron orbital that is reserved for an electron with the same spin as the electron remaining in the ground state. Upon excitation to an excited singlet state, there is some possibility that the electron will reverse spin. When this happens, the electron then drops down to the triplet state, which is at a lower energy level than the excited singlet state. Once in the triplet state, the electron must remain there until its spin is again reversed. Molecular oxygen is unusual in that it is normally in a triplet state.

tRNA (abbr. For transfer rna) : The rna that moves amino acids to the ribosome to be placed in the order prescribed by the messenger rna.

Tuber : An underground stem adapted for storage, typically produced at the end of a stolon.

Tumor (syn. Gall) : Abnormal swelling or localized outgrowth, often roughly spherical, produced by a plant as a result of attack by a fungus, bacterium, nematode, insect or other organism.

Turbidity : The amount of solid particles that are suspended in water and that cause light rays shining through the water to scatter. Thus, turbidity makes the water cloudy or even opaque in extreme cases. (USGS Water Science Glossary of Terms, 2004)

Turgidity : State of being rigid or swollen as a result of internal water pressure.

Turgor : Pressure within a cell resulting from the uptake of water.

Turgor pressure : The positive pressure built up when water is compressed.

Turnover number : The number of reactions per second that the enzyme can catalyze under optimum conditions.

Tylosis (pl. Tyloses) : Balloonlike extrusions of parenchyma cells into lumina of contiguous vessels that partially or completely block them.

Type : The example on which the description of a scientific name is based, and which fixes the application of the name (the type genus of a family, or the type species of a genus).

Tyrosine : Nonessential amino acid.

Ultrastructure : Submicroscopic structure of a macromolecule, cell, or tissue.

Umbel : An indeterminate inflorescence, usually flattopped with peidcels arising from a single point, like an umbrella.

Undifferentiated : Cells that have not developed into specialized tissues. Undifferentiated cells, called callus, are used in tissue culture during the transformation process.

Unicellular : One-celled (see multicellular).

Uniflagellate : Having one flagellum.

Uninucleate : Having one nucleus (see multinucleate).

Unitunicate : Having a single ascus wall (see bitunicate).

Universal code : The theory that the genetic code is used by all forms of life. This is what allows DNA from different organisms to be introduced into a new organism and function correctly.

Uptake : The process of getting a herbicide from the outer surface of a plant to the inside of the plant where the herbicide will either induce its herbicidal activity or be translocated.

Urea : A fertilizer type widely used in agricultural system to supply N. The average N content in urea is about 46%. Urea formula: $(NH_2)_2CO$

Urea : A dry nitrogen material produced by reacting ammonia with carbon dioxide. Urea contains the highest percentage of nitrogen of the commonly used dry fertilizers and is rapidly replacing ammonium nitrate. When surface applied, urea is the most rapidly volatilized of the dry nitrogen materials.

Ureaammonium nitrate : A nonpressure solution of ammonium nitrate, urea and water. Two grades are most common: 2800 (10.7 lbs./gal.) And 3200 (11.1 lbs./gal.).

Urediniospore (also urediospore, uredospore) : The asexual, dikaryotic, often rusty-colored spore of a rust fungus, produced in a structure called a uredinium; the "repeating stage" of a heteroecious rust fungus, i.e. Capable of infecting the host plant on which it is produced.

Uredinium (also uredium; pl. Uredinia) : Fruiting body (sorus) of rust fungi that produces urediniospores.

Utilization : The proportion of current year's forage production that is consumed or destroyed by grazing animals.

Vacuolate cell : A cell derived from a progenitor cell during embryogenesis and having a large vacuole. Vacuolate cells usually are larger than cytoplasmic cells which result from the same cell division.

Vacuole : Generally spherical organelle within a plant cell bound by a membrane and containing dissolved materials such as metabolic precursors, storage materials, or waste products.

Vacuole : A membranebound (tonoplast) space larger than a vesicle which stores material, either dissolved in water or as a crystalline or flocculent mass. Vacuoles are key organelles for storage of compounds and osmoregulation in plants.

Vadose zone : The zone of material between the surface and the water table.

Vapor : A substance in the gaseous state.

Variegation : Pattern of two or more colors in a plant part, as in a green and white leaf.

Variety (syn. Cultivar; adj. Varietal) : Crop plants within a species that have the same genetic composition. Because plants in a variety are usually heterozygous, their offspring will not remain genetically pure (i.e., corn hybrid varieties). OR A plant type within a species, resulting from deliberate manipulation, which has recognizable characteristics (color, shape of flowers, fruits, seeds, height and form).

Vascular : Pertaining to fluid-conducting (xylem and phloem) tissues in plants.

Vascular bundle : Strand of conductive tissue, usually composed of xylem and phloem (in leaves, small bundles are called veins).

Vascular cambium : A layer of meristematic cells located between xylem and phloem tissues which divide to give off cells toward both tissues. This process forms secondary growth in stems and roots in broadleaf plants.

Vascular cylinder (syn. Stele) : The cylinder of vascular tissue in stems or roots.

Vascular wilt disease : A xylem disease that disrupts normal uptake of water and minerals, resulting in wilting and yellowing of foliage.

Vector : A living organism (e.g., insect, mite, bird, higher animal, nematode, parasitic plant, human) able to carry and transmit a pathogen and disseminate disease; (in genetic engineering) a vector or cloning vehicle is a self-replicating

DNA molecule, such as a plasmid or virus, used to introduce a fragment of foreign DNA into a host cell. A plasmid, virus, BAC or YAC used to carry a gene(s) of interest.

Vegetative : Referring to somatic or asexual parts of a plant, which are not involved in sexual reproduction.

Vegetative propagation : Asexual reproduction; in plants, the use of cuttings, bulbs, tubers, and other vegetative plant parts to grow new plants.

Vein : Small vascular bundle in a leaf.

Vein banding : Symptom of virus disease in which regions along veins are either darker green or distinctly more yellow than tissue between veins.

Vein clearing : Disappearance of green color in or around leaf veins.

Vermiform : Worm-shaped.

Vertical resistance (syn. Specific resistance) : Resistance which is effective against some biotypes or races of the pathogen, but not others, usually inherited monogenically and expressed qualitatively (see general resistance, horizontal resistance, race non-specific resistance).

Verticillaster : A false whorl of flowers.

Verticillate : Arranged in whorls. Armitage, A; Herbaceous Perennial Plants

Verticils : A whorl of flowers or leaves.

Vertifolia effect : A term used to describe the loss of general (horizontal, minor gene, multigenic, polygenic) resistance in a cultivar after several generations of selection during which a major gene confers resistance to the dominant race or biotype of the pathogen; first observed in the potato cultivar vertifolia with late blight resistance.

Vesicle : Thin sac in which zoospores are differentiated and released;

The bulbous head terminating the conidiophores of aspergillus;

Structure formed by endomycorrhizal fungi within living cells of the root.

Vessel : Water-conducting structure of xylem tissue with pit openings in end walls.

Vessels : Vessels, or vessel elements, are one of the two major types of waterconducting cells found in plants (tracheids are the others). The elongate cells are dead at maturity, have tapered ends and are arranged end to end and have walls that are partially coated with lignin.

Viability : The measure of the number of individuals in a given phenotypic class that survive relative to another class.

Viable (n. Viability) : The state of being alive; able to germinate, as seeds, fungus spores, sclerotia, etc.; capable of growth.

Virescence : State or condition in which normally white or colored tissues (e.g. Flower petals) become green.

Virion : Complete virus particle.

Viroid : An infectious, nonencapsidated (naked) circular, single-stranded rna.

Viroplasm : Cellular inclusions that are sites of synthesis of viral components and the assembly of virus particles.

Virulence : Degree or measure of pathogenicity; relative capacity to cause disease.

Virulent : Highly pathogenic; having the capacity to cause severe disease (see avirulent).

Viruliferous : Virus-laden, usually applied to insects or nematodes as vectors.

Virus : A submicroscopic, intracellular, obligate parasite consisting of a core of infectious nucleic acid (either rna or DNA) usually surrounded by a protein coat.

Viscin : Sticky substance produced on the seed coat of parasitic flowering plants in the viscaceae (mistletoes) that helps attach the seed to the host plant branch.

Vm : The limiting velocity of an enzymatic reaction. This would occur at an infinite substrate concentration.

Volatile : Liquids or solids with relatively high vapor pressure at normal temperatures. The higher the vapor pressure, the more likely the substance will vaporize into a gas.

Volatility : The ability of a compound to change from a liquid phase to its gaseous phase, dependant on chemical properties of compound and environmental conditions.

Volatilization : The conversion of ammonium into ammonia gas. $NH4 > NH3$(gas).

Volunteer : Self-set plant; plant seeded by chance.

Vulva : Exterior opening of a mature female nematode's reproductive system.

Walling-off : Separation of diseased from healthy tissues by barrier tissues produced by a diseased plant.

Water potential : The chemical potential of water; a measure of the ability of a substance to absorb or release water relative to another substance.

Water sprout : Small, rapidly growing shoot or branch on a large stem, developed from adventitious tissues.

Water table : The upper surface of ground water or that level in the ground where the water is at atmospheric pressure. (sssa).

Waterfilled porosity : The percent of water occupying the empty spaces of soil and is inversely related to the percent of air occupying the empty spaces of soil. Total porosity= airfilled porosity plus waterfilled porosity.

Waterholding capacity : The maximum amount of water a soil can hold against the pull of gravity.

Watershed : The land area that drains water to a particular stream, river, or lake. It is a land feature that can be identified by tracing a line along the highest elevations between two areas on a map, often a ridge. Large watersheds, like the Mississippi River basin contain thousands of smaller watersheds.

Water-soaked : Describing disease symptom of plants or lesions that appear wet, dark, and usually sunken and translucent.

Wavelength : The distance between equivalent positions in the electromagnetic wave formed by a photon as it moves through space. For visible photons, the wavelengths are in the range of 400 to 700 nm.

Weathering : The complex combination of physical, chemical and biological processes that decompose, disintegrate and alter rocks and minerals at or near the Earth's surface. Weathering can be subdivided into chemical weathering, physical weathering, and biological weathering. (Soil and Environmental Science Dictionary). OR The means by which soil, rocks and minerals are changed by physical and chemical processes into other soil components. Weathering is an integral part of soil development.

Western blot : A technique in which proteins are separated using electrophoresis and transferred to a membrane so they can be identified and studied.

Wetland : A transitional area between aquatic and terrestrial ecosystems that is inundated or saturated for long enough periods to produce hydric soils and support hydrophytic vegetation. (SSSA).

White rot (of wood) : Type of wood decay resulting from enzymatic action of fungi; it degrades all components of wood, including lignin, leaving the wood light-colored and spongy.

White rust : Common name usually used for diseases caused by oomycetes in the genus albugo.

Whorl : Circular arrangement of like parts.

Wild type : The phenotype characteristic of the majority of individuals of a species under natural conditions.

Wilt : Drooping of leaves and stems from lack of water (inadequate water supply or excessive transpiration);

Vascular disease that interrupts normal water uptake.

Wilting point : The moisture content of a soil at which most plants wilt.the wilting point is commonly estimated by measuring the 15bar or 1.5 mpa percentage of a soil.

Wind break : A row or other grouping of trees or shrubs used to provide protection against the effects of high velocity winds.

Winterburn : Foliar necrosis, often marginal, of plants that retain their leaves in winter due to water deficiency because they cannot take up water from frozen soils.

Witches' broom : Disease symptom characterized by an abnormal, massed, brushlike development of many weak shoots arising at or close to the same point.

Wood : Secondary xylem.

Woody caudex : The swollen stem base of certain nonwoody perennial plants that survives over the winter and from which new growth is produced. (encarta® world english dictionary).

Wound : Injuries to plant tissues that often breach barriers (cuticle, bark, cell walls) that might otherwise exclude pathogens; some pathogens (e.g. Viruses) can enter plants only through a wound; wounds may occur from natural growth processes, physical and chemical agents, animals (especially insects), and many human agricultural activities, such as pruning.

Xenobiotic : Foreign chemicals; natural or synthetic substances that cannot be utilized by plants for energyyielding processes.

Xilema : Células especializadas de los tejidos vasculares de las plantas, que mueren en la madurez y que conducen agua y minerales.

Xylem : Water and mineral conducting, food-storing, supporting tissue of a plant. OR Nonliving cells connected end to end to form a series of long, thin tubes from root to the shoot, used to supply cells with water and nutrients that are dissolved in water.

Xylem-limited fastidious bacteria (xlb) : Bacterial pathogens of plants found only in the xylem, causing wilt, scorch, and stunting symptoms; vectored by insects that feed on xylem fluid; not sap transmissible; require complex culture media.

Yeast : Unicellular ascomycetous fungus that reproduces asexually by budding .

Yellows : Disease characterized by chlorosis and stunting of the host plant.

Yield : An aggregate of the products resulting from growth or cultivation .

Yield drag : A negative effect on grain yield associated with crop plants that have a specific gene or a specific trait.

Yield lag : A relative reduction in yield observed in some hybrids or varieties compared with the yield observed in the most recently produced hybrids or varieties.

Yield potential : The highest yield a plant (hybrid, variety, etc.) Is capable of producing when grown in ideal conditions.

Zinc Carbonate : Inorganic compounds that are very insoluble in water. If granulated, zinc carbonate is not an effective source of zinc; however, it is effective if very finely ground.

Zinc Chelate : Material blended with dry fertilizers and applied in a row band. Granular zinc chelate is likely to be more effective than granular zinc sulfate because of its mobility.

Zinc Oxide : Inorganic compounds that are very insoluble in water. Zinc oxide is not effective as a source of zinc if granulated, but is effective when finely ground.

Zonate : Targetlike development of tree canker, characterized by successive, perennial rings of callus; any symptom appearing in concentric rings.

Zoosporangium : Sporangium, or spore case, that bears zoospores.

Zoospore : Fungal spore with flagella, capable of locomotion in water.

Zwitterionic : Having both a positive and negative charge.

Zygomycetes : A fungus group, also called the zygomycota, characterized by nonseptate hyphae, sexual zygospores, and sporangiospores produced in a sporangium; common post-harvest pathogens.

Zygospore : Sexual resting spore formed from the union of gametangia in the Zygomycetes.

Zygote : The cell formed when the egg and sperm fuse. The cell has two copies of each chromosome (2n) and will continually divide to develop into an entire organism.

Á-(1,4)-glucan branching enzyme : Catalyzes the transfer of a segment of an α-d-1,4-linked glucan chain to a primary hydroxyl group in a similar glucan chain.

α-amylase : Catalyzes the endohydrolysis of α-d-1,4 glucosidic linkages in polysaccharides containing three or more 1,4-α-linked d-glucose units. The term ?α? indicates the initial anomeric configuration of the free sugar group released and not the configuration of the linkage hydrolysed.

α-glucosidase : Catalyzes the hydrolysis of terminal, nonreducing 1,4-linked α-d-glucose residues with the release of α-d-glucose.

β-amylase : Catalyzes the hydrolysis of β-d-1,4 glucosidic linkages in polysaccharides so as to remove successive maltose units from the nonreducing ends of the chains. The term 'β' denotes the initial anomeric configuration of the free sugar group released and not the configuration of the linkage hydrolysed.

β-amylolysis : The stepwise hydrolysis of alternate glucosidic bonds in starch-type polysaccharides with the liberation of maltose.

Index